# Infinite Group Theory

## From the Past to the Future

# Infinite
# Group Theory
## From the Past to the Future

*editors*

**Paul Baginski**

**Benjamin Fine**

*Fairfield University, USA*

**Anthony M Gaglione**

*United States Naval Academy, USA*

**World Scientific**

NEW JERSEY · LONDON · SINGAPORE · BEIJING · SHANGHAI · HONG KONG · TAIPEI · CHENNAI · TOKYO

*Published by*

World Scientific Publishing Co. Pte. Ltd.

5 Toh Tuck Link, Singapore 596224

*USA office:* 27 Warren Street, Suite 401-402, Hackensack, NJ 07601

*UK office:* 57 Shelton Street, Covent Garden, London WC2H 9HE

**Library of Congress Cataloging-in-Publication Data**

Names: Baginski, Paul, editor. | Fine, Benjamin, 1948–  editor. | Gaglione, Anthony M., editor.
Title: Infinite group theory : from the past to the future / edited by Paul Baginski
   (Fairfield University, USA), Benjamin Fine (Fairfield University, USA),
   Anthony M. Gaglione (United States Naval Academy, USA).
Description: New Jersey : World Scientific, 2017. | Includes bibliographical references.
Identifiers: LCCN 2017038766 | ISBN 9789813204041 (hardcover : alk. paper)
Subjects: LCSH: Infinite groups. | Group theory. | Geometry, Algebraic.
Classification: LCC QA178 .I5227 2017 | DDC 512/.2--dc23
LC record available at https://lccn.loc.gov/2017038766

**British Library Cataloguing-in-Publication Data**
A catalogue record for this book is available from the British Library.

For any available supplementary material, please visit
http://www.worldscientific.com/worldscibooks/10.1142/10354#t=suppl

Typeset by Stallion Press
Email: enquiries@stallionpress.com

# Introduction

Paul Baginski, Benjamin Fine, and Anthony Gaglione

Over the past several decades there has been tremendous progress in infinite group theory and, especially, combinatorial group theory. Many of these advances have centered on four areas, that are of course interrelated: the development of geometric group theory; the development of an algebraic geometry over groups; the integration of group theory and computer science via the study of automatic groups and rewriting systems; and, finally, the creation of group-based cryptography. Much of this material came together in the monumental work that resulted in a proof, by Kharlampovich and Myasnikov and, independently, by Sela, of the Tarski conjectures on free groups.

On April 23–24, 2015 a conference was held to show an appreciation of these achievements and to honor on their 70th birthdays two people, Gerhard Rosenberger and Dennis Spellman, who have made significant contributions to the area. The conference was a two-day conference held Thursday April 23 at Fairfield University in Fairfield, Connecticut, and April 24 at CUNY Graduate Center 34th Street and 5th Avenue in Manhattan. This festschrift is a collection of papers concerning the material at the conference. We think that it is a nice collection and a good contribution to the discipline. Below we say a few things about the honorees.

## Gerhard Rosenberger

Gerhard Rosenberger grew up in a small village in the old East Germany before his family moved to the west during his gymnasium years. He attended the University of Hamburg and after an early fascination with Physics did his doctorate in 1973 in Analytic Number Theory and his habilitation in 1974 in Combinatorial Group Theory at the University of Hamburg. From 1972 to 1976 he was an assistant in Hamburg and from 1976 to 1977 he was a professorial chair representitive in Bielefeld. Then from 1977 to 2010 he was a mathematics professor in Dortmund. Since 2010 he has taught in Hamburg and in Passau. Worldwide he worked for longer terms at nine univerities. He has written more than 200 research articles and 21 books in Mathematics. His published work and journal articles represent many of the areas spoken about in this conference; combinatorial group theory, geometric group theory, group based cryptography and the interplay between number theory and group theory. He has made seminal contributions to each of these areas; the theory of Nielsen transformations, the structure of discrete groups, the theory of one-relator products, the classification of the generalized triangle and generalized tetrahedron groups and the development of group based cryptosystems to name but a few. His book *The Fundamental Theorem of Algebra* joint with Ben Fine won the 2000 Alpha Sigma Nu Book of the Year Award in Science. His coauthors are wordwide from more than 25 different countries. He is a member of the American Math. Society, the London Math. Society, the German Math. Society (DMV) and the Hamburger Math. Gesellschaft. He was the founder of the journal *Groups, Complexity, Cryptology* and, together with V. Shpilrain, the managing editor for many years. He has supervised nineteen PhD students and many diplome students who have also worked on the areas that Dr. Rosenberger pioneered.

### Dennis Spellman

Dennis Spellman received his doctorate in 1971 from Courant Institute, New York University under the direction of Wilhelm Magnus. He is the author of over seventy papers and several books many of which have had a tremendous impact on the field. We say more on this in a bit. He was a proofreader for *Index to Mathematical Problems 1980–1984*, an editor of *Contmeporary Mathematics 421; Combinatorial Group Theory, Discrete Groups and Number Theory*, a coauthor with Seymour Lipschiutz of two volumes in the Schaum's Outline Series and a coauthor with Ben Fine, Anthony Gaglione, Alexei Myasnikov and Gerhard Rosenberger of *The Elementary Theory of Groups*. One outlier deserves mention. Having corrected a manuscript of the engineer Gregory Coxson, Dr. Spellman has exactly one paper (joint with Coxson) in an electrical engineering journal. Aware that this volume is dedicated to more than one person, Dr. Spellman wants it known that one of his proudest accomplishments is having a **Rosenberger number** equal to one.

Dr. Spellman is the elder son of Harold and Ann Spellman and grew up in Philadelphia. He attended Temple University as an undergraduate before going to New York University for his graduate work. His work under Wilhelm Magnus involved the commutator calculus but in 1971 he was seduced by two open problems: the Sims conjecture and the Tarksi conjectures. Encouraged by Verena Huber-Dyson he then devoted most of his career with his long time research collaborator, Anthony Gaglione, (among others) to the intersection of group theory and model theory.

Sim's conjecture asked whether or not for each positive integer $n$, the $n$-th term of the lower central series of a finitely generated nonabelian free group is the normal closure of any fixed set of weight $n$ Hall basic commutators. Working with Anthony Gaglione and David Jackson, Dr. Spellman proved that the Sims question has a positive answer for all weights $n \leq 5$. Dr. Jackson has conjectured that, like the Burnside problem, the question has a positive answer for small $n$ but admits counterexamples for all sufficiently large $n$, perhaps even in weight 6.

The Tarski conjectures were proved by Olga Kharlampovich and Alexei Myasnikov and independently by Zil Sela. Both proofs were monumental pieces of work. Dr. Spellman can claim some small first steps. First, collaborating with Anthony Gaglione, they proved that every finitely presented universally free group (a group that has the same universal theory as a nonabelian free group) is fully residually free. At about the same time, and unaware of each other's work. Vladimir Remeslennikov proved that every finitely generated universally free group is fully residually free. Subsequently it was shown that every finitely generated universally free group is finitely presented. In one of Dr. Spellman's early papers he proved that every 3-generator model of the elementary theory of the nonabelian free groups must already be free. This led to a collaboration with Ben Fine, Anthony Gaglione, Alexei Myasnikov and Gerhard Rosenberger where they obtained the result that for any universally free group in which every 2-generator subgroup is free (the locally hyperbolic case) one already has that every 3-generator subgroup is free. The theorem relied heavily on Dr. Rosenberger's results on Nielsen cancellation.

Late in his career Dr. Spellman was on the faculty of Temple University and collaborated with Seymour Lipschutz and Anthony Gaglione on a series of papers involving generalizations of pregroups.

# Contents

# Chapter 1

# Groups With the Weak Minimal Condition on Non-Permutable Subgroups

Laxmi K. Chataut[1] and Martyn R. Dixon[2]

[1]*Department of Mathematics*
*University of Wisconsin-Marinette*
*Marinette, WI54143-4253, U.S.A.*
*laxmi.chataut@uwc.edu*

[2]*Department of Mathematics*
*University of Alabama*
*Tuscaloosa, AL 35487-0350, U.S.A.*
*mdixon@ua.edu*

*To Gerhard Rosenberger and Dennis Spellman, on their 70th birthdays*

**ABSTRACT.** Let $G$ be a group with the weak minimal condition on non-permutable subgroups. The main results are as follows. If $G$ is locally finite then either $G$ is quasihamiltonian or $G$ is Chernikov; if $G$ is a generalized radical group then $G$ is either quasihamiltonian or soluble-by-finite of finite rank.

## 1. Introduction

In this paper, we shall often be concerned with the well-known class of *generalized radical* groups. Here a group $G$ is called generalized radical if it has an ascending normal (indeed characteristic) series,

$$1 = R_0 \leq R_1 \leq R_2 \leq \cdots \leq R_\alpha \leq R_{\alpha+1} \leq \cdots \leq R_\gamma = G,$$

where $R_{\alpha+1}/R_\alpha$ is either locally nilpotent or locally finite, for each ordinal $\alpha$ and $R_\lambda = \cup_{\beta<\lambda} R_\beta$, for each limit ordinal $\lambda$.

2010 *Mathematics Subject Classification.* Primary: 20E15; Secondary 20F19, 20F22.
*Keywords and phrases.* weak minimal condition, permutable subgroup.
The first author would like to thank the University of Alabama for financial support during his Ph. D. studies. This work formed part of his Ph. D. dissertation.

Other well-studied classes of groups which are of relevance in this paper are the class of groups with the minimal condition and the class of groups with the maximal condition, which we briefly discuss. Shunkov [23] proved that a locally finite group with the minimal condition is Chernikov and it is then an easy consequence to show that every generalized radical group with the minimal condition is Chernikov, a result which can be substantially generalized still further.

The maximal condition is well-known to be equivalent to the condition that every subgroup of $G$ is finitely generated and from this it is easy to deduce that a soluble group with the maximal condition is polycyclic. Consequently a generalized radical group with the maximal condition is polycyclic-by-finite. However, even residually finite groups with the maximal condition do not appear to be well-understood.

Of course, once we leave the realm of generalized radical groups, classification problems often become much more difficult. The well-known Tarski Monsters of Ol'shanskii [17], for example, have both the maximal condition and the minimal condition.

One generalization of both the maximal and minimal conditions that has proved useful is the condition of being minimax. A group $G$ is called *minimax* if it has a finite series each of whose factors have either the maximal condition or the minimal condition.

If $G$ is a soluble minimax group then $G$ has a number of additional properties which have played an important role in the study of structural properties of infinite groups and to which we now turn. A group $G$ is said to have the *weak minimal condition*, denoted by min-$\infty$ if, in every descending chain of subgroups

$$G_1 \geq G_2 \geq \cdots \geq G_n \geq \cdots$$

of $G$, at most finitely many of the indices $|G_i : G_{i+1}|$ are infinite. There is a corresponding weak maximal condition, max-$\infty$, connected with ascending chains. It is clear that if $G$ has the minimal condition then it satisfies the weak minimal condition and likewise if $G$ has the maximal condition then it has the weak maximal condition. These properties were first defined in the papers of D. I. Zaitsev [25–27] and R. Baer [1], where it is shown that in a soluble group the conditions min-$\infty$, max-$\infty$ and minimax are equivalent. Zaitsev [26, Theorem 5] also showed that a locally soluble group with the weak minimal condition is a soluble minimax group.

More general conditions than the minimal condition and the weak minimal condition have been defined. A group $G$ is said to satisfy the *minimal condition* on $\mathcal{P}$-subgroups (min-$\mathcal{P}$) if every descending chain of $\mathcal{P}$-subgroups terminates in finitely many steps and $G$ satisfies the *weak minimal condition* on $\mathcal{P}$-subgroups (min-$\infty$-$\mathcal{P}$) if there is no infinite descending chain

$$H_1 \geq H_2 \geq H_3 \geq \cdots$$

of $\mathcal{P}$-subgroups of $G$ with each $|H_i : H_{i+1}|$ infinite. Equivalently $G$ has min-$\infty$-$\mathcal{P}$ if, for every descending chain $H_1 \geq H_2 \geq H_3 \geq \cdots$ of $\mathcal{P}$-subgroups of $G$, $|H_i : H_{i+1}|$ is infinite only for finitely many $i$. The maximal condition on $\mathcal{P}$-subgroups (max-$\mathcal{P}$) and the weak maximal condition on $\mathcal{P}$-subgroups can be defined for ascending chains.

In Section 2 of this paper we give a brief survey of results concerned with the condition min-$\infty$-$\mathcal{P}$ for various properties or classes, $\mathcal{P}$. Then in Sections 3 and 4 we will present some new results concerned with the weak minimal condition on non-permutable subgroups. Our notation, when not explained, will be that in general usage or can be found in [20], for example.

## 2. Some Results Concerning the Weak Minimal Condition

Generally we only wish here to discuss properties related to normality, in order to keep this section brief, but it is worth pointing out a further paper of Zaitsev [28], who was really instrumental in studying these kinds of topics. In [28] Zaitsev studied the class of groups satisfying the weak minimal condition on non-abelian subgroups. If $\mathcal{P}$ is a group theoretical property or class of groups then we let $\bar{\mathcal{P}}$ denote the class of all groups which are not $\mathcal{P}$-groups together with all trivial groups. Zaitsev proved:

**Theorem 2.1.** *Let $G$ be a locally (almost soluble) group. Then the conditions min-$\infty$ and min-$\infty$-$\overline{ab}$ are equivalent for $G$. In particular, $G$ is a soluble minimax group.*

As usual, a group is called locally (almost soluble) if every finitely generated subgroup is soluble-by-finite: This line of investigation has been considered in a number of papers and we mention here just the papers [19] [5] and [6].

We turn now to a different type of extension of the weak minimal condition. The condition min-sn, the minimal condition on subnormal subgroups, has been studied in several papers and there is a good exposition in [20, Section 5.4]. More generally the class of groups with the weak minimal condition on subnormal subgroups (min-∞-sn) was studied by L. A. Kurdachenko [11] and the following result was proved.

**Theorem 2.2.** *Let $G$ be a group with an ascending normal series whose factors are finite or locally nilpotent. Then $G$ satisfies min-∞-sn (or max-∞-sn) if and only if $G$ is a soluble-by-finite minimax group.*

Groups with the weak minimal (or weak maximal) condition on normal subgroups, denoted by min-∞-n, have also been studied. For locally nilpotent groups it is well-known that the minimal condition on normal subgroups implies the minimal condition (see [20, Theorem 5.27, Corollary 2] for example). However L. A. Kurdachenko has given locally nilpotent examples of groups with min-∞-n in [10] and [12] which are not minimax. However locally nilpotent groups with min-∞-n are hypercentral and soluble.

Dedekind groups, those groups with all subgroups normal, represent a class of groups which have the property min-$\bar{n}$, the minimal condition on non-normal subgroups. More generally, the class of groups with the weak minimal condition on non-normal subgroups (min-∞-$\bar{n}$) has been studied by L. A. Kurdachenko and V. E. Goretsky [13], where they show that a locally (almost soluble) group with min-∞-$\bar{n}$ either is Dedekind or a soluble-by-finite minimax group, thus showing that in this class of groups the conditions min-∞ and min-∞-$\bar{n}$ are equivalent.

To finish this short history of relevant results we note that Kurdachenko and Smith ([14] and [15]) have also studied groups with the weak minimal condition on non-subnormal subgroups, the condition min-∞-$\overline{sn}$. In this case, examples include minimax groups and also the interesting class of groups with all subgroups subnormal, which of course are soluble by a result of Möhres [16]. The main result of [14] runs as follows.

**Theorem 2.3.** *Let $G$ be a generalized radical group and suppose that $G$ satisfies min-∞-$\overline{sn}$. Then either $G$ is a soluble-by-finite minimax group or all subgroups of $G$ are subnormal.*

Locally (soluble-by-finite) groups need not be generalized radical, so in [15] the authors proved that a locally (soluble-by-finite) group satisfying the condition min-∞-$\overline{sn}$ also is a soluble-by-finite minimax group or has all subgroups subnormal.

## 3. Groups with the Weak Minimal Condition on Non-Permutable Subgroups

One generalization of normality that has received some considerable attention recently is the notion of permutability. A subgroup $H$ of a group $G$ is said to be permutable in $G$ if $HK = KH$ for all subgroups $K$ of $G$. Of course, normal subgroups are permutable and O. Ore [18] proved that in a finite group $G$ all permutable subgroups of $G$ are subnormal in $G$. Permutable subgroups have also been called quasinormal subgroups by some authors, and during his investigations of permutable subgroups, S. E. Stonehewer [24] proved that all permutable subgroups of a group $G$ are ascendant in $G$. He also proved that a simple group never contains a proper nontrivial permutable subgroup.

Groups with all subgroups permutable, so called quasihamiltonian groups, were classified quite precisely by Iwasawa [8]. The following theorem summarizes some of the main results and proofs can be found in [22], for example.

**Theorem 3.1.** *Let $G$ be a non-abelian quasihamiltonian group. Then*

(i) *$G$ is locally nilpotent;*
(ii) *$G$ is metabelian;*
(iii) *If $T(G)$ is the torsion subgroup of $G$ and if $G$ is not periodic then $T(G)$ is abelian and $G/T(G)$ is a torsion-free abelian group of rank 1.*

Groups with the minimal condition on non-permutable subgroups were studied in [3]. In that paper the authors obtained the following result. As usual a group is locally graded if every nontrivial finitely generated subgroup has a nontrivial finite image.

**Theorem 3.2.** *Let $G$ be a group which is either locally graded or not periodic. If $G$ satisfies the minimal condition on non-quasinormal subgroups, then either $G$ is Chernikov or $G$ is quasihamiltonian.*

In the remainder of this article we will investigate the structure of groups satisfying the weak minimal condition on non-permutable subgroups, a class of groups which we denote by min-$\infty$-$\overline{\mathrm{qn}}$ and present some new results for this class of groups. Examples of groups with this property include the soluble minimax groups and quasihamiltonian groups. We do not know if this exhausts the generalized radical groups, or even locally graded groups, satisfying min-$\infty$-$\overline{\mathrm{qn}}$.

It is easy to see that if a group $G$ satisfies the weak minimal condition on non-normal subgroups then $G$ has the weak minimal condition on non-permutable subgroups. Similarly, if all the subgroups of infinite index are permutable in $G$, then $G$ has the weak minimal condition on non-permutable subgroups. The structure of such groups is given in [4].

In order to prove our main theorems we need some preliminary results and the rest of this section is devoted to obtaining these. First we note the following easily proved fact.

**Lemma 3.3.** *Let $C \leq B \leq A$ be subgroups of a group $G$ and suppose $A, B, C$ are permutable where $|A : B|$, $|B : C|$ are infinite. Let $x \in G$. Then at least one of $|A\langle x \rangle : B\langle x \rangle|$, $|B\langle x \rangle : C\langle x \rangle|$ is infinite.*

**Proof.** It is easy to see that $|A\langle x \rangle : B\langle x \rangle| = |A : B(A \cap \langle x \rangle)|$ and that $|B(A \cap \langle x \rangle) : B| = |A \cap \langle x \rangle : B \cap \langle x \rangle|$, so

$$|A : B| = |A\langle x \rangle : B\langle x \rangle| \cdot |A \cap \langle x \rangle : B \cap \langle x \rangle|.$$

We may suppose that $|A\langle x \rangle : B\langle x \rangle|$ is finite. Then $|A \cap \langle x \rangle : B \cap \langle x \rangle|$ is infinite, so $B \cap \langle x \rangle = 1$. However $|B\langle x \rangle : C\langle x \rangle||B \cap \langle x \rangle : C \cap \langle x \rangle| = |B : C|$ also and hence $|B\langle x \rangle : C\langle x \rangle|$ infinite. This completes the proof. $\square$

**Corollary 3.4.** *Let $A_1 \geq A_2 \geq A_3 \geq \cdots$ be a descending chain of permutable subgroups of $G$ with $|A_i : A_{i+1}|$ infinite for all $i$ and let $x \in G$. Then there is a subsequence $\{i_j\}_{j \geq 1}$ such that $|A_{i_j}\langle x \rangle : A_{i_{j+1}}\langle x \rangle|$ is infinite for all $j \geq 1$.*

Elements of infinite order often create difficulties. However the following lemma, which is very easy to prove, is often useful.

**Lemma 3.5.** *Let $A_1 \gtrsim A_2 \gtrsim A_3 \gtrsim \cdots$ be a chain of subgroups of a group $G$, where $A_i = B_i \times B_{i+1} \times B_{i+2} \times \cdots$ and $B_i \leq G$ for all $i$. If $x$ is an element of $G$ of infinite order then, $\langle x \rangle \cap A_j = 1$, for some $j$.*

The next lemma is a very important observation.

**Lemma 3.6.** *Let $G$ be a group with the weak minimal condition on non-permutable subgroups and suppose that there are subgroups $X, Y$ with $Y \triangleleft X$ such that $X/Y$ is a direct product of infinitely many nontrival subgroups. Then*

(i) $X$ *is permutable in $G$.*
(ii) $X\langle x \rangle$ *is permutable in $G$, for all $x \in G$.*

**Proof.** (i) It is easy to see that $X/Y = \underset{i\geq 1}{\mathrm{Dr}}\,(C_i/Y)$, where each of the subgroups $C_i/Y$ is itself an infinite direct product of nontrivial subgroups. Clearly, $X = \prod_{i\geq 1} C_i$.

Let $D_i = C_i C_{i+1} C_{i+2} \cdots$ so that $D_1 \gneqq D_2 \gneqq D_3 \gneqq \cdots$ is a descending chain of subgroups of $G$ and $|D_i : D_{i+1}|$ infinite for all $i$, Since $G$ satisfies min-$\infty$-$\overline{\mathrm{qn}}$ there is a positive integer $k$ such that $D_k$ is permutable $G$. Also, we have the descending chain

$$C_1 C_2 \ldots C_k D_{k+2} \gneqq C_1 C_2 \ldots C_k D_{k+3} \gneqq C_1 C_2 \ldots C_k D_{k+4} \gneqq \cdots$$

and the same argument implies the existence of a natural number $l$ such that $C_1 C_2 C_3 \cdots C_k D_{k+l}$ is permutable in $G$. It follows that $X = C_1 \ldots C_k D_{k+l} D_k$ is permutable in $G$.

(ii) By Part (i), $D_k$ is permutable in $G$, for each $k \geq 1$, and we have a descending chain $D_1 \gneqq D_2 \gneqq D_3 \gneqq \cdots$ of permutable subgroups of $G$ with $|D_k : D_{k+1}|$ infinite for all $k$.

Fix $x \in G$. By Corollary 3.4 there is a subsequence $D_{k_1} \gneqq D_{k_2} \gneqq D_{k_3} \gneqq \cdots$ such that $|D_{k_l}\langle x\rangle : D_{k_{l+1}}\langle x\rangle|$ is infinite for all $l \geq 1$. Since $G$ satisfies min-$\infty$-$\overline{\mathrm{qn}}$ there exists a positive integer $m$ such that $D_{k_m}\langle x\rangle$ is permutable in $G$. Since $\prod_{r\neq k_m} D_r$ is permutable in $G$, we deduce that $X\langle x\rangle$ is permutable in $G$, as required. $\square$

In [24], Stonehewer proved the very important lemma that if $G = H\langle x\rangle$, where $H$ is permutable in $G$ and $\langle x\rangle$ is infinite then $x \in N_G(H)$ provided $H \cap \langle x\rangle = 1$. We use this in the next result.

**Lemma 3.7.** *Let $G$ be a group satisfying the weak minimal condition on non-permutable subgroups. Suppose that $G$ contains a subgroup $A$ of the form $B_1 \times B_2 \times B_3 \times \cdots$, with $B_i \neq 1$. Let $x \in G$ be any element of infinite order such that $A \cap \langle x\rangle = 1$. Then $x \in N_G(B_k)$ for all $k$.*

**Proof.** Since $A = B_1 \times B_2 \times B_3 \times \cdots$. We can write $A = B_k \times \underset{i\geq 1}{\mathrm{Dr}}\,C_i$ where $|\underset{i\geq j}{\mathrm{Dr}}\,C_i : \underset{i\geq j+1}{\mathrm{Dr}}\,C_i|$ is infinite. Then $B_k \times C_i$ per $G$, by Lemma 3.6, and so $x \in N_G(B_k \times C_i)$ for all $i$, by [21, 13.2.3]. Hence $x \in N_G(\underset{i\geq 1}{\cap}(B_k \times C_i)) = N_G(B_k)$ $\square$

We continue this section of preliminary results with the following useful lemma. We note that in general even the intersection of two permutable subgroups need not be permutable.

**Lemma 3.8.** *Let $G$ be a group satisfying the weak minimal condition on non-permutable subgroups. Let $C_1 \geq C_2 \geq C_3 \geq \cdots$ be a descending chain of subgroups of $G$ such that $\bigcap_{i \geq 1} C_i = L \lhd C_i$ and suppose that $C_i/L = \underset{j \geq i}{\mathrm{Dr}}(D_j/L)$ where $|C_i : C_{i+1}|$ is infinite. Then $\bigcap_{i \geq 1} C_i \langle x \rangle = L \langle x \rangle$ for all $x \in G$. In particular, $L$ is permutable in $G$.*

**Proof.** We note that, by Lemma 3.6, $C_i$ is permutable in $G$ and hence we can form the subgroups $C_i \langle x \rangle$, for each $x \in G$. Let $d \in \bigcap_{i \geq 1} C_i \langle x \rangle$. Then $d \in C_i \langle x \rangle$ for all $i$ and we may write

$$d = c_1 x^{i_1} = c_2 x^{i_2} = c_3 x^{i_3} = \cdots \tag{1}$$

where $c_i \in C_i$ and $i_k \in \mathbb{Z}$ for each $k$. This implies $c_{k+1}^{-1} c_k = x^{i_{k+1}} x^{-i_k} \in C_1 \cap \langle x \rangle$ for all $k$. If $C_j \cap \langle x \rangle = 1$, for some $j$, then $c_k = c_j$ for all $k \geq j$, so $c_j \in \bigcap_{k \geq j} C_k = L$ and hence $d \in L \langle x \rangle$. Therefore $L \langle x \rangle = \bigcap_{i \geq 1} C_i \langle x \rangle$ in this case.

Hence we may assume that $C_j \cap \langle x \rangle \neq 1$ for all $j$ and, in particular, there exists $l \in \mathbb{Z}$ such that $x^l \in C_1$. Then, for some natural number $k$, we have $x^l L \in \underset{i=1}{\overset{k}{\mathrm{Dr}}} D_i/L$. However there exists $m \in \mathbb{Z}$ such that $x^m \in C_{k+1}$ so that $x^m L \in \underset{i \geq k+1}{\mathrm{Dr}} D_k/L$. This implies that $x^{lm} \in L$, so there exists an integer $\mu$ such that $L \cap \langle x \rangle = \langle x^\mu \rangle$.

Now, in equation (1), we write $i_j = \mu q_j + r_j$ with $0 < r_j < \mu$ and observe that $d_j = c_j x^{q_j \mu} \in C_j$. Equation (1) then becomes

$$d = d_1 x^{r_1} = d_2 x^{r_2} = d_3 x^{r_3} = \cdots$$

Since $0 < r_i < \mu$, we can find a subset $\{l_1, l_2, l_3, \ldots\}$ of $\{1, 2, 3, \ldots\}$ with $l_1 \leq l_2 \leq l_3 \leq \cdots$ such that $r_{l_i} = r_{l_j}$, for all $i, j \geq 1$. We denote this common value by $t$. Therefore

$$d = d_{l_1} x^t = d_{l_2} x^t = d_{l_3} x^t = \cdots$$

and consequently $d_{l_1} \in \bigcap_{i \geq 1} C_{l_i} = L$ and $d = d_{l_1} x^t \in L \langle x \rangle$. Therefore, $\bigcap_{i \geq 1} C_i \langle x \rangle = L \langle x \rangle$. It follows that $L \langle x \rangle$ is a subgroup of $G$, for each $x \in G$ and hence $L$ is permutable in $G$. This completes the proof. $\qquad\square$

We conclude this section with a further elementary lemma which helps us handle elements of finite order.

**Lemma 3.9.** *Let $G$ be a group and let $x, y \in G$, where $y$ has finite order. Suppose that $\{A_k\}_{k \geq 1}$ is a collection of subgroups of $G$ such that $\bigcap_{k \geq 1} A_k = L$ and that $A_k \langle x \rangle \langle y \rangle$ is a subgroup for all $k$.*

(i) *If $x$ has finite order then $L \langle x \rangle \langle y \rangle$ is a subgroup of $G$;*

(ii) *If $x$ has infinite order and $L \triangleleft A_k$, where $A_k / L = \operatorname*{Dr}_{j \geq k} C_j / L$, for certain nontrivial subgroups $C_j$, then $L \langle x \rangle \langle y \rangle$ is a subgroup of $G$.*

**Proof.** (i) It suffices to show that $\bigcap_{k \geq 1} A_k \langle x \rangle \langle y \rangle \subseteq L \langle x \rangle \langle y \rangle$. Suppose that $d \in \bigcap_{k \geq 1} A_k \langle x \rangle \langle y \rangle$. Then $d \in A_k \langle x \rangle \langle y \rangle$ for all $k$ and we have

$$d = a_1 x^{i_1} y^{j_1} = a_2 x^{i_2} y^{j_2} = a_3 x^{i_3} y^{j_3} = \cdots \tag{2}$$

where $a_k \in A_k$ and $i_k$, $j_k$ are non-negative integers for all $k$. Since both $x$ and $y$ have finite order, there is a subset $\{k_1, k_2, k_3, \dots\}$ of $\{1, 2, 3, \dots\}$ with $k_1 \leq k_2 \leq k_3 \leq \cdots$ such that $i_{k_l} = i_{k_{l+1}}$ and $j_{k_l} = j_{k_{l+1}}$, for all $l \geq 1$. Denote these common values by $r, s$ respectively. Then, from (2), we have

$$d = a_{k_1} x^r y^s = a_{k_2} x^r y^s = a_{k_3} x^r y^s = \cdots$$

where $a_{k_i} \in A_{k_i}$. So, $a_{k_l} = a_{k_{l+1}}$, for all $l \geq 1$. From this it follows that $a_{k_1} \in \bigcap_{i \geq 1} A_{k_i} = L$ and hence $d \in L \langle x \rangle \langle y \rangle$, as required.

(ii) Suppose that $d \in \bigcap_{i \geq 1} A_i \langle x \rangle \langle y \rangle$. Then $d \in A_i \langle x \rangle \langle y \rangle$ for all $i$ and hence we may write

$$d = a_1 x^{i_1} y^{j_1} = a_2 x^{i_2} y^{j_2} = a_3 x^{i_3} y^{j_3} = \cdots$$

where $a_i \in A_i$, $i_k \in \mathbb{Z}$, $j_k$ is a positive integer. As in Part (i), we can find a subset $\{k_1, k_2, k_3, \dots\}$ of $\{1, 2, 3, \dots\}$ with $k_1 \leq k_2 \leq k_3 \leq \cdots$ such that $j_{k_l} = j_{k_{l+1}}$, for all $l \geq 1$. Denote this common value by $r$. Then it follows that $dy^{-r} \in \bigcap_{i \geq 1} A_{k_i} \langle x \rangle$. However, $\bigcap_{i \geq 1} A_{k_i} \langle x \rangle = L \langle x \rangle$ by Lemma 3.8. Therefore, $d \in L \langle x \rangle \langle y \rangle$ and hence $L \langle x \rangle \langle y \rangle$ is a subgroup. $\square$

## 4. Main Results

Using some of the preliminary results obtained in Section 3, we prove the following theorem.

**Theorem 4.1.** *Let $G$ be a group satisfying the weak minimal condition on non-permutable subgroups. Suppose that $G$ contains a subgroup $B$ of* the form $B = B_1 \times B_2 \times B_3 \times \cdots$ where each $B_i \neq 1$. Then $G$ is quasihamiltonian.

**Proof.** We may assume that each $B_i$ is itself an infinite direct product, so we may suppose that $B_i$ is permutable in $G$, by Lemma 3.6. We also let $A_i = \operatorname*{Dr}_{j \geq i} B_j$. Then $\bigcap_{i \geq 1} A_i = 1$ and we have a descending chain

$$A_1 \gneqq A_2 \gneqq A_3 \gneqq A_4 \gneqq \cdots$$

of permutable subgroups with $|A_i : A_{i+1}|$ infinite for all $i$.

Fix $x, y \in G$. Then, again by Lemma 3.6, $A_k \langle x \rangle$, $A_k \langle y \rangle$ are permutable in $G$ for all $k \geq 1$. We show that $\langle x \rangle \langle y \rangle = \bigcap_{k \geq 1} A_k \langle x \rangle \langle y \rangle$ and, by Lemma 3.9, the only case remaining to be considered is that when $x, y$ are both elements of infinite order. In this case we may assume, by Lemma 3.5, that $A_k \cap \langle x \rangle = 1 = A_k \cap \langle y \rangle = 1$, for all $k$, and we have $x, y \in N_G(B_k)$, for all $k$, by Lemma 3.7. Therefore, $A_k \lhd A_k \langle x \rangle \langle y \rangle$ for all $k$.

If $A_k \langle x \rangle \cap \langle y \rangle = 1$, for some $k$, then we may assume that $A_k \langle x \rangle \cap \langle y \rangle = 1$, for all $k$. Since $A_k \langle x \rangle$ is permutable then $A_k \langle x \rangle \lhd A_k \langle x \rangle \langle y \rangle$ for all $k$, by [21, 13.2.3],. Therefore, $\bigcap_{k \geq 1} A_k \langle x \rangle \lhd \bigcap_{k \geq 1} A_k \langle x \rangle \langle y \rangle$. However, by Lemma 3.8, $\bigcap_{k \geq 1} A_k \langle x \rangle = \langle x \rangle$. It follows that $\langle x \rangle \lhd \bigcap_{k \geq 1} A_k \langle x \rangle \langle y \rangle$ and hence $\langle x \rangle \langle y \rangle = \langle y \rangle \langle x \rangle$ in this case.

Hence we may suppose that $A_k \langle x \rangle \cap \langle y \rangle \neq 1$, for all $k$. If $\langle x \rangle \cap \langle y \rangle \neq 1$, then $x^r = y^s$, for some $r, s \in \mathbb{Z}$. If $d \in \bigcap_{k \geq 1} A_k \langle x \rangle \langle y \rangle$, we have

$$d = a_1 x^{i_1} y^{j_1} = a_2 x^{i_2} y^{j_2} = a_3 x^{i_3} y^{j_3} = \cdots, \tag{3}$$

for certain $i_k, j_k \in \mathbb{Z}, a_k \in A_k$, Write $j_n = s q_n + r_n$, where $0 \leq r_n < s$. Then Equation (3) becomes

$$d = a_1 x^{l_1} y^{r_1} = a_2 x^{l_2} y^{r_2} = a_3 x^{l_3} y^{r_3} = \cdots, \tag{4}$$

where $r_n < s$ and $l_n = i_n + r q_n$. Since, $0 \leq r_i < s$ there is a subset $\{k_1, k_2, k_3, \ldots\}$ of $\{1, 2, 3, \ldots\}$ with $k_1 \leq k_2 \leq k_3 \leq \cdots$ such that $r_{k_i} = r_{k_j}$, for $i, j \in \mathbb{N}$. Let $r$ denote this common value. We obtain

$$dy^{-r} = a_{k_1} x^{l_{k_1}} = a_{k_2} x^{l_{k_2}} = a_{k_3} x^{l_{k_3}} = \cdots$$

which implies, in particular, that $a_{k_m}^{-1} a_{k_1} = x^{l_{k_m}} x^{-l_{k_1}} \in \langle x \rangle \cap A_{k_1} = 1$, $m \geq 1$. Hence $a_{k_m} = a_{k_1}$ and $x^{l_{k_1}} = x^{l_{k_m}}$, for $m \geq 1$. Thus $a_{k_1} \in \bigcap_{i \geq 1} A_{k_i} = 1$ and hence $d = x^{l_{k_1}} y^r \in \langle x \rangle \langle y \rangle$. Consequently, $\langle x \rangle \langle y \rangle = \bigcap_{k \geq 1} A_k \langle x \rangle \langle y \rangle$ is a subgroup of $G$ in this case.

Finally we suppose that $A_k \langle x \rangle \cap \langle y \rangle \neq 1$ for all $k$ and $\langle x \rangle \cap \langle y \rangle = 1$. Again suppose that $d \in \bigcap_{k \geq 1} A_k \langle x \rangle \langle y \rangle$. Then we have

$$d = x^{i_1} a_1 y^{j_1} = x^{i_2} a_2 y^{j_2} = x^{i_3} a_3 y^{j_3} = \cdots, \tag{5}$$

where $a_k \in A_k, i_k, j_k \in \mathbb{Z}$. Suppose that $a_1 = (b_1, b_2, \ldots b_s, 1, 1, \ldots)$ and $a_{s+1} = (1, 1, \ldots, 1, b_{s+1}, b_{s+2}, \ldots, b_t, 1, \ldots)$, for some $s$ and $t$. Then, $x^{i_1}(b_1, b_2, \ldots, b_s, 1, \ldots) y^{j_1} = x^{i_{s+1}}(1, \ldots, 1, b_{s+1}, b_{s+2}, \ldots, b_t, 1, \ldots) y^{j_{s+1}}$. Hence

$$x^{(i_1 - i_{s+1})}(b_1, b_2, \ldots b_s, 1, \ldots)$$
$$= (1, \ldots, 1, b_{s+1}, b_{s+2}, \ldots, b_t, 1, \ldots) y^{(j_{s+1} - j_1)}.$$

Using Lemma 3.7, we deduce that

$$x^{(i_1 - i_{s+1})} = (b'_1, b'_2, \ldots b'_s, b_{s+1}, \ldots b_t, 1, \ldots) y^{(j_{s+1} - j_1)}, \tag{6}$$

for certain $b'_k \in B_k$. Similarly, repeating the above procedure for large enough $k$, we have

$$x^{(i_k - i_{k+1})} = (1, \ldots, 1, b'_{t+1}, b'_{t+2} \cdots, b'_u, b_{u+1}, \ldots, b_v, 1, \ldots) y^{j_{k+1} - j_k} \tag{7}$$

Combining Equations (6) and (7) and, using Lemma 3.7 we may write,

$$x^{(i_1 - i_{s+1})(i_k - i_{k+1})} = (c_1, c_2, \ldots c_t, 1, 1, \ldots) y^{(j_2 - j_1)(i_k - i_{k+1})}, \tag{8}$$

for certain $c_k \in B_k$. Similarly,

$$x^{(i_k - i_{k+1})(i_1 - i_{s+1})} = (1, \ldots, 1, c_{t+1}, \ldots c_v, 1, \ldots) y^{(j_{k+1} - j_k)(i_1 - i_{s+1})}. \tag{9}$$

Thus from Equations (8) and (9), we deduce

$$(c_1, c_2, \ldots, c_t, c_{t+1}^{-1}, \ldots, c_v^{-1}, 1, \ldots)$$
$$= y^{(j_{k+1} - j_k)(i_1 - i_{s+1}) - (j_{s+1} - j_1)(i_k - i_{k+1})} \tag{10}$$

Since $A_1 \cap \langle y \rangle = 1$ it follows that $c_k = 1$, for all $k \in \mathbb{N}$ and hence

$$x^{(i_1 - i_{s+1})(i_k - i_{k+1})} = y^{(j_{s+1} - j_1)(i_k - i_{k+1})} \in \langle x \rangle \cap \langle y \rangle = 1.$$

Therefore $(i_1 - i_{s+1})(i_k - i_{k+1}) = (j_{s+1} - j_1)(i_k - i_{k+1}) = 0$.

If $i_1 = i_{s+1}$, then from Equation (5) it follows that $a_{s+1}^{-1} a_1 = y^{j_{s+1} - j_1} \in A_1 \cap \langle y \rangle = 1$ and we have $a_1 = a_{s+1}$, contradicting the choice of $a_1$ and $a_{s+1}$.

If $i_1 \neq i_{s+1}$, then for all large $k$, $i_k = i_{k+1}$, which we denote by $i$. Equation (7) gives $x^{-i}d \in \cap_{j \geq k}A_j\langle y \rangle = \langle y \rangle$, by Lemma 3.8. Then $d \in \langle x \rangle \langle y \rangle$ and hence $G$ is quasihamiltonian. $\qquad \square$

Our first main result concerning the structure of a locally finite group satisfying the weak minimal condition on non-permutable subgroups follows, its proof being modeled on that of Theorem 3.2.

**Theorem 4.2.** *Let $G$ be a locally finite group satisfying the weak minimal condition on non-permutable subgroups. Then either $G$ is quasihamiltonian or $G$ is a Chernikov group.*

**Proof.** Suppose that $G$ is not Chernikov. By a theorem of Shunkov [9, Theorem 5.8], $G$ does not satisfy the minimal condition on abelian subgroups. Hence $G$ contains a subgroup of the form $A = \underset{i \geq 1}{\mathrm{Dr}} \langle a_i \rangle$, where each $a_i$ has prime power order and the result now follows, using Theorem 4.1. $\qquad \square$

Our next result generalizes Theorem 4.1.

**Proposition 4.3.** *Let $G$ be a generalized radical group with the weak minimal condition on non-permutable subgroups. Suppose that $G$ contains an abelian subgroup $A$ that has a subgroup $K$ such that $A/K$ is periodic and $\pi(A/K)$ is infinite. Then $K\langle x \rangle \langle y \rangle$ is a subgroup for all $x, y \in G$.*

**Proof.** As usual we can write $A/K = \underset{j \geq 1}{\mathrm{Dr}}\, C_j/K$ where each $C_j/K$ is an infinite direct product of non-trivial groups, for all $j \geq 1$. By Lemma 3.6, each $C_j$ is permutable in $G$. We let $B_m/K = \underset{j \geq m}{\mathrm{Dr}}\, C_j/K$ and note that

$$B_1 \gneqq B_2 \gneqq B_3 \gneqq B_4 \gneqq \cdots$$

is a descending chain of permutable subgroups with $|B_i : B_{i+1}|$ infinite for all $i$. Furthermore $B_k\langle x \rangle$ is a permutable subgroup of $G$, for all $k \geq 1$.

If $x, y \in G$ and at least one of $x, y$ has finite order then we can use Lemma 3.9 to deduce that $K\langle x \rangle \langle y \rangle = \underset{i \geq 1}{\cap} B_i\langle x \rangle \langle y \rangle$. Hence $K\langle x \rangle \langle y \rangle$ is a subgroup of $G$ in these cases so we may assume from now on that $x$ and $y$ are both elements of infinite order.

If $B_i\langle x \rangle \cap \langle y \rangle = 1$ for all $i \geq 1$, then by [21, 13.2.3], $B_i\langle x \rangle \lhd B_i\langle x \rangle \langle y \rangle$ for all $i$. Therefore, $\underset{i \geq 1}{\cap} B_i\langle x \rangle \lhd \underset{i \geq 1}{\cap} B_i\langle x \rangle \langle y \rangle$. Since $\underset{i \geq 1}{\cap} B_i\langle x \rangle = K\langle x \rangle$, by

Lemma 3.8, we have, $K\langle x\rangle \triangleleft \bigcap_{i\geq1} B_i\langle x\rangle\langle y\rangle$, so $K\langle x\rangle\langle y\rangle$ is a subgroup in this case.

Consequently we may suppose that $B_i\langle x\rangle\cap\langle y\rangle \neq 1$ for all $i$ and to treat this case first we suppose that $\langle x\rangle \cap \langle y\rangle \neq 1$.

In this case we have $x^r = y^s$ for some $r, s \in \mathbb{Z}$. If $d \in \bigcap_{i\geq1} B_i\langle x\rangle\langle y\rangle$, we have

$$d = b_1 x^{i_1} y^{j_1} = b_2 x^{i_2} y^{j_2} = b_3 x^{i_3} y^{j_3} = \ldots,$$

where $b_i \in B_i, i_k, j_k \in \mathbb{Z}$. We write $j_k = q_k s + r_k$, $0 \leq r_k < s$ and $l_k = i_k + r q_k$ to reduce this, using $x^r = y^s$, to

$$d = b_1 x^{l_1} y^{r_1} = b_2 x^{l_2} y^{r_2} = b_3 x^{l_3} y^{r_3} = \ldots. \tag{11}$$

Since $r_k < s$, we can find a subset $\{k_1, k_2, k_3, \ldots\}$ of $\{1, 2, 3, \ldots\}$ such that $r_{k_i} = r_{k_j}$, for all $i, j \in \mathbb{N}$. We denote this common value by $r$. Then Equation (11) reduces to

$$d = b_{k_1} x^{l_{k_1}} y^r = b_{k_2} x^{l_{k_2}} y^r = b_{k_3} x^{l_{k_3}} y^r$$

and, consequently $dy^{-r} \in \bigcap_{i\geq1} B_{k_i}\langle x\rangle$. Now Lemma 3.8 implies that $dy^{-r} \in K\langle x\rangle$ and therefore $d \in K\langle x\rangle\langle y\rangle$. Hence $\bigcap_{i\geq1} B_i\langle x\rangle\langle y\rangle = K\langle x\rangle\langle y\rangle$ is a subgroup in this case.

Hence we may suppose that $B_i\langle x\rangle\cap\langle y\rangle \neq 1$, for all $i$ and $\langle x\rangle\cap\langle y\rangle = 1$.

Suppose next that $B_i\cap\langle y\rangle \neq 1$, $B_1\cap\langle x\rangle = 1$ and that $d \in \bigcap_{i\geq1} B_i\langle x\rangle\langle y\rangle$. As usual, write

$$d = x^{i_1} b_1 y^{j_1} = x^{i_2} b_2 y^{j_2} = x^{i_3} b_3 y^{j_3} = \ldots \tag{12}$$

Since $B_1 \cap \langle y\rangle \neq 1$, we have $y^k \in B_1$ for some $k$ and hence $y^k K \in B_1/K$. Therefore $y^k K = (b_1 K, b_2 K, b_3 K, \ldots, b_r K, K, K, \ldots)$, where $b_i \in B_i$. If $\langle y\rangle \cap B_{r+1} \neq 1$, then for some $l$, $y^l \in B_{r+1}$ and hence $y^l K \in B_{r+1}/K$. Thus, $y^l K = (K, K, K, \ldots K, b_{r+1}K, b_{r+2}K, \ldots)$. Consequently,

$$y^{kl} K = (b_1^l K, b_2^l K, \ldots b_r^l K, K, K, \ldots)$$

and

$$y^{lk} K = (K, K \cdots K, b_{r+1}^k K, b_{r+2}^k K, \ldots).$$

It follows that $y^{kl} \in K = \bigcap_{i\geq1} B_i$, and hence $K \cap \langle y\rangle = \langle y^\mu\rangle$ for some $\mu \in \mathbb{Z}$. Thus $y^\mu \in B_i$ for all $i$. We write $j_k = q_k\mu + r_k$ where $0 < r_k < \mu$, for some

$q_k \in \mathbb{Z}$ and $c_k = b_k y^{\mu q_k} \in B_k$. Then, from Equation (12), we have,

$$d = x^{i_1} c_1 y^{r_1} = x^{i_2} c_2 y^{r_2} = x^{i_3} c_3 y^{r_3} = \dots.$$

Since, $0 < r_k < \mu$, we can find a subset $\{l_1, l_2, l_3, \dots\}$ of $\{1, 2, 3, \dots\}$ with $l_1 \le l_2 \le l_3 \le \cdots$ such that $r_{l_i} = r_{l_j}$, for all $i, j \in \mathbb{N}$. Denote this common value by $r$. Then

$$d = x^{i_{l_1}} c_{l_1} y^r = x^{i_{l_2}} c_{l_2} y^r = x^{i_{l_3}} c_{l_3} y^r = \dots$$

which implies that $dy^{-r} \in \bigcap_{i \ge 1} B_i \langle x \rangle$. By Lemma 3.8, $dy^{-r} \in K \langle x \rangle$ so $d \in K \langle x \rangle \langle y \rangle$ and $K \langle x \rangle \langle y \rangle$ is a subgroup of $G$ in this case.

Finally we suppose also that $B_1 \cap \langle x \rangle = 1 = B_1 \cap \langle y \rangle$ so, by [21, 13.2.3], $x, y \in N_G(B_i)$ for all $i$. Hence $x, y \in N_G(K)$ and we can form the groups $B_i \langle x \rangle \langle y \rangle / K$. By Theorem 4.1, $\bigcap_{i \ge 1} B_i \langle x \rangle \langle y \rangle / K = \langle Kx \rangle \langle Ky \rangle$. Therefore $K \langle x \rangle \langle y \rangle / K$ is a subgroup of $B_i \langle x \rangle \langle y \rangle / K$ and hence $K \langle x \rangle \langle y \rangle$ is a subgroup of $G$. □

Until now we have assumed that our groups are generalized radical, but our next result shows that they are radical-by-finite.

**Proposition 4.4.** *Let $G$ be a generalized radical group satisfying the weak minimal condition on non-permutable subgroups. Then $G$ is radical-by-finite.*

**Proof.** Let $R$ be the maximal normal radical subgroup of $G$. If $R \ne G$, then there exists $N \lhd G$ such that $R \le N$ and $N/R$ is either locally finite or locally nilpotent. The choice of $R$ implies that $N/R$ is locally finite. Let $L = G/R$ and suppose that $K$ is the maximal normal locally finite subgroup of $L$. Since $K$ satisfies the weak minimal condition on non-permutable subgroups Theorem 4.2 implies that $K$ is either quasihamiltonian or Chernikov. If $K$ is quasihamiltonian then, $K$ is locally nilpotent by [8] and therefore $K$ is trivial by the choice of $R$. Thus $K$ is Chernikov. Since a radical-by-abelian group is radical it also follows that $K$ must be finite.

If $L$ is infinite then $K \ne L$. Since $L/K$ is a generalized radical group it follows from our work above that the locally finite radical of $L/K$ is trivial. Hence there is a normal locally nilpotent subgroup $M/K$ of $L/K$. Furthermore $M/K$ must be torsion-free and of course $M/K$ is infinite. Moreover $L/C_L(K)$ is finite since $K$ is finite and $K \cap C_M(K) = \zeta(K) = 1$, by the choice of $R$. The isomorphisms

$$C_M(K) \simeq C_M(K)/C_M(K) \cap K \simeq C_M(K)K/K \le M/K$$

imply that $C_M(K)$ is locally nilpotent. Also $L/C_L(K)$ is finite so that $MC_L(K)/C_L(K) \simeq M/M \cap C_L(K)$ is finite and hence $M \cap C_L(K) = C_M(K)$ is infinite. This contradicts the choice of $R$. Hence the result follows. $\qquad \square$

It is now time to obtain our main theorem.

**Theorem 4.5.** *Let $G$ be a generalized radical group satisfying the weak minimal condition on non-permutable subgroups. Then either*

(i) *$G$ is quasihamiltonian, or*
(ii) *$G$ is soluble-by-finite of finite rank.*

**Proof.** By Proposition 4.4, $G$ is radical-by-finite. Let $N$ be a normal radical subgroup of $G$ such that $G/N$ is finite. Consider the abelian subgroups of $G$. If $G$ contains an abelian subgroup $A$ of infinite rank, then $A$ contains a subgroup of the form $A_1 \times A_2 \times A_3 \times \cdots$, with $A_i \neq 1$. It follows, by Theorem 4.1, that $G$ is quasihamiltonian.

If all the abelian subgroups of $G$ have finite rank then, by [2], $G$ also has finite rank. We prove that $G$ is soluble-by-finite in this case. Since quasihamiltonian groups are metabelian, we may assume that $G$ is non-quasihamiltonian. Since $G$ has finite rank then, by [7, Theorem A], there exist normal subgroups $1 \leq T \leq L \leq M \leq G$ such that $T$ is locally finite, $L/T$ is a torsion-free nilpotent group, $M/L$ is a finitely generated torsion-free abelian group and $G/M$ is finite. Clearly $M/T$ is soluble.

Since $T$ satisfies the weak minimal condition on non-permutable subgroups Theorem 4.2 implies that $T$ is either quasihamiltonian or Chernikov. If $T$ is quasihamiltonian then $T$ is locally nilpotent by [8]. Hence $T = \underset{p \in \pi}{\mathrm{Dr}}\, T_p$, where $\pi$ is a set of primes and $T_p \neq 1$, for each $p \in \pi$. If $|\pi|$ is infinite then $G$ is quasihamiltonian by Theorem 4.1. Therefore $|\pi|$ is finite. Since a locally finite $p$-group of finite rank is Chernikov, each $T_p$ is Chernikov and hence so is $T$. Since $T$ is Chernikov, it is either finite or has a non-trivial normal abelian subgroup of finite index. Hence without loss of generality, we may assume that $T$ is finite. This implies that $M/C_M(T)$ is finite. The isomorphism

$$C_M(T)/\zeta(T) = C_M(T)/C_M(T) \cap T \simeq C_M(T)T/T \leq M/T$$

implies that $C_M(T)/\zeta(T)$ is soluble and hence $C_M(T)$ is soluble. Moreover, $G/M$ and $M/C_M(T)$ both are finite and $C_M(T)$ is soluble. Consequently $G$ is soluble-by-finite. This completes the proof. $\qquad \square$

**Theorem 4.6.** *Let $G$ be a generalized radical group satisfying the weak minimal condition on non-permutable subgroups. Suppose that $G$ is not minimax. Then the set of elements of finite order is a subgroup $T$ of $G$. Furthermore, if also $G$ is not quasihamiltonian then $T$ is Chernikov and $G/T$ is torsion-free.*

**Proof.** Since a quasihamiltonian group is locally nilpotent, by [8], we may assume that $G$ is not quasihamiltonian. Then $G$ is soluble-by-finite of finite rank, by Theorem 4.5. Furthermore, if all abelian subgroups of $G$ are minimax then $G$ is a soluble-by-finite minimax group, by [20]. Hence we may assume that $G$ has an abelian non-minimax subgroup $A$. Thus $A$ contains a free abelian subgroup $K$ of the form $\underbrace{\mathbb{Z} \times \mathbb{Z} \times \mathbb{Z} \times \cdots \times \mathbb{Z}}_{n}$, such that $A/K$ is periodic.

Here $\pi(A/K)$ is infinite, otherwise $A$ is minimax, and $A/K = \operatorname*{Dr}_{i \geq 1}(A_i/K)$ where $\bigcap_{i \geq 1} A_i = K$ and each $A_i/K \neq 1$. Hence, by Proposition 4.3, $K\langle x \rangle \langle y \rangle$ is a subgroup for all $x, y \in G$.

Let $K^n = \{k^n | k \in K\}$. Then $K/K^n$ is finite. Moreover, $A/K^n$ is also periodic and we may replace $K$ by $K^n$ in Proposition 4.3 to deduce that $K^n \langle x \rangle \langle y \rangle$ is also a subgroup, for each $n$ and for each $x, y \in G$. Let $H_n = K^n$. It is clear that $\bigcap_{n \geq 1} H_n = 1$. It follows, by Lemma 3.9, that if $x, y$ are two elements of finite order, then $\langle x \rangle \langle y \rangle$ is a subgroup of $G$ and hence the set, $T$, of elements of finite order is a characteristic subgroup of $G$ and $G/T$ is torsion-free.

Clearly $T$ is quasihamiltonian and hence $T$ is locally nilpotent, by [8]. Therefore, $T = \operatorname*{Dr}_{p \in \pi} T_p$, where $\pi$ is a set of primes and $T_p$ is the $p$-component of $T$. If $\pi$ is infinite, then $G$ is quasihamiltonian by Theorem 4.1. Hence $\pi$ is finite and since each $T_p$ is Chernikov, $T$ is Chernikov. This completes the proof. $\qquad\square$

We have mostly concentrated in this paper on the weak minimal conditions. We complete this list of results by stating the following, whose proof appears in [4].

**Theorem 4.7.** *Let $G$ be a generalized radical group satisfying the weak maximal condition on non-permutable subgroups. Then either $G$ is quasihamiltonian or $G$ is soluble-by-finite of finite rank.*

# References

[1] R. Baer, *Polyminimaxgruppen*, Math. Ann. **175** (1968), 1–43.

[2] R. Baer and H. Heineken, *Radical groups of finite abelian subgroup rank*, Illinois J. Math. **16** (1972), 533–580.

[3] M. R. Celentani and A. Leone, *Groups with restrictions on non-quasinormal subgroups*, Boll. Un. Mat. Ital. A (7) **11** (1997), no. 1, 139–147.

[4] L. K. Chataut, *Groups with conditions on non-permutable subgroups*, 2015, Thesis (Ph.D.)–The University of Alabama.

[5] M. R. Dixon, M. J. Evans, and H. Smith, *Groups with some minimal conditions on non-nilpotent subgroups*, J. Group Theory **4** (2001), 207–215.

[6] ———, *Locally soluble-by-finite groups with the weak minimal condition on non-nilpotent subgroups*, J. Algebra **249** (2002), 226–246.

[7] M. R. Dixon, L. A. Kurdachenko, and N. V. Polyakov, *On some ranks of infinite groups*, Ricerche di Matematica **56** (2007), 43–59.

[8] K. Iwasawa, *On the structure of infinite M-groups*, Jap. J. Math. **18** (1943), 709–728.

[9] O. H. Kegel and B. A. F. Wehrfritz, *Locally Finite Groups*, North-Holland Mathematical Library, North-Holland, Amsterdam, London, 1973, Volume 3.

[10] L. A. Kurdachenko, *Groups that satisfy weak minimality and maximality conditions for normal subgroups*, Sibirsk. Mat. Zh. **20** (1979), no. 5, 1068–1076, 1167.

[11] ———, *Groups satisfying weak minimality and maximality conditions for subnormal subgroups*, Mat. Zametki **29** (1981), no. 1, 19–30, 154.

[12] ———, *Locally nilpotent groups with weak minimality and maximality conditions for normal subgroups*, Dokl. Akad. Nauk Ukrain. SSR Ser. A (1985), no. 8, 9–12, 86.

[13] L. A. Kurdachenko and V. E. Goretskii, *Groups with weak minimality and maximality conditions for subgroups that are not normal*, Ukrain. Mat. Zh. **41** (1989), no. 12, 1705–1709, 1728.

[14] L. A. Kurdachenko and H. Smith, *Groups with the weak minimal condition for non-subnormal subgroups*, Ann. Mat. Pura Appl. **173** (1997), 299–312.

[15] ———, *Groups with the weak minimal condition for non-subnormal subgroups. II*, Comment. Math. Univ. Carolin. **46** (2005), no. 4, 601–605.

[16] W. Möhres, *Auflösbarkeit von Gruppen, deren Untergruppen alle subnormal sind*, Arch. Math. (Basel) **54** (1990), no. 3, 232–235.

[17] A. Yu Ol'shanskii, *Groups of bounded exponent with subgroups of prime order*, Algebra i Logika **21** (1982), 553–618 (Russian), English transl. in Algebra and Logic, **21** (1982), 369-418.

[18] O. Ore, *On the application of structure theory to groups*, Bull. Amer. Math. Soc. **44** (1938), no. 12, 801–806.

[19]    A. N. Ostylovskii, *A weak minimality condition for nonnilpotent subgroups*, Algebra i Logika **23** (1984), 439–444 (Russian), English transl. in Algebra and Logic, **23** (1984), 303–306.

[20]    D. J. S. Robinson, *Finiteness Conditions and Generalized Soluble Groups Vols. 1 and 2*, Ergebnisse der Mathematik und ihrer Grenzgebiete, Springer-Verlag, Berlin, Heidelberg, New York, 1972, Band 62 and 63.

[21]    ———, *A Course in the Theory of Groups*, Graduate Texts in Mathematics, vol. 80, Springer Verlag, Berlin, Heidelberg, New York, 1996.

[22]    R. Schmidt, *Subgroup Lattices of Groups*, de Gruyter Expositions in Mathematics, vol. 14, Walter de Gruyter & Co., Berlin, 1994.

[23]    V. P. Shunkov, *The problem of minimality for locally finite groups*, Algebra i Logika **9** (1970), 220–248, English Translation in Algebra and Logic 9 (1970), 137151.

[24]    S. E. Stonehewer, *Permutable subgroups of infinite groups*, Math. Z. **125** (1972), 1–16.

[25]    D. I. Zaitsev, *Groups which satisfy a weak minimality condition*, Dokl. Akad. Nauk SSSR **178** (1968), 780–782 (Russian), English transl. in Soviet Math. Dokl., **9** (1968), 194–197.

[26]    ———, *Groups which satisfy a weak minimality condition*, Ukrain. Mat. Zh. **20** (1968), 472–482 (Russian), English transl. in Ukrainian Math. J., **20** (1968), 408–416.

[27]    ———, *The groups which satisfy a weak minimality condition*, Mat. Sb. (N.S.) **78 (120)** (1969), 323–331.

[28]    ———, *Groups satisfying the weak minimal condition for non-abelian subgroups*, Ukrain. Mat. Zh. **23** (1971), 661–665 (Russian), English transl. in Ukrainian Math. J., **23** (1971), 543–546.

# Chapter 2

# A Survey: Shamir Threshold Scheme and Its Enhancements

Chi Sing Chum[1], Benjamin Fine[2], and Xiaowen Zhang[1,3]

[1] *Computer Science Dept., Graduate Center, CUNY*
*365 Fifth Ave., New York, NY 10016, U.S.A.*
*cchum@gradcenter.cuny.edu*

[2] *Mathematics Dept., Fairfield University*
*1073 North Benson Road, Fairfield, CT 06824, U.S.A.*
*fine@fairfield.edu*

[3] *Computer Science Dept., College of Staten Island, CUNY*
*2800 Victory Blvd, Staten Island, NY 10314, U.S.A.*
*xiaowen.zhang@csi.cuny.edu*

**ABSTRACT.** This paper serves as an introduction to secret sharing scheme, and it provides the fundamental understandings to the scheme from various aspects. We first review the basics of a Shamir threshold scheme, and discuss various enhancements so that the scheme can be proactive and verifiable. We then show how a Shamir scheme can be extended to realize any general access structure. We also point out the relationship between a Shamir scheme and other topics such as error correction code, ramp scheme, information disposal algorithm and multiparty computation. Finally, we briefly discuss other platforms for its implementation.

## 1. Introduction

A **secret sharing scheme** is a method that distributes a secret among a group of participants by giving a share of the secret to each. The secret can be recovered only if a sufficient number of participants combines their shares.

Formally we have the following. We have a secret $K$ and a group of $n$ participants. This group is called the **access control group**. A **dealer**

**Keywords**: Secret sharing, threshold scheme, access structure, Reed-Solomon code, ramp scheme, information disposal algorithm, multiparty computation.

allocates shares to each participant under given conditions. If a sufficient number of participants combine their shares, then the secret can be recovered. If $t \leq n$, then an $(t, n)$-**threshold scheme** is one with $n$ total participants and in which any $t$ participants can combine their shares and recover the secret, but not fewer than $t$. The number $t$ is called the **threshold**. It is a **secure secret sharing scheme**; if given less than the threshold, then there is no chance to recover the secret. If a measure is placed on the set of secrets, and on the set of shares, security can be made precise by saying that when given less than the threshold, all secrets are equally likely, but when given the threshold, there is a unique secret. Secret sharing is an old idea but was formalized mathematically in independent papers in 1979 by Adi Shamir [26] and George Blakley [2].

Shamir [26] proposed a beautiful $(t, n)$ threshold scheme, based on polynomial interpolation, that has many desirable properties. We describe this in Section 3. It is now a standard method for solving the $(t, n)$ secret sharing problem, although there are modifications for different situations that we will discuss in this paper. Blakley [2] in his original paper proposed a geometric solution based on hyperplanes that is less space efficient for computer storage than Shamir's. In Blakley's scheme, the distributed shares are larger than the secret, whereas in Shamir's scheme, they are the same size.

The protection of a private key in an encryption protocol provides strong motivation for the ideas of secret sharing. Based on Kerchhoffs' principle [18], only the private key in an encryption scheme is the secret and not the encryption method itself. When we examine the problem of maintaining sensitive information, we will consider two issues: availability and secrecy. If only one person keeps the entire secret, then there is a risk that the person might lose the secret or the person might not be available when the secret is needed. Hence, it is often wise to allow several people to have access to the secret. On the other hand, the higher the number of people who can access the secret, the higher the chance the secret will be leaked. A secret sharing scheme is designed to solve these issues by splitting a secret into multiple shares and distributing these shares among a group of participants. The secret can only be recovered when the participants of an authorized subset join together to combine their shares.

A secret sharing scheme is a cryptographic primitive with many applications, such as in security protocols, multiparty computation (MPC), Pretty Good Privacy (PGP) key recovering, visual cryptography, threshold cryptography, threshold signature, etc.

The remainder of this paper is organized as follows. In Section 2, we give a brief review on entropy which is related to secret sharing schemes. In Section 3, we discuss the principles of share distribution and secret recovery of a Shamir threshold scheme and its properties. We further talk about different enhancements which make the original threshold scheme proactive or verifiable. In Section 4, we further show how to extend a Shamir threshold scheme to realize any general access structure. In Sections 5 to 8, we discuss the relationship between a Shamir threshold scheme and Reed-Solomon code, ramp scheme, information disposal algorithm, and multiparty computation, respectively. In Section 9, we give an alternative to the Shamir threshold scheme. In Section 10, we discuss another platform for its implementation. We conclude the paper in Section 11.

## 2. Entropy

In information theory, developed by Shannon [27, 28], entropy is a measure of information or uncertainty. Also see [4, 14, 30] for the details. Let $X$ be a random variable with possible outcomes $\mathcal{X}$ and probability distribution $p(x)$, where $p(x) \geq 0$, $\sum_{x \in \mathcal{X}} p(x) = 1$. Then, the entropy of $X$ is defined as

$$Ent(X) = - \sum_{x \in \mathcal{X}} p(x) \log_2 p(x). \tag{1}$$

In probabilistic terms this is the expected value of $- \log_2 p(x)$. We assume $p(x) \log_2 p(x) = 0$, if $p(x) = 0$. This is justified because

$$\lim_{p(x) \to 0} p(x) \log_2 p(x) = 0. \tag{2}$$

Example: Let $X$ be a random variable of the event of an unbiased fair coin flipping with the possible outcomes of $\mathcal{X} = \{\text{Head, Tail}\}$, with $p(X = \text{Head}) = p(X = \text{Tail}) = 1/2$, then:

$$Ent(X) = -p(X = \text{Head}) \log_2 p(X = \text{Head})$$
$$- p(X = \text{Tail}) \log_2 p(X = \text{Tail}) = \frac{1}{2} + \frac{1}{2} = 1. \tag{3}$$

If the coin is biased with $p(X = \text{Head}) = 1$ and $p(X = \text{Tail}) = 0$, then $Ent(X) = 0$. In this case there is no uncertainty. We can use $Ent(X) = 0$ to infer that $\exists x_i \in \mathcal{X}$ such that $p(x_i) = 1$ and $p(x_j) = 0$ for $j \neq i$.   □

Let $X$ and $Y$ be two random variables. The joint entropy $H(X,Y)$ is defined as:

$$H(X,Y) = -\sum_{x \in \mathcal{X}}\sum_{y \in \mathcal{Y}} p(x,y) \log_2 p(x,y). \qquad (4)$$

Again, as in the case of a single random variable, this is the expected value of $-\log_2(p(x,y))$

The conditional entropy $H(X|Y)$ is defined as:

$$H(X|Y) = \sum_{y \in \mathcal{Y}} p(y) H(X|Y = y)$$

$$= -\sum_{y \in \mathcal{Y}} p(y) \left( \sum_{x \in \mathcal{X}} p(x|y) \log_2 p(x|y) \right)$$

$$= -\sum_{y \in \mathcal{Y}}\sum_{x \in \mathcal{X}} p(y)p(x|y) \log_2 p(x|y). \qquad (5)$$

However, if $X$ and $Y$ are independent, then

$$H(X|Y) = -\sum_{y \in \mathcal{Y}} p(y) \left( \sum_{x \in \mathcal{X}} p(x|y) \log_2 p(x|y) \right)$$

$$= \sum_{y \in \mathcal{Y}} p(y) \left( -\sum_{x \in \mathcal{X}} p(x) \log_2 p(x) \right)$$

$$= 1 \cdot H(X) = H(X). \qquad (6)$$

## 3. Shamir $(t, n)$ Threshold Scheme

Given a secret $K$ in general, a $(t, n)$ secret sharing threshold scheme is a cryptographic primitive in which a secret is split into pieces (shares) and distributed among $n$ participants $p_1, p_2, \ldots, p_n$ so that any group of $t$ or more participants, with $(t \le n)$, can recover the secret. Meanwhile, any group of $t - 1$ or fewer participants cannot recover the secret. By sharing a secret in this way, the availability and reliability issues can be solved. Distributing share and recovering secrets [3, 14, 30] will be discussed as follows.

The general idea of a Shamir $(t, n)$ threshold scheme is the following. Let $F$ be any field and $(x_1, y_1), \ldots, (x_n, y_n)$ be $n$ points in $F^2$ with distinct $x_i$. We say that a polynomial $P(x)$ of degree less than or equal to $n - 1$ over $F$ **interpolates** these points if $P(x_i) = y_i$ for $i = 1, \ldots, n$. The relevant theoretical result that we need is the following. We can see Atkinson [1] for a reference and for a proof.

**Theorem 3.1.** *Let $F$ be any field and $x_1, \ldots, x_n$ be $n$ distinct elements of $F$ and $y_1, \ldots, y_n$ any elements of $F$. Then there exists a **unique** polynomial of degree $\leq n - 1$ that interpolates the $n$ points $(x_i, y_i), i = 1, \ldots, n$.*

Using this theorem, a Shamir $(t, n)$ threshold scheme is roughly this. We choose a field $F$. The secret is $K \in F$ and we choose a polynomial $P(x)$ of degree at most $t - 1$ with $K$ as its constant term. We choose distinct $x_1, \ldots, x_n$ with no $x_i = 0$ and distribute to each of the $n$ participants a point $(x_i, P(x_i)), i = 1, \ldots, n$. By the theorem above, any $t$ people can determine the interpolating polynomial $P(x)$ and hence recover the secret $K$. Given an infinite field and people fewer than $t$ there are infinite polynomials of degree $t$ that can interpolate the given points. Thus finding the correct polynomial has a probability of zero.

We now present a more explicit version of the Shamir scheme using the finite field $\mathbb{Z}_q$ where $q$ is a large prime. By using a finite field, Shamir was able to place a finite measure on the set of plaintexts and ciphertexts and showed that with this scheme, if there are fewer than $t$ people, all secrets are equally likely.

**Distributing share:** Let $K$ be the secret. The dealer generates a polynomial $P(x)$ of degree at most $t-1$ over $\mathbb{Z}_q$, where $q$ is a prime number $> n$ as follows:

$$P(x) = a_0 + a_1 x + \ldots + a_{t-1} x^{t-1} \pmod{q} \tag{7}$$

where $a_0 = K$ is the secret, $a_1, \ldots, a_{t-1} \in \mathbb{Z}_q$ and are generated randomly.

The dealer arbitrarily chooses different $x_i \in \mathbb{Z}_q - \{0\}$, $i = 1, 2, \ldots, n$. Usually, $x_i = i$ will be chosen for simplicity. The values $x_1, x_2, \ldots, x_n$ are stored in a public area. The dealer calculates $y_i = P(x_i) \pmod{q}, i = 1, 2, \ldots, n$, and distributes to the $n$ participants via a secure channel so that each participant $p_i$ gets one share $y_i$. For the rest of the paper, we will not repeat the criteria of the generation of the coefficient $a_i$ of the polynomial $P(x)$ and the calculation of the shares $P(x_i)$.

**Recovering secret (i):** When any $t$ participants join together, we have the following system of $t$ equations. For simplicity, we assume $p_1, p_2, \ldots, p_t$

join together.

$$y_1 = P(x_1) = a_0 + a_1 x_1 + \ldots + a_{t-1} x_1^{t-1} \pmod{q},$$
$$y_2 = P(x_2) = a_0 + a_1 x_2 + \ldots + a_{t-1} x_2^{t-1} \pmod{q},$$

$$\ldots ,$$

$$y_t = P(x_t) = a_0 + a_1 x_t + \ldots + a_{t-1} x_t^{t-1} \pmod{q}. \tag{8}$$

In matrix representation, it will be:

$$\begin{bmatrix} 1 & x_1 & \cdots & x_1^{t-1} \\ 1 & x_2 & \cdots & x_2^{t-1} \\ \vdots & \vdots & \cdots & \vdots \\ 1 & x_t & \cdots & x_t^{t-1} \end{bmatrix} \begin{bmatrix} a_0 \\ a_1 \\ \vdots \\ a_{t-1} \end{bmatrix} = \begin{bmatrix} y_1 \\ y_2 \\ \vdots \\ y_t \end{bmatrix} \pmod{q}. \tag{9}$$

Let $M$ be the above $t \times t$ Vandermonde matrix. Its determinant is

$$\det(M) = \prod_{1 \leq j < k \leq t} (x_k - x_j) \pmod{q}. \tag{10}$$

Since we choose different points for the participants, i.e., different $x_i$'s, $\det(M) \neq 0$, and this guarantees a unique solution. We can solve the system of equations by Gaussian elimination or Crammer's rule. Hence, the secret can be recovered.

**Recovering secret (ii):** Another method is to use Lagrange interpolation. We can construct the polynomial of degree at most $t - 1$ by any $t$ different points $(x_1, y_1), \ldots, (x_t, y_t)$ as

$$P(x) = \sum_{i=1}^{t} y_i l_i(x), \text{ where } l_i(x) = \prod_{j=1, j \neq i}^{t} \frac{x - x_j}{x_i - x_j} \pmod{q}. \tag{11}$$

So, the secret $a_0$ will be

$$a_0 = P(0) = \sum_{i=1}^{t} y_i \prod_{j=1, j \neq i}^{t} \frac{-x_j}{x_i - x_j} \pmod{q}. \tag{12}$$

### 3.1. *Access structure*

In a $(t, n)$ threshold scheme, any group of $t$ or more participants forms an authorized subset, since we assume it has the monotone property. A group of participants, which can recover the secret when they join together, is called an **authorized subset**. On the other hand, any group of participants that cannot recover the secret is called an unauthorized subset. An **access structure** $\mathcal{A}$ is a set of all authorized subsets.

Given any access structure $\mathcal{A}$, $A \in \mathcal{A}$ is called a minimal authorized subset if $A' \subsetneq A$ then $A' \notin \mathcal{A}$.

We use $\mathcal{A}_0$ to denote the set of the minimal authorized subsets of $\mathcal{A}$. In a $(t, n)$ threshold scheme, let $P$ be the set of the participants:

$$\mathcal{A} = \{A | A \subseteq P \text{ and } |A| \geq t\}, \tag{13}$$

$$\mathcal{A}_0 = \{A | A \subseteq P \text{ and } |A| = t\}. \tag{14}$$

In secret sharing, we first define the access structure. Then, we realize the access structure by a secret sharing scheme.

### 3.2. *Perfect and ideal scheme*

A Shamir $(t, n)$ threshold scheme allows no partial information to be given out even up to $t - 1$ participants joined together [9, 29]. In other words, any group of up to $t - 1$ participants cannot get more information about the secret than any outsider. A secret sharing scheme with this property is called a **perfect scheme**.

In terms of entropy in information theory, we have

$$H(S|A) = 0, \text{ if } A \in \mathcal{A} \text{ (correctness)}, \tag{15}$$

$$H(S|A) = H(S), \text{ if } A \notin \mathcal{A} \text{ (privacy)}. \tag{16}$$

Equation (15) says that for an authorized subset $A$ the entropy is equal to zero (i.e., no uncertainty) and the secret $S$ can be determined/recovered. Equation (16) says that for an unauthorized subset $A$, the entropy remains unchanged and no information about the secret $S$ is leaked out even if the participants pool all their shares together.

Based on the information theory, the length of any share must at least be as long as the secret itself in order to have perfect secrecy. The argument is that up to $t-1$ participants have zero information about the secret under perfect sharing scheme, but when one extra participant joins the group, the secret can be recovered. That means any participant has his share at least as long as the secret.

Following [30], the information rate for participant $p_i, i = 1, \ldots, n$, is defined as

$$\rho_i = \frac{\log_2 |\mathbb{K}|}{\log_2 |S_i|}, \tag{17}$$

where $\mathbb{K}$ is the key space, $S_i \subseteq S$ is the set of shares that $p_i$ has. The information rate of the scheme is defined as

$$\rho = \min \{\rho_i : 1 \le i \le n\}. \tag{18}$$

For a perfect scheme, the information rate will be less than or equal to 1. If the shares and the secret come from the same domain, we call it an **ideal scheme**. In this case, the shares and the secret have the same size, i.e., the information rate is equal to 1.

### 3.3. *Proactive scheme*

In a secret sharing scheme, we need to consider the possibility that a smart adversary may find out all the shares in an authorized set to discover the secret eventually if given a very long time to gather the necessary information. This means that if the adversary can successfully break into $t$ servers in a $(t, n)$ threshold scheme, he can steal the secret. In order to prevent this from happening, we may try to reset the shares. We re-fresh and re-distribute all the shares to all the participants periodically. After finishing this phase, the old shares are erased safely and the secret remains unchanged. By doing this, an adversary has to get enough information about the shares within any two periodic resets in order to break into the system. This would make it more difficult to achieve.

Based on the Shamir scheme, Herzberg, Jarecki, Krawczyk, and Yung [13] derived a proactive scheme, which uses the following method to reset the shares.

Let $P(x)$ be an arbitrary polynomial of degree at most $t - 1$ over $\mathbb{Z}_q$, same as in the Shamir scheme,

$$P(x) = a_0 + a_1 x + \ldots + a_{t-1} x^{t-1} \pmod{q}, \tag{19}$$

where $q$ is a prime number, $a_0$(secret) $a_1, \ldots, a_{t-1} \in \mathbb{Z}_q$. For simplicity, let $P(1), \ldots, P(n)$ be the shares of the participants $p_1, \ldots, p_n$. The dealer generates another polynomial $Q(x)$ of degree at most $t - 1$ over $\mathbb{Z}_q$ without a constant term,

$$Q(x) = b_1 x + \ldots + b_{t-1} x^{t-1} \pmod{q}, \tag{20}$$

where $b_1, \ldots, b_{t-1} \in \mathbb{Z}_q$. The dealer sends out $Q(1), \ldots, Q(n)$ to the participants $p_1, \ldots, p_n$, respectively. Each participant $p_i$ will update/renew his share as $S(i) = P(i) + Q(i)$ and destroy his old share $P(i)$ safely. Here

$$S(x) = P(i) + Q(i) = a_0 + c_1 x + \ldots + c_{t-1} x^{t-1} \pmod{q}, \tag{21}$$

where $c_i = a_i + b_i \pmod{q}$ for $i = 1, \ldots, t-1$. The scheme remains a $(t, n)$ threshold scheme with the same original secret $a_0$.

The above technique can be extended so that each participant $p_i$, by turn, generates a polynomial $P_i(x)$ of degree at most $t - 1$ without a constant term and sends values of $P_i(1), \ldots, P_i(i-1), P_i(i+1), \ldots, P_i(n)$ to participants $p_1, \ldots, p_{i-1}, p_{i+1}, \ldots, p_n$, respectively. That means participant $p_j$ will get $P_i(j)$ from participant $p_i$. After the above exchange process, each participant $p_i$ resets his new shares as follows:

$$\text{newshare} = \text{oldshare} + P_1(i) + \ldots + P_n(i). \tag{22}$$

After the calculation of the new shares, all participants will destroy their old shares safely. In other words, all the participants can engage in the share renewing process. This method can eliminate all the work done by the dealer and be more secure.

### 3.4. *Verifiable scheme*

Shamir's original sharing scheme assumes the dealer and all the participants are honest. However, in reality, we need to consider the situation that the dealer or some of the participants might be malicious. In this case, we need to set up a verifiable scheme so that the shares of the participants can be verified. In order to make this possible, additional information is required for the participants to verify their shares' consistency.

Feldman [8] presented a simple verifiable scheme that is based on the Shamir scheme. It is based on the homomorphic properties of the exponentiation function $x^{a+b} = x^a \cdot x^b$.

The idea is to find a cyclic group $G$ of order $q$, where $q$ is a prime. Since it is cyclic a generator of $G$, say $g$, exists. As other cryptographic protocols, we assume the parameters of $G$ are carefully chosen so that the discrete logarithm problem is hard to solve in $G$.

Let $p, q$ be primes such that $q|p-1$, $g \in Z_p^*$ of order $q$. A polynomial over $Z_q$ of degree at most $t-1$ as a Shamir $(t, n)$ threshold scheme is generated as

$$P(x) = a_0 + a_1 x + \ldots + a_{t-1}x^{t-1} \pmod{q}, \tag{23}$$

where $a_0, a_1, \ldots, a_{t-1} \in Z_q$.

The dealer sends out $P(i)$ to participant $i$ as before. In addition, he broadcasts in a public channel the commitments $g^{a_0} \pmod{p}$, $g^{a_1} \pmod{p}, \ldots, g^{a_{t-1}} \pmod{p}$ for the participants to verify.

Each participant $P_i$ can verify if the following equation is true.

$$g^{P(i)} = (g^{a_0})(g^{a_1})^i(g^{a_2})^{i^2} \ldots (g^{a_{t-1}})^{i^{t-1}} \pmod{p}, i = 1, \ldots, n. \qquad (24)$$

Based on the homomorphic properties of the exponentiation, the above condition will hold true if the dealer sends out consistent information. If this is the case, we conclude that the dealer is honest, and the scheme is verifiable. Later, when the participants return their shares for secret recovering, the dealer can verify their shares' validity by the same method.

Feldman's scheme is not a perfect scheme since partial information, $g^{a_0} \pmod{p}$, is leaked out. However, we assume it is difficult to get the secret $a_0$ from $g^{a_0} \pmod{p}$ if the discrete logarithm problem is hard to solve under $G$.

## 3.5. *Enhancements by one-way function and RSA*

In order to make secret sharing schemes practical, researchers have proposed to apply one-way functions [20], hash functions [17, 31] and RSA [7, 12, 23] cryptosystems in Shamir threshold scheme. These enhancements add proactive-ness, verifiability and other desired features to the Shamir scheme.

### 3.5.1. *Applying one-way function in Shamir scheme*

Liu *et al.* [20] enhanced the Shamir $(t, n)$ threshold scheme by applying a one-way function. Their scheme works as follows.

**Scheme setup:** Suppose $f : \mathbb{Z}_q \to \mathbb{Z}_q$ is a collision-free one-way function, where $\mathbb{Z}_q$ is a finite field and $q > n$ is a large prime. (a) The dealer $\mathcal{D}$ randomly chooses $n$ distinct elements $s_1, \ldots, s_n$ in $\mathbb{Z}_q$ as shares for $n$ participants, sends $s_i$ to $p_i$ via a secure channel. (b) $\mathcal{D}$ randomly chooses an element $\alpha \in \mathbb{Z}_p$ and a polynomial $P(x)$ of degree $t - 1$, such that $P(0) = K$ is the secret. Dealer computes $y_i = P(f(\alpha + s_i))$, $i = 1, 2, \ldots, n$. (c) $\mathcal{D}$ publishes $f$, $\alpha$ and the sequence $(y_1, y_2, \ldots, y_n)$ in a public area (such as a bulletin board). All evaluations for $P(x)$ and $f(x)$ are reduced by $\bmod q$.

**Secret recovery:** Any $t$ participants, say $p_1, p_2, \ldots, p_t$, can recover the secret $K$. Every $p_i$ gets $\alpha$ and their corresponding $y_i$ from the public area. With his private share $s_i$ (only known to him), $p_i$ computes $x_i = f(\alpha + s_i)$ and presents $x_i$, the masked share, to a trusted agent $\mathcal{T_A}$. After collecting $t$ pairs of $(x_i, y_i)$, $i = 1, \ldots, t$, $\mathcal{T_A}$ uses Lagrange interpolation method to recover $P(x)$, hence the secret $K = P(0)$.

The collision-free property of the one-way function $f$ guarantees that $x_i = f(\alpha + s_i)$ will be distinct for distinct $s_i$, therefore $\mathcal{T}_A$ will surely get $t$ distinct points to recover the polynomial $P(x)$. One-way function $f$ also keeps share $s_i$ private, a participant $p_i$ only needs to present his masked share $x_i$. When the secret $K$ needs to be replaced by a new secret $K'$, $\mathcal{D}$ chooses element $\alpha'$ ($\alpha' \neq \alpha$) and a new polynomial $P'(x)$ of degree $(t-1)$ such that $K' = P'(0)$, and new $y'_i = P'(f(\alpha' + s_i))$, $s_i$ remains the same and can be used an unlimited number of times.

The scheme can be made verifiable simply adding a verifying message $v_i = f(x_i)$ in the public area for every participant $p_i$. $\mathcal{T}_A$ or a participant can verify the validity of any participants by this. When a new participant, say $p_{n+1}$, is admitted to the scheme, $\mathcal{D}$ only needs to generate $s_{n+1}$ and appends $y_{n+1} = f(\alpha + s_{n+1})$ to the $y_i$ sequence. When a participant $p_i$ needs to be removed from the scheme, $\mathcal{D}$ generates another polynomial $P'(x)$ of the same degree and let $P'(0) = K$, and update the $y_i$ sequence with the new $P'(x)$.

### 3.5.2. *Using one-way functions and RSA in a Shamir scheme*

Fei and Wang [7] enhanced the Shamir $(t, n)$ threshold scheme by applying one-way function and RSA cryptosystem. Their scheme works as follows.

**Scheme setup**: Suppose $q > n$ is a big prime, $g$ is a primitive element of finite field $\mathbb{Z}_q$, $u, w$ are two RSA prime numbers and $m = uw$, and $f$ is a one-way function. (a) Dealer $\mathcal{D}$ chooses a polynomial $P(x)$ of degree $t - 1$ over $\mathbb{Z}_q$, such that $K = P(0)$ is the secret to be shared among $n$ participants $p_1, p_2, \ldots, p_n$. (b) $\mathcal{D}$ chooses an $e$, such that $\gcd(e, \phi(m)) = 1$, and computes $d = e^{-1} \bmod \phi(m)$ (here $\phi$ is Euler's totient function), and publishes $e$. (c) $\mathcal{D}$ computes $s_i = P(g^i)$, $v_i = (f(s_i))^d \bmod m$, and sends $s_i$ and $v_i$ to participant $p_i$ as his share and verifying message.

**Secret recovery**: When a trusted agent $\mathcal{T}_A$ receives $t$ points $(g^1, s_1)$, $(g^2, s_2), \ldots, (g^t, s_t)$ from any $t$ participants, $\mathcal{T}_A$ uses Lagrange interpolation method to reconstruct the polynomial $P(x)$, and hence the secret $K = P(0)$. Participant $p_i$ can be verified by $v_i^e = f(s_i) \bmod m$.

## 4. Extension to Any General Access Structure

Ito, Saito and Nishizeki [15, 16] showed how to extend a Shamir threshold scheme to a multiple assignment scheme to realize any general access structure which fulfills the following monotone property:

$$A' \in \mathcal{A} \text{ and } A' \subseteq A'' \subseteq P \Longrightarrow A'' \in \mathcal{A}, \tag{25}$$

$$B' \in \beta \text{ and } B'' \subseteq B' \Longrightarrow B'' \subseteq \beta \tag{26}$$

where $P$ is the set of the participants, $\mathcal{A}$ is the access structure. $\beta = 2^P - \mathcal{A}$ will be the set of all unauthorized subsets.

Following the notations in [15, 16], we give a brief discussion here. For details, please refer to [15, 16]. The family of maximal sets in $\mathcal{A}$ is defined as

$$\partial^+ \mathcal{A} = \{A \subset \mathcal{A} : A \not\subseteq A' \, \forall A' \in \mathcal{A} - \{A\}\}. \tag{27}$$

Let $S$ be the set of shares. A multiple assignment scheme assigns a subset $S_i \subseteq S$ to participant $p_i \in P$ as follows:

$$g : P \rightarrow 2^S \text{ or } g(p_i) = S_i, \forall i = 1, \ldots, n. \tag{28}$$

Define

$$\mathcal{A}(S, g, k) = \{Q \subseteq P \mid |\bigcup_{p \in Q} g(p)| \geq k\}. \tag{29}$$

That means if the number of distinct shares of the union of the participants in a subset $Q$ of $P$ is more than the threshold $k$, it is an authorized subset.

For any access structure $\mathcal{A} \subseteq 2^P$ satisfying the monotone property, there exists a set of shares $S$, an assignment function $g : P \rightarrow 2^S$ and a non-negative integer $k$ such that $\mathcal{A}(S, g, k) = \mathcal{A}$.

Proof: Let $\beta = 2^P - \mathcal{A}$. We determine $\partial^+ \beta$ and set up a $(k, k)$ threshold scheme, where $k = |\partial^+ \beta|$.

Construct a set of shares $S$ so that $|S| = k$. We have $\partial^+ \beta = \{\beta_1, \ldots, \beta_k\}$ and $S = \{s_1, \ldots, s_k\}$. There exists a one-to-one correspondence between $S$ and $\partial^+ \beta$, say $s_1 \leftrightarrow \beta_1$, $s_2 \leftrightarrow \beta_2$, $\ldots, s_k \leftrightarrow \beta_k$. That means $S = \{S_i, \beta_i \in \partial^+ \beta, i = 1, \ldots, k\}$. We also define $g : P \rightarrow 2^S$ as follows:

$$g(p) = \{S_i, \beta_i \in \partial^+ \beta, p \notin \beta_i, i = 1, \ldots, k\}. \tag{30}$$

(i) $\mathcal{A} \subseteq \mathcal{A}(S, g, k)$.

Assume there exists $Q \in \mathcal{A}$ such that $Q \notin \mathcal{A}(S, g, k)$, then $|\bigcup_{p \in Q} g(p)| < k$ and hence $\bigcup_{p \in Q} g(p) \neq S$. There exists $s_i \in S - \bigcup_{p \in Q} g(p)$ for some $i$. So, for every $p \in Q, s_i \notin g(p)$ and therefore $p \in \beta_i$. Hence $Q \subseteq \beta_i \in \partial^+ \beta$.

By monotone property, $Q \in \beta$. This contradicts $Q \in \mathcal{A}$, since $\beta = 2^P - \mathcal{A}$.

(ii) $\mathcal{A}(S, g, k) \subseteq \mathcal{A}$.

Assume there exists $Q \in \mathcal{A}(S, g, k)$, but $Q \notin \mathcal{A}$. Since $Q \notin \mathcal{A}$, there exists $\beta_i \in \partial^+\beta$ such that $Q \in \beta$. By the definition of the function $g$, $s_i \notin g(p)$ for all $p \in Q$.

So, $s_i \notin \underset{p \in Q}{\cup} g(p)$ and hence $Q \notin \mathcal{A}(S, g, k)$. This contradicts the assumption.

Example: Let $P = \{p_1, p_2, p_3\}$ be the set of participants. Suppose $\mathcal{A} = \{\{p_1, p_2\}, \{p_1, p_3\}, \{p_1, p_2, p_3\}\}$, then $\beta = \{\{p_1\}, \{p_2\}, \{p_3\}, \{p_2, p_3\}\}$, then $\partial^+\beta = \{\{p_1\}, \{p_2, p_3\}\}$.

Since $|\partial^+\beta| = 2$, we set up a $(2, 2)$ threshold scheme with $S = \{s_1, s_2\}$ being the set of shares. $s_1$ will be assigned to participant(s) $p_2, p_3$ $[P - \{p_1\}]$; $s_2$ will be assigned to participant(s) $p_1$ $[P - \{p_2, p_3\}]$. It can be easily verified that all the following are unauthorized subsets $\{p_2\}, \{p_3\}, \{p_2, p_3\}$ (with $s_1$ only), $\{p_1\}$ (with $s_2$ only).

On the other hand, $\{p_1, p_2\}, \{p_1, p_3\}, \{p_1, p_2, p_3\}$ will have shares $s_1$ and $s_2$ to recover the secret.

## 5. Relation with Reed-Solomon Code

Here we discuss briefly error correction code, in particular, Reed-Solomon code. Then, we talk about the relationship or similarity between Reed-Solomon code and Shamir threshold scheme. Please refer to the textbooks for details in error correction code, for instance, [14, 21].

A $[m, q]$ code $C$ is a mapping from a vector space of dimension $m$ over a finite field $F$ into a vector space of dimension $q$ (here $q > m$) over the same field, i.e.,

$$C : F^m \rightarrow F^q; m < q. \tag{31}$$

That means an information word $a = (a_0, \ldots, a_{m-1}) \in F^m$ is mapped to a codeword $c = (c_0, \ldots, c_{q-1}) \in F^q$. There are $q - m$ extra symbols to detect or correct the errors occurred during the transmission. We call $q$ and $m$ the length and the dimension of the code $C$, respectively.

The Hamming distance between two codewords $c_1, c_2 \in C$ is defined as the number of the differences between the corresponding positions in $c_1$ and $c_2$. For example, let $c_1 = (0, 0, 1, 1)$, $c_2 = (1, 0, 1, 0)$. Since the first and fourth positions are different, the Hamming distance $d(c_1, c_2) = 2$. The minimum distance of $C$, $d$, is defined as

$$d = min\{d(c_1, c_2) | c_1, c_2 \in C, \text{and } c_1 \neq c_2\}. \tag{32}$$

$d$ is important that it tells us the minimum of errors that will convert a codeword $c_1$ to another codeword $c_2$.

A code $C$ can detect and correct up to $t_1$ and $t_2$ errors, respectively, if $t_1 \leq d - 1$ and $2t_2 + 1 \leq d$. The error detection is based on the fact that fewer than $d$ errors cannot convert a codeword to another codeword. The error correction is based on the nearest neighbor decoding principle. The received invalid word $c'$ will be converted to the codeword $c$ such that $d(c', c)$ is the smallest.

Reed-Solomon code, which is one type of error correcting codes with many applications such as compact disc (CD), spacecraft etc., was invented by Irving Reed and Gus Solomon in 1959 [25].

Let $F$ be a field with $q$ elements. There exists a primitive element $\alpha$ such that the $q$ elements in $F$ can be represented as $\{0, \alpha, \alpha^2, \ldots, \alpha^{q-1} = 1\}$.

Given an information word $a = (a_0, \ldots, a_{m-1})$, we set up a polynomial $P(x) = a_0 + a_1 x + \ldots + a_{m-1} x^{m-1}$, where $a_i \in F$. And the Reed-Solomon code is the mapping of the information word $a = (a_0, \ldots, a_{m-1})$ to a codeword $c = (P(0), P(\alpha), P(\alpha^2), \ldots, P(\alpha^{q-2}), P(1))$ as follows:

$$P(0) = a_0,$$

$$P(\alpha) = a_0 + a_1 \alpha + a_2 \alpha^2 + \ldots + a_{m-1} \alpha^{m-1},$$

$$P(\alpha^2) = a_0 + a_1 \alpha^2 + a_2 (\alpha^2)^2 + \ldots + a_{m-1} (\alpha^2)^{m-1},$$

$$\cdots \quad \cdots \quad \cdots \quad \cdots \tag{33}$$

$$P(\alpha^{q-2}) = a_0 + a_1 \alpha^{q-2} + a_2 (\alpha^{q-2})^2 + \ldots + a_{m-1} (\alpha^{q-2})^{m-1},$$

$$P(1) = a_0 + a_1 + a_2 + \ldots + a_{m-1}.$$

Any $m$ correct equations without error from Eq. (33) will determine $a$ correctly. On the other hand, any $m$ equations from Eq. (33) with one or more errors will determine $a$ incorrectly.

Suppose $t$ errors occur during the transmission. There will be $\binom{q-t}{m}$ and $\binom{t+m-1}{m}$ sets of $m$ equations that will give correct and incorrect results, respectively. By taking the majority vote for determination of the information word $a$, we can get the correct result if

$$\binom{q-t}{m} > \binom{t+m-1}{m}. \tag{34}$$

That is $t < \frac{q-m+1}{2}$. Please refer to [25] for details.

McEliece and Sarwate [22] pointed out that Shamir scheme is closely related to Reed-Solomon code. Suppose $s$ pieces of $P_i$ (Eq. (33)) are transmitted and $t$ out of these $s$ pieces are in error. Replacing $q$ by $s$ plus rearrangement and modifications in Eq. (34), we can recover $a = (a_0, a_1, \ldots, a_{m-1})$ as long as $s - 2t \geq m$. This is exactly a $(m, s)$ threshold scheme with $t = 0$, and $a_0$ of $a$ is the secret and $F = Z_q$ ($q$ is a prime), $\alpha^i = i$. Recall that the original Shamir threshold scheme assumes the dealer and the participants are honest and $P(1), \ldots, P(s)$ are the shares of the participants.

## 6. Shamir Ramp Scheme

Recall that in the Shamir $(t, n)$ threshold scheme, $n$ shares $P(x_1), \ldots, P(x_n)$ are distributed to $n$ participants $p_1, \ldots, p_n$ so that any $t$ out of these $n$ participants when joined together can recover the secret. Let $q$ be a large prime, $x_1, \ldots, x_n \in Z_q - \{0\}$ are all different to each other ($x_i \neq x_j$ if $i \neq j, 1 \leq i, j \leq n$) and chosen arbitrarily. $a_0, \ldots, a_{t-1} \in Z_q$ are chosen randomly. For simplicity, suppose $p_1, \ldots, p_t$ join together and let $y_1 = P(x_1), y_2 = P(x_2), \ldots,$ etc. We have the following $t$ independent equations. [Note: If $y_i$ is not available, let $y_i'$ be its assumed value.]

$$y_1 = a_0 + a_1 x_1 + a_2 x_1^2 + \ldots + a_{t-1} x_1^{t-1} \pmod{q}; \tag{35}$$

$$\ldots$$

$$y_t = a_0 + a_1 x_t + a_2 x_t^2 + \ldots + a_{t-1} x_t^{t-1} \pmod{q}. \tag{36}$$

From Eq. (35), rewrite $a_{t-1}$ in terms of $a_0, \ldots, a_{t-2}$ and substitute this into other equations, we reduce $t$ equations in $t$ unknowns into $(t-1)$ equations in $(t-1)$ unknowns. Continuing this way, we can reduce the system of $t$ independent equations to one equation with one unknown $a_0$. We can solve for $a_0$, which is the secret.

If only $t - 1$ participants, say $p_1, \ldots, p_{t-1}$, join together, the last equation will have 2 unknowns left, namely, $y_t$ and $a_0$. Any assumed or guessed value of the secret $a_0' \in Z_q$ will lead to a corresponding valid share of the missing participant $y_t' \in Z_q$, and vice versa. In other words, we can find a unique polynomial $P'(x)$ such that it will pass through all these $t - 1$ points and the assumed secrets $a_0'$. $P'(0) = a_0', P'(1) = y_1, \ldots, P'(t - 2) = y_{t-2}, P'(t-1) = y_{t-1}$. Since we cannot rule out any possibility, the scheme is perfect. The secret $a_0$ and the shares $y_i (i = 1, \ldots, n)$ are elements of $Z_q$, so it is ideal. From Eq. (17), it is obvious that the information rate is 2.

Suppose $(a_0, a_1)$ is the secret. If $(t-2)$ participants $p_1, \ldots, p_{t-2}$ join together, we have 2 equations left:

$$y'_{t-1} = \text{ in terms of } a_1 \text{ and } a_0,$$

$$y'_t = \text{ in terms of } a_1 \text{ and } a_0.$$

Any guessed values of the secret $(a'_0, a'_1)$ will lead to valid shares $y'_{t-1} \in Z_q$ and $y'_t \in Z_q$ of missing participants, and vice versa. So no partial information is given out here. The scheme is perfect.

Now, assume $(t-1)$ participants $p_1, \ldots, p_{t-1}$ join together. We have one equation left:

$$y'_t = \text{ in terms of } a_0. \tag{37}$$

As before, any guessed value of the share $y'_t \in Z_q$ gives a unique $a'_0 \in Z_q$. However, once $a'_0$ is determined, all the $a'_1, \ldots, a'_{t-1}$ are determined. We can thus eliminate the possibilities from $|Z_q| \times |Z_q|$ to $|Z_q|$. Hence, partial information is given out.

The above can be summarized by Shamir ramp scheme. For more details, please refer to [30].

A Shamir $(t_1, t_2, n)$ ramp scheme, where $t_1 < t_2 \leq n$, is one in which $n$ shares of information are distributed to $n$ participants so that

(i) if $t_2$ or more participants join together, the secret can be recovered.
(ii) if up to $t_1$ participants join together, the secret cannot be recovered and no partial information about the secret is leaked out.
(iii) if $t$ $(t_1 < t < t_2)$ participants join together, the secret cannot be recovered. However, partial information will be leaked out. The larger the $t$, the more information will be leaked out.

For a Shamir $(t_1, t_2, n)$ ramp scheme, let $l = t_2 - t_1$ be the gap. The bigger the gap $l$, the more efficient the size of the share would be, but the lesser the secrecy the scheme will provide (see Figure 1-Right).

One implementation for a ramp scheme is also by polynomial evaluation and interpolation. Let $s = (a_0, a_1, \ldots, a_{l-1}) \in Z_q^l$. We create a polynomial of degree of at most $t_2 - 1$ as follows:

$$P(x) = a_0 + a_1 x + \ldots + a_{l-1} x^{l-1} + a_l x^l + \ldots + a_{t_2-1} x^{t_2-1} (\text{mod } q) \tag{38}$$

where $a_i \in Z_q$ will be generated randomly, $i = l, \ldots, t_2-1$. $x_i \in Z_q - \{0\}$ will be chosen arbitrarily and $P(x_i)$ will be evaluated and sent to $P_i, i = 1, \ldots, n$ as his/her share. The information rate is equal to $l$.

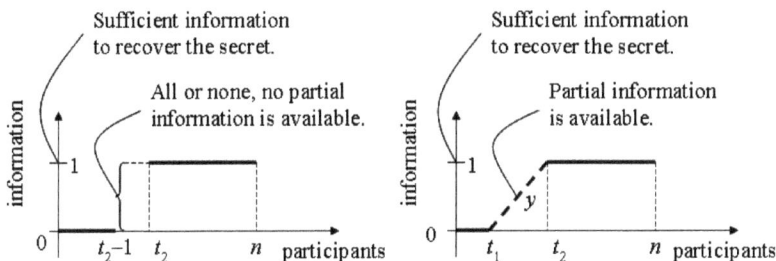

Figure 1.   A Shamir $(t_2, n)$ scheme is a $(t_2 - 1, t_2, n)$ ramp scheme.

Let us fix $t_2$ and $n$. That means any $t_2$ out of $n$ participants can recover the secret. One special case is as follows: A $(t_2 - 1, t_2, n)$ ramp scheme is just the same as a $(t_2, n)$ threshold scheme. The information rate is equal to 1 but perfect secrecy is provided. The secret will be the constant term of the polynomial. Figure 1-Left is to illustrate this.

## 7.  Information Disposal Algorithm and Making Secret Short

Rabin [24] proposed the information disposal algorithm (IDA) in 1989. IDA is a scheme to distribute a piece of information into $n$ participants such that any $t$ of these participants can recover the original information while up to $(t - 1)$ participants cannot. One implementation is also by polynomial interpolation, same as the Shamir threshold scheme. In a Shamir threshold scheme, the constant term will be the secret. However, in IDA, the secret will be split into all the coefficients. In other words, the secret will be represented by the whole polynomial. This gives the optimal rate of information, but even one participant has some partial information.

A $(0, t_2, n)$ ramp scheme is an information dispersal algorithm. The information rate is optimal. But no secrecy is provided. Any participant has some partial information. The secret is made up of all the coefficients of the polynomial, as Figure 2 illustrated.

Krawczyk [19] showed a method to make the secret short and provides secrecy at the same time. Suppose we have a secure encryption $(ENC_K)$ and decryption $(DEC_K)$ scheme and a symmetric key $K$ will be chosen randomly from the key space $\mathbb{K}$.

(a)  We first encrypt the secret $S$ to give a ciphertext $C$, i.e. $ENC_K(S) = C$. Then we use IDA to split $C$ into $C_1, \ldots, C_n$ shares and distribute them

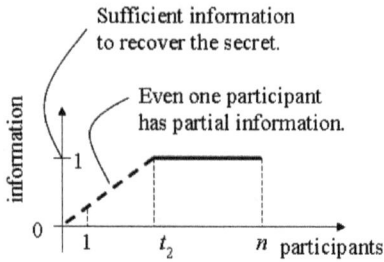

Figure 2.    A $(t_2, n)$ IDA is a $(0, t_2, n)$ ramp scheme.

to participants $p_1, \ldots, p_n$ so that each participant $p_i$ gets one share $C_i$, $i = 1, \ldots, n$.

(b) We use a perfect secret sharing scheme, say a Shamir $(t, n)$ threshold scheme, to safeguard the key $K$. Each participant $p_i$ gets one share of the key $K_i$, $i = 1, \ldots, n$.

In this way any $t$ participants can recover the key $K$ and the ciphertext $C$. Then use $K$ to get back the original secret $S$ by $DEC_K(C) = S$.

The information rate is optimal. IDA helps to make the size of the share short. But, it does not provide secrecy. So, we need a secure encryption and decryption scheme to protect it. In turn, we need a perfect secret sharing scheme to safeguard the key.

## 8. Secure Multiparty Computation

Secure multiparty computation (MPC), a subfield of cryptography, was first introduced in Yao's seminal two millionaire's problem [32]. The goal is to create methods for parties to jointly compute a function over their inputs while keeping those inputs private. In MPC $n$ parties $p_1, p_2, \ldots, p_n$ join together to compute a public function $f(x_1, x_2, \ldots, x_n)$, where $x_i$ is the private input held by party $p_i$, $i = 1, \ldots, n$. After the computation, each $p_i$ will know the correct function result, the value of $f(x_1, x_2, \ldots, x_n)$, but he or she will not know the inputs of the other parties. For more MPC materials, please refer to [6].

For security reason, instead of storing a secret in a single server, we split it as shares and store in different servers. That is why secret sharing schemes are important in multiparty computation. We also want to have the computations based on the shares of the parties instead of the secrets. Let $p_1, \ldots, p_n$ be the parties and $p_i$ holds $A(i)$ and $B(i)$ as shares for the secrets $a_0$ and $b_0$, respectively. We want to calculate $c_0 = a_0 + b_0$ based on $(A(i), B(i))$, $i = 1, \ldots, n$.

Since Shamir threshold scheme is linear, we can proceed as follows:

$$A(x) = a_0 + a_1 x + \ldots, a_{t-1} x^{t-1}, \ a_i \in Z_q, \tag{39}$$

$$B(x) = b_0 + b_1 x + \ldots, b_{t-1} x^{t-1}, \ b_i \in Z_q, \text{ and} \tag{40}$$

$$C(x) = A(x) + B(x) = c_0 + c_1 x + \ldots, c_{t-1} x^{t-1}, \text{ where}$$
$$c_i = a_i + b_i, 0 \le i \le t - 1. \tag{41}$$

Any $t$ parties (say $1, \ldots, t$) can join together to calculate $C(i) = A(i) + B(i)$, $1 \le i \le t$, and then recover $c_0$ which is equal to $a_0 + b_0$, the sum of the original secrets.

But for multiplication, it is different. Here,

$$D(x) = A(x)B(x) = a_0 b_0 + \ldots. \tag{42}$$

$D(x)$ will be a polynomial of degree $(t - 1) + (t - 1) = 2t - 2$. So, we need $2t - 1$ parties to pull their shares to recover $a_0 b_0$, which is the product of the original secrets $a_0$ and $b_0$. Obviously, $2t - 1$ cannot be greater than $n$. So Shamir threshold scheme is multiplicative provided that $n \ge 2t - 1$. Also, a linear secret sharing scheme (LSSS) is strongly multiplicative if any subset $A \subseteq P$, such that $P - A$ is not qualified, and the product $a_0 b_0$ can be computed only from the values of $A$. In a Shamir $(t, n)$ threshold scheme, the maximum size of an unauthorized subset is $t - 1$. So, a Shamir $(t, n)$ threshold scheme will be strongly multiplicative if $n - (t - 1) \ge 2t - 1$, i.e., $3t - 2 \le n$.

## 9. Private Information Retrieval and Shamir Scheme

Private information retrieval (PIR) deals with the privacy of a user when he queries a public database. It was first introduced by Chor *et al.* [5] in 1995. It is formalized as follows: given a database $x$ which consists of $n$ bits, $x = x_1 \ldots x_n$, a user wants to inquire the $i$th bit without letting the database know any information about $i$. A trivial solution is to let the user download the entire database. In this case, the communication complexity, which is the number of bits transferred between the user and the database, is $n$. Chor *et al.* proved that this trivial solution turned out to be optimal for a single database in the information theoretic setting. However, Chor *et al.* further showed that if we had more than one non-colluding servers with each having a complete database, we could reduce the communication complexity and preserve the perfect privacy as well.

In PIR, a user sends out queries to a group of non-colluding databases, and then combines the answers from the databases to come up with the results. The answers from the databases act like shares from the participants, and based on that, the desired information, somewhat like the secret, can be obtained. In the literature, there are papers discussing the applications of secret sharing schemes to PIR. For example, Goldberg [10] proposed a Byzantine-robust PIR based on the Shamir secret sharing scheme.

## 10.  Practical Applications

Many companies start to store their data outside their premises in cloud storage provided by various cloud providers, for instance, Amazon, Google, etc. The advantages to use cloud storage mainly include shorter setup time, lower implementation cost, easier scaling up/down, cheaper ongoing cost (pay-as-you-go). Big data has 3Vs characteristics, i.e., the velocity — the data go in and out or change very fast, the variety — different types of data (structured, semi-structured, and unstructured), and the volume - exponentially growing huge volume of data. This has been the trend for the last decade and will remain this way at least in the foreseeable future. Both cloud storage/computing and big data give rise to many big challenges to the existing data center infrastructure. They affect almost all areas to a certain extent. Here let us discuss some applications based on Shamir's secret sharing scheme and its variants.

Big data: In order to provide the data availability for the users, the traditional approach is to replicate one or more copies of data in different locations so that when one operating node goes down, the system can switch to another node so that the service will not be interrupted and is transparent to the users. However, under big data scenarios, this method is not feasible anymore. We need another efficient approach. By applying information dispersal algorithm, a large file can be separated into several smaller segments and a subset of these segments can combine to reconstruct the original file. This solves the problem of single point failure and as we saw before, the storage needed is the optimal.

Cloud storage/computing: Even if we trust a company, the data would turn out to be stored outside the premises. Privacy is a big concern to cloud storage/computing.

## 11. Other Platforms

Since many cryptographic protocols are based on the assumed hardness of certain mathematical problems, there is always a strong motivation to continue looking for harder problems especially after knowing that a powerful quantum computer could break RSA easily.

Since 1990, there are new proposals coming up, by using multivariate polynomials, braid group cryptography, etc. For example, Habeeb, Kahrobaei and Shpilrain [11] proposed an $(n, n)$ secret splitting scheme construction based on non-abelian groups using $n$ secure channels. The $(n, n)$ scheme combined with the Shamir's idea can be further generalized to a $(t, n)$ threshold scheme. Under this $(t, n)$ threshold scheme, the shares of the secret are sent out to the participants over the open channels as integers in the form of tuples of words. The participants then use group-theoretic techniques to recover the integers as their shares. Then following polynomial interpolation as in Shamir's threshold scheme, any $t$ participants can recover the polynomial and the secret.

As we mentioned earlier, Ito, Saito and Nishizeki [15, 16] showed how to extend a threshold scheme to a multiple assignment scheme to realize any general access structure, so this provides a new direction to set up any secret sharing scheme based on another platform, non-abelian groups.

## 12. Conclusions and Future Research

Based on a Shamir threshold scheme, many properties of secret sharing schemes can be easily demonstrated. It has a simple access structure. It is perfect and ideal. The shares distribution and secret recovery are through polynomial evaluation and polynomial interpolation, which are easy to follow. It can be further implemented as proactive or verifiable. A Shamir threshold scheme can be used as a building block to realize any general access structure. It is also closely related to Reed-Solomon code, a ramp scheme, an information dispersal algorithm and multiparty computation.

Even though the Shamir scheme was introduced more than 30 years ago, we can still use it as a building block for other cryptographic primitives and/or protocols. It has many applications in different areas such as big data and cloud storage/computing. It still remains an important active research area in the future and is worth more attention.

Another direction for research is to set up secret sharing schemes based on other alternative platforms as briefly mentioned in this paper, should this be proved more effective.

# References

[1]   K. Atkinson. *An Introduction to Numerical Analysis.* Wiley, 2nd edition, 1989.

[2]   G.R. Blakley. Safeguarding cryptographic keys. In *Proc. of the National Computer Conference, American Federation of Information Processing Societies Proceedings 48,* pages 313–317, 1979.

[3]   D. Bogdanov.   Foundations and properties of Shamir's secret sharing scheme. University of Tartu, available online http://www.cs.ut.ee/~peeter_l/teaching/seminar07k/bogdanov.pdf, 2007.

[4]   R.M. Capocelli, A. De Santis, L. Gargano, and U. Vaccaro. On the size of shares for secret sharing schemes. *Journal of Cryptology,* 6(3):157–167, 1993.

[5]   B. Chor, O. Goldreich, E. Kushilevitz, and M. Sudan. Private information retrieval. In *36th Annual IEEE Symposium on Foundations of Computer Science,* pages 41–50, 1995.

[6]   R. Cramer, I. Damgård, and J.B. Nielsen. Multiparty Computation, an Introduction.   Lecture Notes. Available http://www.brics.dk/~jbn/smc.pdf, 2009.

[7]   R. Fei and L. Wang. Cheat-proof secret sharing schemes based on rsa and one-way function. *Journal of Software,* 14(1):146–150, 2003. (In Chinese).

[8]   P. Feldman. A practical scheme for non-interactive verifiable secret sharing. In *Proc. of the 28th IEEE Symposium on the Foundations of Computer Science,* pages 427–437, 1987.

[9]   H. Ghodosi and R. Safavi-Naini.   Remarks on the multiple assignment secret sharing scheme.   In *in Proceedings of ICICS'97 –International Conference on Information and Communications Security,* pages 72–80. SpringerVerlag, 1997.

[10]  I. Goldberg. Improving the robustness of private information retrieval. In *Proc. of IEEE S&P 2007,* pages 131–148, May 2007. Oakland, California.

[11]  M. Habeeb, D. Kahrobaei, and V. Shpilrain.   A secret sharing scheme based on group presentations and the word problem. In *Contemporary Mathematics, Volume 582 - Computational and Combinatorial Group Theory and Cryptography (American Mathematical Society),* pages 143–150, 2012.

[12]  J. He, L. Li, and X. Li.   Verifiable multi-secret sharing scheme. *Acta Electronica Sinica,* 31(1):45–47, 2003. (In Chinese).

[13]  A. Herzberg, S. Jarecki, H. Krawczyk, and M. Yung.   Proactive secret sharing. In *Proc. of CRYPTO 1995,* volume 963 of *LNCS,* 1995.

[14]  W.C. Huffman and V. Pless. *Fundamentals of Error-Correcting Codes.* Cambridge University Press, 2003.

[15]  M. Ito, A. Saio, and T. Nishizeki. Multiple assignment scheme for sharing secret. *J. Cryptology,* 6:15–20, 1993.

[16]  M. Ito, A. Saito, and T. Nishizeki. Secret sharing scheme realizing general access structure. In *Proc. of IEEE GLOBECOM 1987,* pages 99–102, 1987.

[17]  W. Ji, S. Oh, S. Kim, and D. Won. New on-line secret sharing scheme using hash function. *Acta Electronica Sinica,* 31(1):45–47, 2003. (In Chinese).

[18]   J. Katz and Y. Lindell. *Introduction to Modern Cryptography*. Chapman and Hall/CRC, 2007.

[19]   H. Krawczyk. Secret sharing made short. In *CRYPTO 1993*, volume 773 of *LNCS*.

[20]   H. Liu, M. Hu, B. Fang, and Y. Yang. A dynamic secret sharing scheme based on one-way function. *Journal of Software*, 13(5):1009–12, 2002. (In Chinese).

[21]   F.J. MacWilliams and N.J.A. Sloane. *The Theory of Error-Correcting Codes*. North Holland Publishing Co., 1977.

[22]   R.J. McEliece and D.V. Sarwate. On sharing secrets and reed-solomon codes. *Communications of the ACM*, 24(9):583–584, 1981.

[23]   L. Pang and Y. Wang. $(t, n)$ threshold secret sharing scheme based on rsa cryptosystem. *Acta Electronica Sinica*, 31(1):45–47, 2003. (In Chinese).

[24]   M.O. Rabin. Efficient dispersal of information for security, load balancing, and fault tolerance. *Journal of the ACM*, 36(2):335–348, 1989.

[25]   I.S. Reed and G. Solomon. Polynomial codes over certain finite fields. *J. of the Society for Industrial and Applied Mathematics*, 8(2):300–304, 1960.

[26]   A. Shamir. How to share a secret. *Communications of the ACM*, 22(11):612–613, 1979.

[27]   C.E. Shannon. A mathematical theory of communication. *Bell Systems Technical Journal*, 27:379–423, 623–656, 1948.

[28]   C.E. Shannon. Communication theory of secrecy systems. *Bell Systems Technical Journal*, 28:656–715, 1949.

[29]   D. Stinson. An explication of secret sharing schemes. *Design, Codes and Cryptology*, 2:357–390, 1992.

[30]   D. Stinson. *Cryptography, Theory and Practice*. Chapman and Hall/CRC, 3rd edition, 2005.

[31]   H. Yang and G. Lin. Security research of secret sharing schemes based on hash functions. *Computer Engineering and Design*, 27(24):4718–19, 2006. (In Chinese).

[32]   A.C. Yao. Protocols for secure computations (extended abstract). In *the 21st Annual IEEE Symposium on the Foundations of Computer Science*, pages 160–164, 1982.

# Chapter 3

# The Zappa-Szep Product of Left-Orderable Groups

Fabienne Chouraqui

*University of Haifa at Oranim, Israel.*
*fabienne.chouraqui@gmail.com*
*fchoura@sci.haifa.ac.il*

**ABSTRACT.** It is well-known that the direct product of left-orderable groups is left-orderable and that, under a certain condition, the semi-direct product of left-orderable groups is left-orderable. We extend this result and show that, under a similar condition, the Zappa-Szep product of left-orderable groups is left-orderable. Moreover, we find conditions that ensure the existence of a partial left and right invariant ordering (bi-order) in the Zappa-Szep product of bi-orderable groups and prove some properties satisfied.

## 1. Introduction

Let $G$ be a group with subgroups $H$ and $K$ such that $G = HK$ and $H \cap K = \{1\}$. Then $G$ is isomorphic to the Zappa-Szep product of $H$ and $K$, denoted by $H \bowtie K$. If both subgroups $H$ and $K$ are normal, $G$ is their direct product $H \times K$ and if $H$ only is normal, $G$ is their semi-direct product $H \rtimes K$. The Zappa-Szep product of groups is a generalisation of the direct and the semi-direct product which requires the embedding of neither of the factors to be normal in the product. We recall that a group $G$ is *left-orderable* if there exists a total ordering $\prec$ of its elements which is invariant under left multiplication, that is $g \prec h$ implies $fg \prec fh$ for all $f$, $g$, $h$ in $G$. It is well-known that the direct product of left-orderable groups is left-orderable and that, under a certain condition, the semi-direct product of left-orderable groups is left-orderable [9], [8][p. 27]. We extend this result and show that, under a similar condition, the Zappa-Szep product of left-orderable groups is left-orderable. Before we state our first result, we introduce some notations.

**Definition.** Let $H$, $K$ be groups. Let $\alpha$ be the homomorphism defined by $\alpha : K \to \text{Sym}(H)$, $k \mapsto \alpha_k$, where $k \in K$ and $\text{Sym}(H)$ denotes the group of bijections of $H$. Let $\beta : H \to \text{Sym}(K)$, $h \mapsto \beta_h$, be such that $\beta_{h_1 h_2} = \beta_{h_2} \circ \beta_{h_1}$ and $\beta_1 = Id_K$ ($\beta$ an anti-homomorphism). Assume $(\alpha, \beta)$ satisfies: $\alpha_k(h_1 h_2) = \alpha_k(h_1) \alpha_{\beta_{h_1}(k)}(h_2)$ and $\beta_h(k_1 k_2) = \beta_{\alpha_{k_2}(h)}(k_1) \beta_h(k_2)$. *The Zappa-Szep product $H \bowtie K$ of $H$ and $K$ with respect to the pair $(\alpha, \beta)$ is the set $H \times K$ endowed with the following product:*

$$(h_1, k_1)(h_2, k_2) = (h_1 \, \alpha_{k_1}(h_2), \, \beta_{h_2}(k_1) \, k_2).$$

The identity is $(1, 1)$ and the inverse of an element $(h, k)$ is

$$(h, k)^{-1} = (\alpha_{k^{-1}}(h^{-1}), \, \beta_{h^{-1}}(k^{-1})).$$

In fact, $H \times \{1\}$ and $K \times \{1\}$ are subgroups of $H \bowtie K$, isomorphic to $H$ and $K$, respectively.

The Zappa-Szep product is also called crossed product, bi-crossed product, knit product, two-sided semidirect product. We refer to [3], [1], [13], [14] for details. Note that every element in $H \bowtie K$ is uniquely represented by $hk$, with $h \in H$ and $k \in K$, and $kh$ is equal to $\alpha_k(h) \beta_h(k)$ in $H \bowtie K$.

**Theorem 1.** *Let $G$ be the Zappa-Szep product of the groups $H$ and $K$ with respect to $(\alpha, \beta)$. Assume $H$ and $K$ are left-orderable groups. Let $P_H$ and $P_K$ denote the positive cones (of a left order) of $H$ and $K$ respectively. Assume the following condition $(*)$ is satisfied:*

$$\alpha_k(P_H) \subseteq P_H, \quad \forall k \in K.$$

*Then, there exists a total left order $\prec$ on $G = H \bowtie K$, with positive cone $P$, such that (the embedding of) $K$ is a convex subgroup with respect to $\prec$ and $P_K = P \cap K$.*

We introduce some definitions and refer to [2], [6], [7], [8], [9], [10], [11], [12]. Let $G$ be a left-orderable group with a strict total left order $\prec$. An element $g$, $g \in G$, is called *positive* if $1 \prec g$ and the set of positive elements $P$ is called the *positive cone of* $\prec$. The positive cone $P$ satisfies:

(1) $P$ is a semigroup, that is $P \cdot P \subseteq P$
(2) $G$ is partitioned, that is $G = P \cup P^{-1} \cup \{1\}$ and $P \cap P^{-1} = \emptyset$.

Conversely, if there exists a subset $P$ of $G$ that satisfies (1) and (2), then $P$ determines a unique total left order $\prec$ defined by $g \prec h$ if and only if $g^{-1}h \in P$. If $P$ satisfies only (1), the left order obtained is partial. A subgroup $N$ of a left-orderable group $G$ is called *convex* if for any $x, y, z \in G$ such that $x, z \in N$ and $x \prec y \prec z$, we have $y \in N$.

**Proof.** *of Theorem 1.* Let $P \subseteq G$ be defined by: $g = hk \in P$ if $h \in P_H$ or if $h = 1$, $k \in P_K$. We show there exists a total left order $\prec$ on $G = H \bowtie K$ with positive cone $P$. First, we prove $P$ is a semigroup. Let $g = hk \in P$ and $g' = h'k' \in P$. If $h' = 1$ and $h = 1$, then $k, k' \in P_K$ and $kk' \in P_K$, since $P_K$ is a semigroup, so $gg' \in P$. If $h' = 1$ and $h \neq 1$, then $h \in P_H$ and so $gg' = hkk' \in P$. If $h' \neq 1$, then $h' \in P_H$ and $gg' = hkh'k' = h\,\alpha_k(h')\beta_{h'}(k)\,k'$. From the assumption $(*)$, $\alpha_k(h') \in P_H$, so if $h = 1$, then $gg' = \alpha_k(h')\beta_{h'}(k)\,k' \in P$ and if $h \neq 1$, $h\,\alpha_k(h') \in P_H$ also, since $P_H$ is a semigroup, so $gg' \in P$. Next, we prove that given $g = hk \neq 1$ in $G = H \bowtie K$, either $g$ belongs to $P$ or $g^{-1} = \alpha_{k^{-1}}(h^{-1})\,\beta_{h^{-1}}(k^{-1})$ belongs to $P$. Assume $h = 1$. If $k \in P_K$, then $g \in P$, otherwise $k^{-1} \in P_K$, since $P_K$ partitions $K$, and then $g^{-1} = k^{-1} \in P$. Furthermore, $g$ cannot belong to $P \cap P^{-1}$, since it would contradict $P_K \cap P_K^{-1} = \emptyset$. Assume $h \neq 1$. If $h \in P_H$, then $g \in P$, otherwise $h^{-1} \in P_H$, since $P_H$ partitions $H$, and from $(*)$, $\alpha_{k^{-1}}(h^{-1}) \in P_H$. So, $g^{-1} = \alpha_{k^{-1}}(h^{-1})\,\beta_{h^{-1}}(k^{-1})$ belongs to $P$. Assume $g \in P \cap P^{-1}$, then $h \in P_H$ and $\alpha_{k^{-1}}(h^{-1}) \in P_H$. It holds that $\alpha_{k^{-1}}(h^{-1}) = (\alpha_{\beta_{h^{-1}}(k^{-1})}(h))^{-1}$. Indeed, on one side $\alpha_{k^{-1}}(1) = 1$ and on the other side $\alpha_{k^{-1}}(1) = \alpha_{k^{-1}}(h^{-1}h) = \alpha_{k^{-1}}(h^{-1})\alpha_{\beta_{h^{-1}}(k^{-1})}(h)$. From $(*)$, $\alpha_{\beta_{h^{-1}}(k^{-1})}(h)$ belongs to $P_H$, so $\alpha_{k^{-1}}(h^{-1}) = (\alpha_{\beta_{h^{-1}}(k^{-1})}(h))^{-1}$ belongs to $P_H \cap P_H^{-1}$ and this is a contradiction. So, $P$ satisfies the conditions (1) and (2) and it determines uniquely a total left order $\prec$.

Let $g = hk \in G$ and $k' \in K$. Assume $1 \prec g \prec k'$ and assume by contradiction that $h \neq 1$. From $1 \prec g$, we have $h \in P_H$. From $hk \prec k'$, we have $k'^{-1}hk \prec 1$. But, $k'^{-1}hk = \alpha_{k'^{-1}}(h)\,\beta_h(k'^{-1})\,k \succ 1$, since, from $(*)$, $\alpha_{k'^{-1}}(h)$ belongs to $P_H$. So, $h = 1$. $\qquad\square$

In the following example, we remark it is sometimes useful to consider a semi-direct product as a special case of Zappa-Szep product in order to show it is left-orderable. Indeed, this permits to check condition $(*)$ either on a positive cone of $H$ or on a positive cone of $K$. Let $H = \mathrm{Gp}\langle y, z \mid y^3 = z^3\rangle$. Since $H = \mathbb{Z} *_{3\mathbb{Z}} \mathbb{Z}$ and $\mathbb{Z}$ is (left) orderable, $H$ is left-orderable [8][p.178]. Let $K = \mathrm{Gp}\langle a, b \mid aba = bab\rangle$, the braid group on 3 strands, $K$ is left-orderable [5]. Let $\alpha : K \to \mathrm{Sym}(H)$ be the trivial homomorphism and let $\beta : H \to \mathrm{Sym}(K)$, the anti-homomorphism defined by $\beta_y = \beta_z = (a, b)$. With respect to $(\alpha, \beta)$, let $G = H \bowtie K$ in this specific order; $G$ is presented by $\mathrm{Gp}\langle y, z, a, b \mid aba = bab, y^3 = z^3, ay = yb, by = ya, az = zb, bz = za\rangle$.

We show $G$ is left-orderable, using Theorem 1. Each element $h \in H$ admits a normal formal $h = y^{\epsilon_1} z^{\mu_1} y^{\epsilon_2} z^{\mu_2} \ldots y^{\epsilon_m} z^{\mu_m} \Delta^{n_h}$, where $\Delta = y^3 = z^3$ is central in $H$, $-1 \le \epsilon_i, \mu_i \le 1$ and $n_h \in \mathbb{Z}$. We define $exp(h) = \epsilon_1 + \mu_1 + .. + \epsilon_m + \mu_m + 3n_h$ and $P_H$ to be the set of elements $h \in H \setminus \{1\}$ such that $exp(h) > 0$, or if $exp(h) = 0$, $n_h > 0$, or if $exp(h) = n_h = 0$, $\epsilon_1 \ne 0$. The set $P_H$ is a positive cone (the proof appears in the appendix) and $P_H$ satisfies trivially the condition $(*)$ from Theorem 1. So, $G$ is left-orderable.

Note that a group is left-orderable if and only if it is right-orderable. This is well illustrated here with the existence of a symmetric version of this result. Indeed, if $H$ and $K$ are right-orderable groups, with $Q_H$ and $Q_K$ positive cones of right orders of $H$ and $K$ satisfying $\beta_h(Q_K) \subseteq Q_K$, $\forall h \in H$. Then there exists a total right order $<$ on $G = H \bowtie K$, with positive cone $Q$ (defined by $g = hk \in Q$ if $k \in Q_K$ or if $k = 1$, $h \in Q_H$), such that (the embedding of ) $H$ is a convex subgroup with respect to $<$. We say a group $G$ is *partially (totally) bi-orderable* if there exists a partial (total) ordering $\ll$ of its elements which is invariant under left and right multiplication, that is $g \ll h$ implies $fgk \ll fhk$ for all $f, g, h, k$ in $G$. In particular, a set $P$ determines a partial bi-order if and only if $P$ is a semigroup, and satisfies $gPg^{-1} \subseteq P$ for all $g \in G$; $P$ determines a total bi-order if additionally $G = P \cup P^{-1} \cup \{1\}$ and $P \cap P^{-1} = \emptyset$. A natural question is when the Zappa-Szep product of bi-orderable groups is a bi-orderable group. In the following theorem, we give conditions that ensure the existence of a partial bi-order.

**Theorem 2.** *Let $G$ be the Zappa-Szep product of the groups $H$ and $K$ with respect to $(\alpha, \beta)$. Assume $H$ and $K$ are partially bi-orderable groups. Let $P_H$ and $P_K$ denote the positive cones (of a partial bi-order) of $H$ and $K$ respectively. Assume the following conditions are satisfied:*

$(*)$ $\alpha_k(P_H) \subseteq P_H$, $\forall k \in P_K$, and $\beta_h(P_K) \subseteq P_K$, $\forall h \in P_H$
$(**)$ $k P_H k^{-1} \subseteq P_H$, $\forall k \in K$, and $h P_K h^{-1} \subseteq P_K$, $\forall h \in H$

*Then, there exists a partial bi-order $\ll$ on $G = H \bowtie K$. Furthermore, if $h, h' \in H$ satisfy $h \ll h'$ then $\alpha_k(h) \ll \alpha_k(h')$ and $\beta_h(k) \ll \beta_{h'}(k)$, $\forall k \in K$ and if $k, k' \in K$ satisfy $k \ll k'$ then $\alpha_k(h) \ll \alpha_{k'}(h)$ and $\beta_h(k) \ll \beta_h(k')$, $\forall h \in H$.*

**Proof.** Let $P \subseteq G$ be defined by: $g = hk \in P$ if $h \in P_H$ and $k \in P_K$; if $h = 1$, $k \in P_K$ or if $k = 1$, $h \in P_H$. We show $P$ is a semigroup and $gPg^{-1} \subseteq P$ for all $g \in G$. Let $g = hk \in P$ and $g' = h'k' \in P$, then $gg' = hkh'k' = h \alpha_k(h') \beta_{h'}(k) k'$. From the assumption $(*)$, $\alpha_k(h') \in P_H$ and $\beta_{h'}(k) \in P_K$, so $h\,\alpha_k(h') \in P_H$ and $\beta_{h'}(k)\,k' \in P_K$, since $P_H$ and $P_K$ are semigroups, so $gg' \in P$. Next, let $g = hk \in G$ and $h'k' \in P$, we show $gh'k'g^{-1} \in P$: $gh'k'g^{-1} = hkh'k'k^{-1}h^{-1} = \left(h\left(kh'k^{-1}\right)h^{-1}\right)\left(h\left(kk'k^{-1}\right)h^{-1}\right)$. From $(**)$, $h' \in P_H$ implies $kh'k^{-1} \in P_H$ and so $h\left(kh'k^{-1}\right)h^{-1} \in P_H$, since $P_H$ is the positive cone of a bi-order. The element $h\left(kk'k^{-1}\right)h^{-1} \in P_K$, first from $kk'k^{-1} \in P_K$, and next from $(**)$.

Let $h, h' \in H$ satisfy $h \ll h'$, and let $k \in K$, then $kh \ll kh'$, that is $\alpha_k(h)\beta_h(k) \ll \alpha_k(h')\beta_{h'}(k)$. Since $\ll$ is a bi-order, this implies $1 \ll (\alpha_k(h))^{-1}\alpha_k(h')\,\beta_{h'}(k)(\beta_h(k))^{-1}$ and from the definition of $\ll$, we have $(\alpha_k(h))^{-1}\alpha_k(h') \in P_H$ and $\beta_{h'}(k)(\beta_h(k))^{-1} \in P_K$, that is $\alpha_k(h) \ll \alpha_k(h')$ and $\beta_h(k) \ll \beta_{h'}(k)$. In case $h = 1\ h' \neq 1$, this means $\alpha_k(P_H) \subseteq P_H$, $\forall k \in K$ and $h' \gg 1$ implies $\beta_{h'}(k) \gg k$, $\forall k \in K$; in case $h \neq 1\ h' = 1$, this means that $h \ll 1$ implies $\alpha_k(h) \ll 1$ and $\beta_h(k) \ll k$, $\forall k \in K$. The same proof works for the symmetric statement. $\qquad\square$

It would be interesting to know if there exists a natural extension of total bi-orders of $H$ and $K$ that defines a total bi-order on $H \bowtie K$ and $H \bowtie K$ is not a direct nor a semi-direct product.

## 2. Appendix

We recall each element $h \in H$ admits a normal formal $h = y^{\epsilon_1} z^{\mu_1} y^{\epsilon_2} z^{\mu_2} \ldots y^{\epsilon_m} z^{\mu_m} \Delta^{n_h}$, where $\Delta = y^3 = z^3$ is a central element in $H$, $-1 \leq \epsilon_i, \mu_i \leq 1$ and $n_h \in \mathbb{Z}$. We define $exp(h) = \epsilon_1 + \mu_1 + .. + \epsilon_m + \mu_m + 3n_h$ and $P_H$ to be the set of elements $h \in H \setminus \{1\}$ such that $exp(h) > 0$ (class (A)), or if $exp(h) = 0$, $n_h > 0$ (class (B)), or if $exp(h) = n_h = 0$, $\epsilon_1 \neq 0$ (class (C)). We show $P_H$ is a positive cone. First, we show $P_H$ is a semigroup. Let $h = y^{\epsilon_1} z^{\mu_1} \ldots y^{\epsilon_m} z^{\mu_m} \Delta^{n_h}$, $h' = y^{\epsilon'_1} z^{\mu'_1} \ldots y^{\epsilon'_m} z^{\mu'_m} \Delta^{n_{h'}}$, with $-1 \leq \epsilon_i, \mu_i, \epsilon'_i, \mu'_i \leq 1$, be elements of $P_H$. We need to check several cases in order to show $hh' \in P_H$.

Case 1: If $h$ is in class (A) and $h'$ is in class (A) or (B) or (C), then $hh'$ is also in (A).

Case 2: If $h, h'$ are in class (B), then $hh' \in P_H$, since $exp(hh') = 0$ and $n_{hh'} \geq n_h + n_{h'} - 1 > 0$. Indeed, there are only two cases in which $n_{hh'}$ decreases: if $h$ ends with $y^{-1}$ and $h'$ begins with $y^{-1}$, or if $h$ ends with $z^{-1}$ and $h'$ begins with $z^{-1}$, and in both cases $n_{hh'} = n_h + n_{h'} - 1$.

Case 3: Assume $h, h'$ are in class (C). Note that if $h = y^{\epsilon_1} z^{\mu_1} \ldots y^{\epsilon_m} z^{\mu_m} \Delta^{n_h}$, $-1 \leq \epsilon_i, \mu_i \leq 1$, satisfies $exp(h) = n_h = 0$, then the length of $h$ is necessarily even. Furthermore, if $h$ is in class (C), then $h = y^{\pm 1} z^{\mu_1} y^{\epsilon_2} z^{\mu_2} \ldots y^{\epsilon_m} z^{\pm 1}$ and the same holds for $h'$. So, $hh'$ is also in (C).

Case 4: If $h$ is in class (B) and $h'$ is in class (C), then $h = y^{\epsilon_1} z^{\mu_1} \ldots y^{\epsilon_m} z^{\mu_m} \Delta^{n_h}$, $n_h > 0$ and $h' = y^{\pm 1} z^{\mu'_1} \ldots y^{\epsilon'_m} z^{\pm 1}$. It holds that $n_h - 1 \leq n_{hh'} \leq n_h + 1$ and we need only to check the case $n_{hh'} = n_h - 1$. If $n_h - 1 > 0$, then $hh'$ is in class (B). If $n_h - 1 = 0$, then $h = y^{\epsilon_1} z^{\mu_1} \ldots z^{\mu_{m-1}} y^{-1} \Delta$, $h' = y^{-1} z^{\mu'_1} \ldots y^{\epsilon'_m} z^{\pm 1}$ and $hh' = y^{\epsilon_1} z^{\mu_1} \ldots z^{\mu_{m-1}} y z^{\mu'_1} \ldots y^{\epsilon'_m} z^{\pm 1}$. We show $hh'$ is in class (C), that is $\epsilon_1 \neq 0$. On one-hand, $exp(h) = \epsilon_1 + \mu_1 + \ldots + \mu_{m-1} - 1 + 3$ and on the second-hand $exp(h) = 0$. By induction on $m$, it holds that $\epsilon_1 + \mu_1 + \ldots + \epsilon_{m-1} + \mu_{m-1} = -2$, with $-1 \leq \epsilon_i, \mu_i \leq 1$, implies $\epsilon_1 \neq 0$.

Case 5: If $h$ is in class (C) and $h'$ is in class (B), then $h = y^{\pm 1} z^{\mu_1} \ldots y^{\epsilon_m} z^{\pm 1}$, and $h' = y^{\epsilon'_1} z^{\mu'_1} \ldots y^{\epsilon'_m} z^{\mu'_m} \Delta^{n_{h'}}$, $n_{h'} > 0$. It holds that $n_{h'} - 1 \leq n_{hh'} \leq n_{h'} + 1$ and in case $n_{hh'} = n_{h'} - 1 = 0$, $hh'$ is necessarily in class (C), as $hh'$ begins with $y^{\pm 1}$.

Next, we show $P_H$ partitions $H$, that is given $h \in H \setminus \{1\}$, either $h \in P_H$ or $h^{-1} \in P_H$. If $exp(h) > 0$, then $h \in P_H$, otherwise assume $exp(h) \leq 0$. If $exp(h) < 0$, then $h^{-1} \in P_H$, since $exp(h^{-1}) = -exp(h)$. So, assume $exp(h) = 0$. If $n_h > 0$, then $h \in P_H$, otherwise assume $n_h \leq 0$. If $n_h < 0$, then $h^{-1} \in P_H$, since $n_{h^{-1}} = -n_h$. So, assume $n_h = 0$. If $h = y^{\epsilon_1} z^{\mu_1} \ldots y^{\epsilon_m} z^{\mu_m} \Delta^{n_h}$, $-1 \leq \epsilon_i, \mu_i \leq 1$, satisfies $exp(h) = n_h = 0$, then there are two possibilities: either $h = y^{\pm 1} z^{\mu_1} y^{\epsilon_2} z^{\mu_2} \ldots y^{\epsilon_m} z^{\pm 1}$ or $h = z^{\pm 1} y^{\epsilon_2} z^{\mu_2} \ldots y^{\epsilon_{m-1}} z^{\mu_{m-1}} y^{\pm 1}$. If $h = y^{\pm 1} z^{\mu_1} y^{\epsilon_2} z^{\mu_2} \ldots y^{\epsilon_m} z^{\pm 1}$, then $h \in P_H$. Otherwise, if $h = z^{\pm 1} y^{\epsilon_2} z^{\mu_2} \ldots y^{\epsilon_{m-1}} z^{\mu_{m-1}} y^{\pm 1}$, then $h^{-1} = y^{\mp 1} z^{-\mu_{m-1}} y^{-\epsilon_{m-1}} z^{-\mu_2} y^{-\epsilon_2} \ldots z^{\mp 1}$, that is $h^{-1} \in P_H$.

## References

[1]  A.L. Agore and G. Militaru, *Classifying complements of groups. Applications*, to appear Ann. Inst. Fourier, Grenoble.

[2]  R. Botto Mura and A.H. Rhemtulla, *Orderable groups*, Lecture Notes in Pure and Applied Mathematics **27**, Marcel Dekker Inc., New York 1977.

[3]  M.G. Brin, *On the Zappa-Szep Product*, Comm. Algebra **33** n.2, (2005), 393–424.

[4]  P.F. Conrad, *Right-ordered groups*, Michigan Math. J. **6** (1959), 267–275.

[5]  P. Dehornoy, I. Dynnikov, D. Rolfsen and B. Wiest, *Ordering braids*, Mathematical Surveys and Monographs 148, American Mathematical Society, Providence, RI, 2008.

[6] B. Deroin, A.Navas, C. Rivas, *Groups, Orders and Dynamics*, Preprint, ArXiv 1408:5805.

[7] F.W. Levi, *Ordered groups*, Proc. Indian Acad. Sci. A. **16** (1942), 256–263.

[8] V.M. Kopytov, N. Ya Medvedev, *Right-ordered groups*, Siberian school of Algebra and Logic, Consultants bureau, New York 1996.

[9] V.M. Kopytov, *Free lattice-ordered groups*, Algebra Logic **18**, n.4 (1979), 259–270.

[10] P.A. Linnell, *The space of left orders of a group is either finite or uncountable*, Bull. London Math. Soc. **43** (2011), no. 1, 200–202.

[11] D. Rolfsen, *Low dimensional topology and ordering groups*, Mathematica Slovaca. **64**, n.3 (2014), 579–600.

[12] A.S. Sikora, *Topology on the spaces of orderings of groups*, Bull. London Math. Soc. **36** (2004), n.4, 519–526.

[13] J. Szep, *On the structure of groups which can be represented as the product of two subgroups*, Acta Sci. Math. Szeged **12** (1950), 57–61.

[14] G. Zappa, *Sulla costruzione dei gruppi prodotto di due dati sottogruppi permutabili traloro*, Atti Secondo Congresso Un. Mat. Ital., Bologna; Edizioni Cremonense, Rome, (1942), 119 125.

# Chapter 4

# Totally Disconnected Groups From Baumslag-Solitar Groups

Murray Elder[1] and George Willis[2]

[1] *School of Mathematical and Physical Sciences*
*University of Technology Sydney, Ultimo NSW 2007, Australia*
*Murray.Elder@uts.edu.au*

[2] *School of Mathematical and Physical Sciences*
*The University of Newcastle, Callaghan NSW 2308, Australia*
*George.Willis@newcastle.edu.au*

**ABSTRACT.** For each Baumslag-Solitar group $\mathrm{BS}(m, n)$ ($m, n \in \mathbb{Z} \setminus \{0\}$), a totally disconnected, locally compact group, $G_{m,n}$, is constructed so that $\mathrm{BS}(m, n)$ is identified with a dense subgroup of $G_{m,n}$. The scale function on $G_{m,n}$, a structural invariant for the topological group, is seen to distinguish the parameters $m$ and $n$ to the extent that the set of scale values is

$$\left\{ \left( \frac{\mathrm{lcm}(m, n)}{|m|} \right)^{\rho}, \left( \frac{\mathrm{lcm}(m, n)}{|n|} \right)^{\rho} \mid \rho \in \mathbb{N} \right\}.$$

It is also shown that $G_{m,n}$ has flat rank 1 when $|m| \neq |n|$ and 0 otherwise, and that $G_{m,n}$ has a compact, open subgroup isomorphic to the product $\prod \{(\mathbb{Z}_p, +) \mid p \text{ is a prime divisor of the scale}\}$.

## 1. Introduction

In this article we investigate a family of totally disconnected, locally compact groups $G_{m,n}$ built from Baumslag-Solitar groups, which are

Date: May 4, 2016.

2010 *Mathematics Subject Classification.* 22D05; 20F65.

*Keywords and phrase.* Totally disconnected locally compact group; commensurated subgroup; Baumslag-Solitar group; scale; minimizing subgroup; flat rank; scale-multiplicative subsemigroup.

Research supported by the Australian Research Council projects DP0984342, DP120100996 and FT110100178.

the groups $\mathrm{BS}(m,n)$ with presentations $\langle a,t \mid ta^mt^{-1} = a^n \rangle$ for non-zero integers $m$ and $n$.

Recall that a topological group $G$ is a group that is also a topological space such that the map $(x,y) \mapsto xy^{-1} : G \times G \to G$ is continuous. A topological space is *Hausdorff* if for any two points $x,y$ there are disjoint open sets $U,V$, with $x \in U$ and $y \in V$, *totally disconnected* if the connected component containing any point is just the point itself, or equivalently that for any two points $x,y$ there are disjoint open sets $U,V$, with $x \in U$ and $y \in V$ and whose union is the whole space, and *locally compact* if each point has a compact neighbourhood. The study of locally compact groups naturally splits into the study of connected and totally disconnected groups, since for every locally compact group $G$, the connected component $G_0$ of the identity is a closed normal subgroup, and $G/G_0$ is totally disconnected.

The subgroup $\langle a \rangle$ is *commensurated* by $\mathrm{BS}(m,n)$. We follow a construction described in [11, 16, 18, 20, 21] that uses a commensurated subgroup of an abstract group to construct a totally disconnected, locally compact group that is the completion of the abstract group with respect to a topology defined on it in terms of cosets of the commensurated subgroup. The group thus obtained from $\mathrm{BS}(m,n)$ is denoted $G_{m,n}$.

In [9] Gal and Januszkiewicz prove that Baumslag-Solitar groups are a-T-menable, by embedding them into the topological group of automorphisms of their corresponding *Bass-Serre* tree. The closure of $\mathrm{BS}(m,n)$ in this group coincides with the group $G_{m,n}$ we construct.

A key invariant of a totally disconnected locally compact group is the set of positive integers corresponding to the *scales* of its elements, which we define in Section 2. The scale function and the related notion of minimizing subgroups were introduced by the second author [23, 24] and fill a similar role in the theory of totally disconnected groups to that played by eigenvalues and triangularising bases in linear algebra and the theory of matrix groups.

We prove that the set of scales for the group $G_{m,n}$ for $m,n \neq 0$ is

$$\left\{ \left( \frac{\mathrm{lcm}(m,n)}{|m|} \right)^{\rho}, \left( \frac{\mathrm{lcm}(m,n)}{|n|} \right)^{\rho} \mid \rho \in \mathbb{N} \right\}.$$

The scale thus partially distinguishes the parameters in the definition of $\mathrm{BS}(m,n)$. In particular, for each pair of relatively prime integers $m,n > 0$ we obtain a distinct nondiscrete totally disconnected group. We offer two alternative proofs of this result: a combinatorial approach is given in

Section 6; while in Section 8 the *tidying procedure* developed by the second author is used.

Another key invariant of totally disconnected locally compact groups is their *flat rank* (defined in Section 7). Here we show that $G_{m,n}$ has flat rank 1 for all $|m| \neq |n|$.

The article is organised as follows. In Sections 2–4 we define scale, Baumslag-Solitar groups and commensurated subgroups, and list properties that will be used later. In Section 5 we outline the construction of a totally disconnected group starting with an abstract group having a commensurated subgroup, which we apply to Baumslag-Solitar groups. In Section 6 we compute the scale function for $G_{m,n}$. In Section 7 we compute a formula for the *modular function* for $G_{m,n}$, and compute the flat rank and maximal scale-multiplicative subsemigroups for $G_{m,n}$. In Section 8 we give more detail on the local structure of $G_{m,n}$, describing its compact open subgroups explicitly, and recompute the modular and scale functions with this description.

Note that throughout this paper $\mathbb{N}$ will denote the set of nonnegative integers (including 0).

The authors wish to thank Mathieu Carette and Yves de Cornulier for suggestions and corrections to earlier drafts.

## 2. The Scale Function

Let $G$ be a totally disconnected locally compact group. By van Dantzig's Theorem (see [22] or [12] Theorem 7.7), every neighourhood of the identity contains a compact open subgroup, $V$ say. An automorphism of $G$ is a group automorphism $\alpha : G \to G$ that is also a topological homeomorphism, meaning $\alpha$ and $\alpha^{-1}$ are continuous. If $\alpha$ is an automorphism and $V$ is a compact open subgroup, the set $\alpha^{-1}(V)$ is compact and open. The cosets of the subgroup $\alpha^{-1}(V) \cap V$ form an open cover of $V$, and since $V$ is compact, the index $[V : \alpha^{-1}(V) \cap V]$ is finite. Define the *scale* of $\alpha$, denoted $s(\alpha)$, to be the minimum such index over all compact open subgroups $V$ of $G$. Then $s(\alpha)$ is the minimum of a set of positive integers and $s : \text{Aut}(G) \to \mathbb{Z}^+$ is a well defined function. A subgroup $V$ for which $s(\alpha)$ is attained is called *minimizing* for $\alpha$.

In the case that $\alpha = \alpha_x : g \mapsto xgx^{-1}$ is an inner automorphism, the scale function induces a function from the group to $\mathbb{Z}^+$, also denoted by $s$, which enjoys the following properties.

**Proposition 2.1 ([23, 24]).** *Let $s : G \to \mathbb{Z}^+$ be the scale function on $G$. Then*

(i)  *$s$ is continuous;*
(ii)  *for each $x \in G$ and $n \in \mathbb{N}$, $s(x^n) = s(x)^n$;*
(iii)  *if $V$ is minimising for $x$ then $V$ is minimising for $x^i$ for all $i \in \mathbb{Z}$;*
(iv)  *$s$ is invariant under conjugation, that is, $s(x) = s(yxy^{-1})$ for any $x, y \in G$;*
(v)  *$s(x) = s(x^{-1}) = 1$ if and only if there is a compact open subgroup $V$ with $x^{-1}Vx = V$.*

Note that if $G$ has the discrete topology then the scale of any element is 1, since $V = \{1\}$ is open and compact (and obviously for each $x \in G, x^{-1}\{1\}x = \{1\}$).

The continuity of the scale function implies the following.

**Corollary 2.2.** *If $B$ is a dense subset of a totally disconnected locally compact group $G$, then $\{s(b) \mid b \in B\} = \{s(g) \mid g \in G\}$.*

**Proof.** If $g \in G$ has scale $s(g) = n$, then the inverse image $U$ of the open set $\{n\}$ in $\mathbb{Z}^+$ under the scale function is an open set in $G$. Since $B$ is dense in $G$, every open set contains points from $B$, so there is a point $b \in U \cap B$ with $s(b) = n$. □

We will make use of an asymptotic formula of Möller which makes it possible to use arbitrary compact open subgroups to calculate the scale function.

**Theorem 2.3 ([14] Theorem 7.7).** *For any compact open subgroup $V$,*

$$s(x) = \lim_{k \to \infty} \left[ V : V \cap x^{-k}Vx^k \right]^{\frac{1}{k}}.$$

## 3. Baumslag-Solitar Groups

In this section we collect some facts that will be useful in computing scales later on. Let $\mathrm{BS}(m, n)$ be the group with presentation $\langle a, t \mid ta^m t^{-1} = a^n \rangle$ for non-zero integers $m$ and $n$.

Define $\rho(w)$ to be the number of $t$ letters minus the number of $t^{-1}$ letters in the word $w \in \{a^{\pm 1}, t^{\pm 1}\}^*$. Recall Britton's lemma [13], which states that if a freely reduced nonempty word in the generators $a^{\pm 1}, t^{\pm 1}$ for $\mathrm{BS}(m, n)$ is equal to the identity element, it must contain a subword of the form $ta^{cm}t^{-1}$ or $t^{-1}a^{cn}t$ for some $c \in \mathbb{Z}$ (such a subword is called a *pinch*). Replacing $ta^{cm}t^{-1}$ or $t^{-1}a^{cn}t$ by $a^{cn}$ or $a^{cm}$ in a word is called *removing a pinch*.

**Lemma 3.1.** *If $w, u \in \{a^{\pm 1}, t^{\pm 1}\}^*$ represent the same element in $\mathrm{BS}(m, n)$ then $\rho(w) = \rho(u)$.*

**Proof.** If $w$ or $u$ contains a pinch, removing them does not change the respective values of $\rho$, so remove them (since each word has a finite number of $t$ letters, this process will terminate). Since $wu^{-1}$ equals the identity in $\mathrm{BS}(m, n)$ it must contain a pinch by Britton's lemma. Assuming there are no pinches in $w$ or $u^{-1}$, the pinch must consist of one $t$ letter in one subword and one $t^{-1}$ letter in the other. Removing all pinches in $wu^{-1}$ until we obtain the empty word gives the result. □

It follows that $\rho$ is an invariant of group elements, and is called the *t-exponent sum* for $w$. The first author has made extensive use of the $t$-exponent sum to prove facts about Baumslag-Solitar groups [5–8].

**Lemma 3.2.** *Each element in $\mathrm{BS}(1, n)$ can be represented uniquely in the form $t^{-p}a^q t^r$ with $p, q, r \in \mathbb{Z}$, $p, r \geq 0$ and $n$ dividing $q$ only if $p = 0$ or $r = 0$.*

**Proof.** Let $w \in \{a^{\pm 1}, t^{\pm 1}\}^*$ be a word representing an element of $\mathrm{BS}(1, n)$. Applying the moves $ta^{\pm 1} \to a^{\pm n}t$, $a^{\pm 1}t^{-1} \to t^{-1}a^{\pm n}$, and free cancellation, $w$ is equal in the group to a word of the form $t^{-p}a^q t^r$ with $p, q, r \in \mathbb{Z}$, $p, r \geq 0$. If $n$ divides $q$ and $p, r > 0$ then the word contains a pinch $t^{-1}a^{nc}t$ which can be replaced by $a^c$. Repeating this gives the result. □

Note that $\mathrm{BS}(1, n)$ is isomorphic to the linear group, generated by

$$a = \begin{bmatrix} 1 & 1 \\ 0 & 1 \end{bmatrix} \quad \text{and} \quad t = \begin{bmatrix} n & 0 \\ 0 & 1 \end{bmatrix}.$$

This group is isomorphic to

$$G = \left\{ \begin{bmatrix} n^\rho & z \\ 0 & 1 \end{bmatrix} \mid \rho \in \mathbb{Z}, \ z \in \mathbb{Z}[1/n] \right\} \tag{1}$$

and Lemma 3.2 thus represents a matrix in this group as

$$\begin{bmatrix} n^{-p} & 0 \\ 0 & 1 \end{bmatrix} \begin{bmatrix} 1 & q \\ 0 & 1 \end{bmatrix} \begin{bmatrix} n^r & 0 \\ 0 & 1 \end{bmatrix}.$$

**Lemma 3.3.** *The subgroup $\langle a \rangle$ has a nontrivial subgroup that is normal in* $\mathrm{BS}(m, n)$ *if and only if* $|m| = |n|$.

**Proof.** If $|m| = |n|$, then $\langle a^m \rangle$ is normal, in fact central, in $\mathrm{BS}(m, n)$. If $|m| \neq |n|$, assume without loss of generality that $|m| < |n|$, and suppose $K \leq \langle a \rangle$ is a nontrivial normal subgroup of $\mathrm{BS}(m, n)$. Then $K = \langle a^k \rangle$ for some positive integer $k$. Let $s$ be the largest integer such that $k = q|n|^s + r$ for some $0 \leq q < |n|$ and $0 \leq r < |n|^s$. If $r = 0$ then $t^{-s}a^k t^s = a^{q(\pm m)^s}$ which is not in $K$ since $|qm^s| < |qn^s| = k$, and if $r > 0$ then $t^{-s}a^k t^s$ is not in $\langle a \rangle$.                                                                □

## 4. Commensurated Subgroups

Define a relation $\sim$ on the set of subgroups of an abstract group $G$ by $H \sim K$ if $H \cap K$ is finite index in both $H$ and $K$. We say that $H$ and $K$ are *commensurable* if $H \sim K$. One may verify that being commensurable is an equivalence relation.[1]

A subgroup $H$ is *commensurated* by $G$ if for each $g \in G$ the subgroups $H$ and $gHg^{-1}$ are commensurable. If $x, y \in G$, $xHx^{-1} \sim H$ and $H \sim yHy^{-1}$ then $xyHy^{-1}x^{-1} \sim H$ (since $H \sim K$ implies $xHx^{-1} \sim xKx^{-1}$). It follows that if $G$ is generated by a set $X$, then $H$ is commensurated by $G$ if and only if $xHx^{-1} \sim H$ for all $x \in X$. This gives a fast way to check for commensurated subgroups in finitely generated groups.

One example of a commensurated subgroup is $\mathrm{SL}(k, \mathbb{Z})$ in $\mathrm{SL}(k, \mathbb{Q})$, and in this important case the construction about to be given in Section 5 yields the embedding of $\mathrm{SL}(k, \mathbb{Q})$ into $\mathrm{SL}(k, \mathbb{A}_f)$, where $\mathbb{A}_f$ denotes the ring of finite adeles. Restricting to the subgroup $G$ of $\mathrm{SL}(2, \mathbb{Q})$ appearing in Equation (1), this construction yields an embedding of $G$ into a group isomorphic to

$$\left\{ \begin{bmatrix} n^\rho & z \\ 0 & 1 \end{bmatrix} \mid \rho \in \mathbb{Z}, \ z \in \mathbb{Q}_{p_1} \times \cdots \times \mathbb{Q}_{p_l} \right\}, \tag{2}$$

where $p_1, \ldots, p_l$ are the prime divisors of $n$. In the case of Baumslag-Solitar groups, the cyclic subgroup $\langle a \rangle$ is commensurated because $t\langle a \rangle t^{-1} \cap \langle a \rangle = \langle a^n \rangle$ is finite index in both $t\langle a \rangle t^{-1}$ and $\langle a \rangle$, and similarly $t^{-1}\langle a \rangle t \cap \langle a \rangle = \langle a^m \rangle$ is finite index in both $t^{-1}\langle a \rangle t$ and $\langle a \rangle$. Recalling the isomorphism

---

[1] Clearly $\sim$ is reflexive and symmetric. That $\sim$ is transitive follows because, for subgroups $H$, $K$ and $L$, $[H : H \cap L] \leq [H : H \cap K \cap L] = [H : H \cap K][H \cap K : H \cap K \cap L]$ and $[H \cap K : H \cap K \cap L] \leq [K : K \cap L]$.

between $BS(1, n)$ and $G$, applying the construction to $BS(m, n)$ will thus extend this embedding of matrix groups.

## 5. Construction of the Totally Disconnected Group

Let $G$ be an abstract group with commensurated subgroup $H$. The action of $G$ on $G/H$ given by $g'.gH = (g'g)H$, $(g, g' \in G)$, determines a homomorphism $\pi : G \to \mathrm{Sym}(G/H)$. For each $x \in \mathrm{Sym}(G/H)$ and each finite subset $F$ of $G/H$, define

$$\mathcal{N}(x, F) = \{y \in \mathrm{Sym}(G/H) \mid y(gH) = x(gH) \; \forall (gH) \in F\}.$$

Note that if $y \in \mathcal{N}(x, F)$ then $\mathcal{N}(x, F) = \mathcal{N}(y, F)$. If $\mathcal{N}(x_1, F_1) \cap \mathcal{N}(x_2, F_2)$ is nonempty then it contains some element $y$, so $N(x_1, F_1) = N(y, F_1)$ and $N(x_2, F_2) = N(y, F_2)$. Then

$$\mathcal{N}(x_1, F_1) \cap \mathcal{N}(x_2, F_2) = \mathcal{N}(y, F_1) \cap \mathcal{N}(y, F_2) = \mathcal{N}(y, F_1 \cup F_2).$$

It follows that $\{\mathcal{N}(x, F) \mid x \in G, \; F \subseteq G/H \text{ finite}\}$ forms a basis for a topology on $\mathrm{Sym}(G/H)$.

**Lemma 5.1.** *The topology defined on* $\mathrm{Sym}(G/H)$ *is Hausdorff. Hence the subspace topology on* $\pi(G)$ *induces a Hausdorff topology on* $G/\ker \pi$.

**Proof.** If $x, y \in \mathrm{Sym}(G/H)$ are distinct, then the neighbourhoods $\mathcal{N}(x, \{gH\})$ and $\mathcal{N}(y, \{gH\})$ are disjoint for some $g \in G$. Hence $\mathrm{Sym}(G/H)$ is a Hausdorff topological group. The second claim is justified by the First Isomorphism Theorem. $\qquad\square$

Note that $\ker \pi$ is a subgroup of $H$ and that $H/\ker \pi$ is commensurated by $G/\ker \pi$. The topology on $G/\ker \pi$ may then be defined equivalently by considering the injection of $G/\ker \pi$ into $\mathrm{Sym}(G/\ker \pi)/(H/\ker \pi)$. From now on it is assumed that the kernel is trivial.

**Lemma 5.2.** *The topology defined is totally disconnected.*

**Proof.** If $x, y \in \mathrm{Sym}(G/H)$ are distinct, there is a coset $gH$ with $x(gH) \neq y(gH)$. Then $\mathcal{N}(x, \{gH\})$ is an open set containing $x$, and its complement,

$$\bigcup \{\mathcal{N}(z, \{gH\}) \mid z(gH) \neq x(gH)\},$$

is open and contains $y$. $\qquad\square$

It is a standard result that $\mathrm{Sym}(G/H)$ is a topological group with the topology defined. In keeping with the expository nature of this article we include a proof of this fact.

**Lemma 5.3.** *The map* $* : (x, y) \mapsto xy^{-1}$ *is continuous.*

**Proof.** The map is continuous at $(x, y)$ if for any open set $V$ containing $xy^{-1}$ there is an open set $U \subseteq \mathrm{Sym}(G/H) \times \mathrm{Sym}(G/H)$ containing $(x, y)$ whose image is in $V$.

Since $V$ is open and contains $xy^{-1}$, it contains a set $\mathcal{N}(xy^{-1}, F)$ for some finite $F$. Take $U = \mathcal{N}(x, y^{-1}F) \times \mathcal{N}(y, y^{-1}F)$. If $(a, b) \in U$, then $b(y^{-1}(gH)) = y(y^{-1}(gH)) = (gH)$ and $b^{-1}(gH) = y^{-1}(gH)$ for all $gH \in F$ and $ab^{-1}(gH) = a(y^{-1}(gH)) = xy^{-1}(gH)$ for all $gH \in F$. Hence $ab^{-1} \in \mathcal{N}(xy, F) \subseteq V$. $\qquad \square$

It follows that $\pi(G)$ is also a (Hausdorff and totally disconnected) topological group, with the subspace topology induced from $\mathrm{Sym}(G/H)$. Define $G/\!/H$ to be the closure of $\pi(G)$ in $\mathrm{Sym}(G/H)$, and define $\tilde{H}$ to be the closure of $\pi(H)$. Then

$$\tilde{H} = \{x \in G/\!/H \mid x(H) = H\} = G/\!/H \cap \mathcal{N}(e, \{H\})$$

is an open subgroup of $G/\!/H$.

It is shown next that $G/\!/H$ is locally compact. The following fact is used in this argument and also later in the calculation of the scale.

**Lemma 5.4.** *Suppose that $V$ is an open subgroup of $G/\!/H$. Then, setting $U = V \cap \pi(G)$, each $U$-orbit $U.(gH) \subseteq G/H$ is equal to $V.(gH)$.*

**Proof.** Since $V$ is open and $\pi(G)$ is dense in $G/\!/H$, $U$ is a dense subgroup of $V$. Hence the intersection $\mathcal{N}(x, \{gH\}) \cap U$ is nonempty for any $x \in V$ and there is $u \in U$ such that $x(gH) = u(gH)$. $\qquad \square$

When applied to the orbits $U.xH$ and $V.xH$, the Orbit-Stabiliser theorem implies the following.

**Corollary 5.5.** *If $V$ and $U$ are as in Lemma 5.4 and $x \in G/\!/H$, then*

$$[V : x^{-1}Vx \cap V] = [U : x^{-1}Ux \cap U].$$

**Lemma 5.6.** *The subspace topology defined on $G/\!/H$ is locally compact.*

**Proof.** Lemma 5.4 implies that $H$ and $\tilde{H}$ act on $G/H$ by permuting cosets in the blocks $H(gH)$ ($g \in G$) so that, making a natural identification,

$$\tilde{H} \leq \prod_{H(gH) \subset G/H} \mathrm{Sym}(H(gH)). \tag{3}$$

It may be checked that $\tilde{H}$ is a closed subgroup under this identification and that the topology on $\tilde{H}$ is equivalent to the subspace topology for the product topology on $\prod_{H(gH) \subset G/H} \mathrm{Sym}(H(gH))$. Since commensurability of $H$ implies that each block $H(gH)$ is finite, it follows by Tychonov's theorem (see for example [15] Theorem 6.50) that $\tilde{H}$ is compact, and is thus a compact, open neighbourhood of $e$. □

The subgroup $\langle a \rangle$ is commensurated by the group $\mathrm{BS}(m,n)$ for $m, n \neq 0$ and the kernel of the map $\pi : \mathrm{BS}(m,n) \rightarrow \mathrm{Sym}(\mathrm{BS}(m,n)/\langle a \rangle)$ is a subgroup of $\langle a \rangle$. Then $\ker \pi$ is trivial if $|m| \neq |n|$, by Lemma 3.3, and $\mathrm{BS}(m,n)$ embeds as a dense subgroup of the totally disconnected, locally compact group $\mathrm{BS}(m,n)/\!\!/\langle a \rangle$. This group is denoted by $G_{m,n}$ in the sequel.

If $|m| = |n|$, we have $a^i(gH) = gH$ for all $g \in \mathrm{BS}(m, \pm m)$ if and only if $m$ divides $i$ and so $\ker(\pi) = \langle a^m \rangle$. Factoring $\mathrm{BS}(m, \pm m)$ by $\ker(\pi)$ then makes the commensurated subgroup $\langle a \rangle / \langle a^m \rangle$ finite, whence $\tilde{H}$ is finite and discrete. Hence

$$(\mathrm{BS}(m, \pm m)/\langle a^m \rangle)/\!\!/(\langle a \rangle / \langle a^m \rangle) \cong \mathrm{BS}(m, \pm m)/\langle a^m \rangle$$

(which we will denote as $G_{m, \pm m}$) and is discrete. Since the construction yields nothing new in this case, the focus of the rest of the article is on the case $|m| \neq |n|$.

### 5.1. *An alternative construction of $G_{m,n}$*

Gal and Januszkiewicz [9] embed $\mathrm{BS}(m,n)$ into a topological group as follows. Recall the *Bass-Serre tree* for a graph of groups [1, 19]. In the case of $\mathrm{BS}(m,n)$ the tree has a vertex for each coset of $\langle a \rangle$, and $(u\langle a \rangle, v\langle a \rangle)$ is a directed edge labeled $t^{\pm 1}$ if $v\langle a \rangle = ua^i t^{\pm 1}\langle a \rangle$. As an example, part of the Bass-Serre tree for the group $\mathrm{BS}(2,3)$ is shown in Figure 1. Note that each vertex will have degree 5 in this example (two outgoing edges labeled $t$ and three labeled $t^{-1}$).

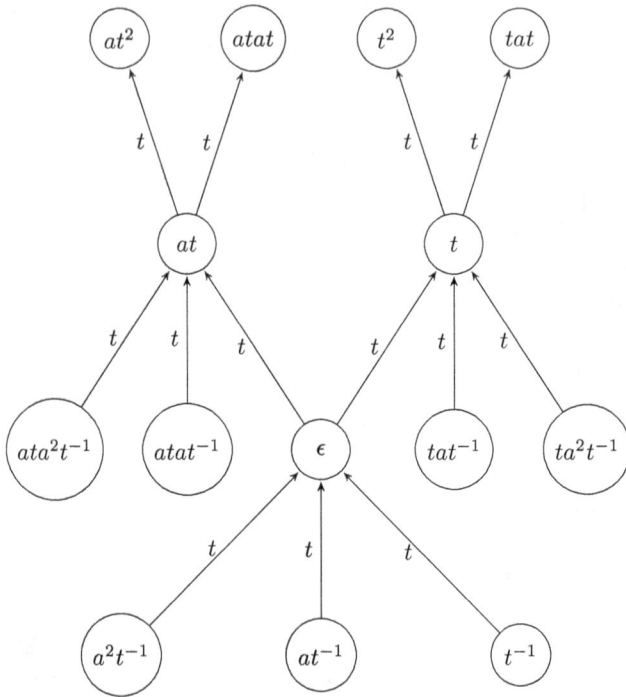

Figure 1.    Part of the Bass-Serre tree for BS(2, 3). Edges labeled $t^{-1}$ correspond to traveling in reverse direction along edges labeled $t$.

Let $T_{m,n}$ denote the Bass-Serre tree for BS($m, n$) with vertex set $V(T_{m,n})$, and Aut($T_{m,n}$) the group of automorphisms of $T_{m,n}$. Let

$$\{N(x, F) \mid x \in \text{Aut}(T_{m,n}), F \subset_{\text{finite}} V(T_{m,n})\}$$

be a base of neighborhoods for a topology on Aut($m, n$), where

$$N(x, F) = \{\beta \in \text{Aut}(T_{m,n}) \mid x.v = \beta.v \text{ for all } v \in F\}.$$

Since vertices correspond to (left) cosets of $\langle a \rangle$ in BS($m, n$) then Aut($T_{m,n}$) $\subseteq$ Sym(BS($m, n$)$/\langle a \rangle$).

To see that the closure of BS($m, n$) in Aut($T_{m,n}$) coincides with $G_{m,n}$ defined above, if $x \in G_{m,n}$ then either $x \in$ BS($m, n$) or every open set in Sym(BS($m, n$)$/\langle a \rangle$) containing $x$ contains an element of BS($m, n$). So for every finite set of cosets (vertices of $T_{m,n}$) $x$ agrees with some element of BS($m, n$) and so preserves adjacencies of the Bass-Serre tree.

## 6. Computing Scales for $G_{m,n}$

In this section the scales of elements of $G_{m,n}$ are computed when $|m| \neq |n|$. For convenience we will abuse notation and identify elements and subsets of $BS(m,n)$ with their images under the embedding $\pi$.

Since $BS(m,n)$ is dense in $G_{m,n}$, Corollary 2.2 shows that in order to compute scales in $G_{m,n}$, it suffices to compute the scale of elements in $BS(m,n)$. If $V$ is a compact open subgroup of $G_{m,n}$ then Corollary 5.5 shows that the index $[V : x^{-1}Vx \cap V]$ is the same as the index $[U : x^{-1}Ux \cap U]$ where $U$ is the intersection of $V$ with $BS(m,n)$. It follows that scales in $G_{m,n}$ can be computed by working entirely in $BS(m,n)$.

We start by considering the case when one of $m, n$ is a proper divisor of the other. Note that this includes the case that one of $m, n$ is $\pm 1$, in which case the scale could be computed more easily by working directly with the topological matrix group in Equation (2), or with its dense subgroup described in Equation (1).

Suppose that $n = mr$ with $|r| > 1$. Then

$$t^{-1}\langle a^m \rangle t \cap \langle a^m \rangle = \langle a^m \rangle,$$
$$t^{-1}\langle a^{mr^j} \rangle t \cap \langle a^m \rangle = \langle a^{mr^{j-1}} \rangle, \quad \text{and}$$
$$t\langle a^{mr^i} \rangle t^{-1} \cap \langle a^m \rangle = \langle a^{mr^{i+1}} \rangle$$

for any $i \geq 0, j > 0$. These facts may encoded in a graph. Define $\Lambda$ to be the labeled directed graph having nodes $N = \{mr^i \mid i \in \mathbb{N}\}$, and directed edges

$$E = \left\{ (m, m), (mr^j, mr^{j-1}), (mr^i, mr^{i+1}) \mid j > 0, i \geq 0 \right\},$$

where the loop $(m, m)$ and edges $(mr^j, mr^{j-1})$ are labeled $t$, and edges $(mr^i, mr^{i+1})$ are labeled $t^{-1}$. Then $(x, y)$ is an edge labeled $t^\epsilon$ if and only if

$$t^{-\epsilon}\langle a^x \rangle t^\epsilon \cap \langle a^m \rangle = \langle a^y \rangle.$$

A picture of part of $\Lambda$ is shown in Figure 2.

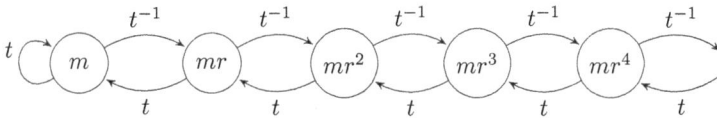

Figure 2. Part of the graph $\Lambda$. An edge from $x$ to $y$ labeled $t^\epsilon$ means that $t^{-\epsilon}\langle a^x \rangle t^\epsilon \cap \langle a^m \rangle = \langle a^y \rangle$.

Recall Britton's lemma and the notion of a pinch from Section 3.

**Lemma 6.1.** *Let $u = a^\eta t^\epsilon v$ with $\eta \in \mathbb{Z}$ and $\epsilon = \pm 1$ be a freely reduced word in $\mathrm{BS}(m, n)$ with no pinches. Then*

$$u^{-1} \langle a^i \rangle u \cap \langle a \rangle = v^{-1} \left( t^{-\epsilon} \langle a^i \rangle t^\epsilon \cap \langle a \rangle \right) v \cap \langle a \rangle .$$

**Proof.** If the word $u^{-1}(a^i)^j u = v^{-1} t^{-\epsilon} a^{-\eta} (a^i)^j a^\eta t^\epsilon v$ is in $\langle a \rangle$ then it must contain a pinch. There is no pinch within $t^\epsilon v$ (or $v^{-1} t^{-\epsilon}$) so the pinch must be $t^{-\epsilon} (a^i)^j t^\epsilon$, so this subword must be in $\langle a \rangle$. □

**Lemma 6.2.** *Let $w$ be a freely reduced word with no pinches, and let $p(w)$ be the word in the free monoid generated by $t$ and $t^{-1}$ obtained by removing all $a^{\pm 1}$ letters from $w$. Consider the path in $\Lambda$ starting at the node $mr^i$, whose edge labels follow the sequence $p(w)$. Then*

$$w^{-1} \langle a^{mr^i} \rangle w \cap \langle a^m \rangle = \langle a^{mr^k} \rangle$$

*where $mr^k$ is the label of the node at the end of this path.*

**Proof.** If $w$ has no $t^{\pm 1}$ letters then the statement is clearly true. Suppose for induction the statement is true for words with $q$ $t^{\pm 1}$ letters. Let $u$ be a freely reduced word with no pinches with $q + 1$ $t^{\pm 1}$ letters, and write $u = a^\eta t^\epsilon v$, where $\epsilon = \pm 1$ and $\eta \in \mathbb{Z}$. Then by Lemma 6.1 we have

$$u^{-1} \langle a^{mr^i} \rangle u \cap \langle a^m \rangle = (v^{-1} t^{-\epsilon} \langle a^{mr^i} \rangle t^\epsilon v \cap \langle a \rangle) \cap \langle a^m \rangle$$

$$= (v^{-1} (t^{-\epsilon} \langle a^{mr^i} \rangle t^\epsilon \cap \langle a \rangle) v \cap \langle a \rangle) \cap \langle a^m \rangle$$

$$= (v^{-1} \langle a^{mr^j} \rangle v \cap \langle a \rangle) \cap \langle a^m \rangle$$

where $(mr^i, mr^j)$ is a directed edge labeled $t^\epsilon$ in $\Lambda$. The path starting at $mr^i$ labeled by $p(u)$ consists of this edge to $mr^j$, followed by a path labeled $p(v)$. By inductive assumption

$$(v^{-1} \langle a^{mr^j} \rangle v \cap \langle a \rangle) \cap \langle a^m \rangle = v^{-1} \langle a^{mr^j} \rangle v \cap \langle a^m \rangle = \langle a^{mr^k} \rangle$$

since $mr^k$ is the endpoint of the path $p(v)$ starting at $mr^j$. The result follows. □

Consider the following example which shows how the previous lemma can be used to compute the scale. Let $w = t^4 a t^{-2} a$. Then by tracing the path $p(w) = t^4 t^{-2}$ starting at $m$ through $\Lambda$ we have $w^{-1} \langle a^m \rangle w \cap \langle a^m \rangle = \langle a^{mr^2} \rangle$. Since scale is invariant under conjugation, we could instead consider $u = t^{-2} a t^4 a$, in which case the path $p(u) = t^{-2} t^4$ starting at $m$ ends at

$m$, so $u^{-1}\langle a^m\rangle u \cap \langle a^m\rangle = \langle a^m\rangle$, and we see that (the closure of) $\langle a^m\rangle$ is minimising for $u$, and so $s(w) = s(u) = 1$.

This pre-conjugating step is the key to proving Proposition 6.4 below. We will need the following fact.

**Lemma 6.3.** *Let $x \in \mathrm{BS}(m,n)$. Then $x$ is conjugate to a word $w \in \{a^{\pm 1}, t^{\pm 1}\}^*$ such that $ww$ is freely reduced and contains no pinches.*

**Proof.** Let $y \in \{a^{\pm 1}, t^{\pm 1}\}^*$ be a word equal to the element $x$ in $\mathrm{BS}(m,n)$, and consider the following four moves:

(1) if $y$ is not freely reduced, then removing a cancelling pair reduces the length of $y$, and gives a word $z$ equal to $y$ and shorter in length.
(2) if $y$ is freely reduced and $yy$ is not freely reduced, then we must have $y = c^{\pm 1} z c^{\mp 1}$, so $z$ is conjugate to $y$ and shorter in length.
(3) if $y$ contains a pinch, then removing gives a word $z$ equal to $y$ with fewer $t$ letters.
(4) if $y$ has no pinches and $yy$ contains a pinch, then we must have $y = a^i t^{-1} v t a^j$ with $i + j = km$, or $y = a^i t v t^{-1} a^j$ with $i + j = kn$. Conjugating by $a^i t$ [or $a^i t^{-1}$ respectively] then removing the pinch we obtain a word $z = v a^{nk}$ [or $z = v a^{mk}$] conjugate to $y$ with fewer $t$ letters.

Define an order $\prec$ on words $y, z \in \{a^{\pm 1}, t^{\pm 1}\}^*$ by $z \prec y$ if $y$ has more $t$ letters than $z$, or they have the same number of $t$ letters but $y$ is longer than $z$. Then since each of the four moves produces a word that is shorter in this order, after a finite number of moves the word $w$ with the required properties is obtained. □

**Proposition 6.4.** *If $x \in \mathrm{BS}(m, mr)$ for $|r| > 1$ is equal to a word of $t$-exponent sum $\rho$, then $s(x) = 1$ if $\rho \geq 0$ and $s(x) = |r|^{|\rho|}$ if $\rho < 0$.*

**Proof.** Suppose $x$ has nonnegative $t$-exponent sum $\rho$ and, applying Lemma 6.3, choose $w$ conjugate to $x$ and with $ww$ freely reduced and containing no pinches. We consider two cases.

If $w$ contains no $t^{-1}$ letters, then the path starting at $m$ labeled by $p(w) = t^\rho$ ends at $m$, so Lemma 6.2 we have $w^{-1}\langle a^m\rangle w \cap \langle a^m\rangle = \langle a^m\rangle$ and $s(w) = s(x) = 1$.

Otherwise $w$ contains a $t^{-1}$ letter, and since $\rho \geq 0$ it is cyclically conjugate to a word $u = t^{-1} w_1 t a^\eta$, which is freely reduced and contains no pinches since it is a subword of $ww$. Suppose $u$ contains $k$ $t$ letters, and consider the word $v = t^{-k}ut^k$. Note that $k > \rho$. This word is freely reduced and contains no pinches since $u$ starts with $t^{-1}$ and ends with $ta^n$.

Now consider the path in $\Lambda$ starting at $m$ labeled $p(v)$. It first travels $k$ steps to the right, then since $p(u)$ has exactly $k$ $t$ letters, it travels left for each $t$ and right for each $t^{-1}$ in $p(u)$ (that is, we have ensured that it never enters the loop at $m$). Since the subpath labeled $p(u)$ started at $mr^k$ and the $t$-exponent sum of $u$ is $\rho \geq 0$, the subpath ends at $mr^{k-\rho}$. Finally the path $p(v)$ travels $k - \rho$ edges left and the $\rho$ edges around the loop at $m$, and so $u^{-1}\langle a^m \rangle u \cap \langle a^m \rangle = \langle a^m \rangle$ and $s(u) = s(w) = s(x) = 1$.

To compute the scale of $x$ of negative $t$-exponent $\rho$, choose $w$ conjugate to $x$ and with $ww$ freely reduced and containing no pinches. If $w$ contains no $t$ letters then $w^{-1}$ contains no $t^{-1}$ letters, so by the argument above the closure of $\langle a^m \rangle$ is minimising for $w^{-1}$, so is minimising for $w$ by Proposition 2.1(iii). Applying Lemma 6.2, the path starting at $m$ labeled by $p(w) = t^\rho$ ends at $mr^{|\rho|}$ so $w^{-1}\langle a^m \rangle w \cap \langle a^m \rangle = \langle a^{mr^{|\rho|}} \rangle$ which has index $|r|^{|\rho|}$ in $\langle a^m \rangle$, so $s(w) = s(x) = |r|^{|\rho|}$.

If $w$ contains $t$ letters, then $w$ is cylically conjugate to $u = t^{-1}u_1ta^n$, and so conjugate to $v = t^{-k}ut^k$, where $k$ is the number of $t$ letters in $u$, and $u, v$ are freely reduced and contain no pinches.

Since $v^{-1} = t^{-k}(a^{-n}t^{-1}u_1^{-1}t)t^k$, the argument for the second case above shows that $\langle a^m \rangle$ is minimising for $v^{-1}$, so it is minimising for $v$. Using Lemma 6.2 once again, the path in $\Lambda$ starting at $m$ labeled $p(v)$ travels $k$ steps to the right, then since $p(u)$ has exactly $k$ $t$ letters, it travels left for each $t$ and right for each $t^{-1}$ in $p(u)$ (once again, we have ensured that it never enters the loop at $m$).

Since the subpath labeled $p(u)$ started at $mr^k$ and never enters the loop, and the $t$-exponent sum of $u$ is $\rho < 0$, the subpath ends at $mr^{k+|\rho|}$. Finally the path $p(v)$ travels $k$ edges left to end at $mr^{|\rho|}$, and so $v^{-1}\langle a^m \rangle v \cap \langle a^m \rangle = \langle a^{mr^{|\rho|}} \rangle$ which has index $|r|^{|\rho|}$ in $\langle a^m \rangle$ so $s(v) = s(x) = |r|^{|\rho|}$. $\qquad \square$

We now turn to the case where neither $m$ or $n$ is a divisor of the other. In this case we make use of Möller's formula (Theorem 2.3), choosing $V$ to be closure of $\langle a \rangle$. For this, it is necessary to compute $w^{-k}\langle a \rangle w^k \cap \langle a \rangle$ for arbitrary natural numbers $k$.

**Corollary 6.5.** *Let $x \in BS(m, n)$ with neither $m$ or $n$ is a divisor of the other. Then $x$ is conjugate to a word $w \in \{a^{\pm 1}, t^{\pm 1}\}^*$ such that $w^k$ is freely reduced and contains no pinches for all $k \in \mathbb{N}$.*

**Proof.** By Lemma 6.3, $x$ is conjugate to $w$ such that $ww$ is freely reduced and contains no pinches. If $w^k$ contains a pair $yy^{-1}$ for some $y \in \{a^{\pm 1}, t^{\pm 1}\}$, then $yy^{-1}$ must lie either in $w$ or $ww$. If $w^k$ contains a pinch, and the pinch is of the form $t^\epsilon u w^i v t^{-\epsilon}$ where $v, u$ are a prefix and suffix of $w$, then $w^i$ must consist of $a^{\pm 1}$ letters, meaning $i = 0$, so $ww$ contains the pinch. $\quad\square$

Define a directed graph $\Omega$ as follows. Let $l = \mathrm{lcm}(m, n)$. The nodes of $\Omega$ are labeled by integers from the set

$$N = \left\{ 1, m\left(\frac{l}{n}\right)^i, l\left(\frac{l}{n}\right)^i, n\left(\frac{l}{m}\right)^i, l\left(\frac{l}{m}\right)^i \mid i \in \mathbb{N} \right\}.$$

These integers are distinct because neither $m$ nor $n$ divides the other.

We define directed labeled edges in $\Omega$ in a similar way to edges in $\Lambda$: $(x, y)$ is an edge from $x$ to $y$ labeled $t^\epsilon$ for $\epsilon = \pm 1$ if $t^{-\epsilon}\langle a^x \rangle t^\epsilon \cap \langle a \rangle = \langle a^y \rangle$. Note that this time we intersect with $\langle a \rangle$ rather than $\langle a^m \rangle$.

**Lemma 6.6.** *The edge set $E$ of $\Omega$ consists of the following edges, for all positive integers $i, j$:*

$(1, m)$,

$\left( m, m\left(\frac{l}{n}\right) \right)$,

$\left( m\left(\frac{l}{n}\right)^i, m\left(\frac{l}{n}\right)^{i+1} \right)$,

and

$(n, m)$,

$\left( n\left(\frac{l}{m}\right)^i, l\left(\frac{l}{m}\right)^{i-1} \right)$,

$\left( l\left(\frac{l}{n}\right)^i, m\left(\frac{l}{n}\right)^{i+1} \right)$,

$\left( l\left(\frac{l}{n}\right)^i \left(\frac{l}{m}\right)^j, l\left(\frac{l}{n}\right)^{i-1} \left(\frac{l}{m}\right)^{j+1} \right)$,

$(1, n)$,

$\left( n, n\left(\frac{l}{m}\right) \right)$,

$\left( n\left(\frac{l}{m}\right)^i, n\left(\frac{l}{m}\right)^{i+1} \right)$,

$(m, n)$,

$\left( m\left(\frac{l}{n}\right)^i, l\left(\frac{l}{n}\right)^{i-1} \right)$,

$\left( l\left(\frac{l}{m}\right)^i, n\left(\frac{l}{m}\right)^{i+1} \right)$,

$\left( l\left(\frac{l}{n}\right)^i \left(\frac{l}{m}\right)^j, l\left(\frac{l}{n}\right)^{i+1} \left(\frac{l}{m}\right)^{j-1} \right)$

*where all edges in the left column are labeled $t$, and all edges in the right column are labeled $t^{-1}$.*

**Proof.** Since $1 \in N$ and

$$t^{-1}\langle a \rangle t \cap \langle a \rangle = \langle a^m \rangle$$
$$t\langle a \rangle t^{-1} \cap \langle a \rangle = \langle a^n \rangle$$

we have $m, n \in N$ and $(1, m), (1, n) \in E$ with $(1, m)$ labeled $t$ and $(1, n)$ labeled $t^{-1}$.

Since

$$t^{-1}\langle a^m \rangle t = \{ t^{-1}a^{|m|c}t \ : \ c \in \mathbb{Z} \}$$
$$= \left\{ t^{-1}a^{|m|(q\frac{l}{|m|}+r)}t \ : \ q, r \in \mathbb{Z}, 0 \le r < \frac{l}{|m|} \right\}$$
$$= \left\{ t^{-1}a^{ql}tt^{-1}a^{|m|r}t \ : \ q, r \in \mathbb{Z}, 0 \le r < \frac{l}{|m|} \right\}$$
$$= \left\{ a^{q\frac{l}{n}m}t^{-1}a^{|m|r}t \ : \ q, r \in \mathbb{Z}, 0 \le r < \frac{l}{|m|} \right\}$$

has intersection $\langle a^{\frac{l}{n}m} \rangle$ with $\langle a \rangle$, we have the edge $(m, m\left(\frac{l}{n}\right))$ labeled $t$. A similar argument gives the edge $(n, n\left(\frac{l}{m}\right))$ labeled $t^{-1}$.

More generally for any $i \in \mathbb{N}$ we have

$$t^{-1}\langle a^{m\left(\frac{l}{n}\right)^i} \rangle t = \left\{ t^{-1}a^{|m|\left(\frac{l}{|n|}\right)^i c}t \ : \ c \in \mathbb{Z} \right\}$$
$$= \left\{ t^{-1}a^{|m|\left(\frac{l}{|n|}\right)^i(q\frac{l}{|m|}+r)}t \ : \ q, r \in \mathbb{Z}, 0 \le r < \frac{l}{|m|} \right\}$$
$$= \left\{ t^{-1}a^{ql\left(\frac{l}{|n|}\right)^i}tt^{-1}a^{|m|\left(\frac{l}{|n|}\right)^i r}t \ : \ q, r \in \mathbb{Z}, 0 \le r < \frac{l}{|m|} \right\}$$
$$= \left\{ a^{q\frac{l}{n}\left(\frac{l}{|n|}\right)^i m}t^{-1}a^{|m|\left(\frac{l}{|n|}\right)^i r}t \ : \ q, r \in \mathbb{Z}, 0 \le r < \frac{l}{|m|} \right\}$$

which has intersection $\langle a^{\frac{l}{n}\left(\frac{l}{|n|}\right)^i m} \rangle = \langle a^{\left(\frac{l}{n}\right)^{i+1} m} \rangle$ with $\langle a \rangle$, so we have edge

$$\left( m\left(\frac{l}{n}\right)^i, m\left(\frac{l}{n}\right)^{i+1} \right)$$

labeled $t$ for all $i \in \mathbb{N}$. A similar argument gives the edges

$$\left( n\left(\frac{l}{m}\right)^i, n\left(\frac{l}{m}\right)^{i+1} \right)$$

labeled $t^{-1}$.

In Figure 3 we have drawn part of the graph $\Omega$, with these edges draw vertically down the left and right sides of the picture.

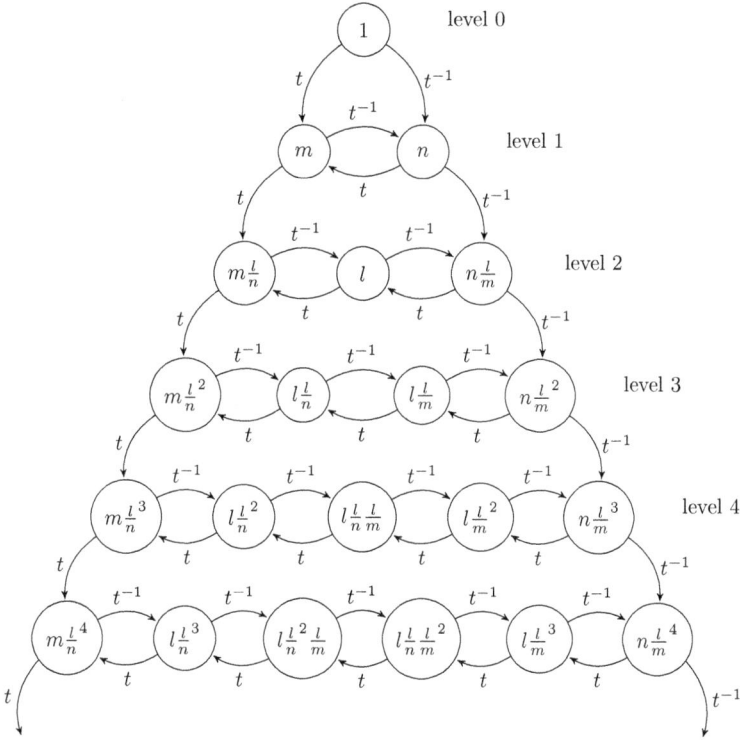

Figure 3.    Part of the graph $\Omega$. An edge from $x$ to $y$ labeled $t^\epsilon$ means that $t^{-\epsilon}\langle a^x\rangle t^\epsilon \cap \langle a\rangle = \langle a^y\rangle$.

The remaining edges (which are drawn horizontally in the picture) all follow from these facts where $c \in \mathbb{N}$:

$$t\langle a^{mc}\rangle t^{-1} = \langle a^{nc}\rangle,$$
$$t^{-1}\langle a^{lc}\rangle t = \langle a^{n\frac{l}{m}c}\rangle,$$
$$t^{-1}\langle a^{nc}\rangle t = \langle a^{mc}\rangle,$$
$$t^{-1}\langle a^{lc}\rangle t = \langle a^{m\frac{l}{n}c}\rangle.$$

Since we have described an edge labeled $t$ and $t^{-1}$ starting from each node in the set $N$, there are no edges other than these.    $\square$

**Corollary 6.7.** *The graph* $\Omega$ *is connected. The length of the shortest directed path from* $m\left(\frac{l}{n}\right)^i$ *to* $n\left(\frac{l}{m}\right)^i$ *is* $i+1$ *for all* $i \in \mathbb{N}$, *and also* $i+1$ *from* $n\left(\frac{l}{m}\right)^i$ *to* $m\left(\frac{l}{n}\right)^i$.

**Proof.** Each node in $N$ can be reached by a path starting at 1 and labeled $t^i t^{-j}$ for $i, j \in \mathbb{N}$, $i \geq j$, as follows.

- The path $t^i$ starting at 1 ends at $m \left(\frac{l}{n}\right)^{i-1}$, for $i \geq 1$.
- The path $t^i t^{-j}$ starting at 1 ends at $l \left(\frac{l}{m}\right)^{j-1} \left(\frac{l}{n}\right)^{i-j-1}$, for $i \geq 2, j \geq 1$, and $j < i$.
- The path $t^i t^{-i}$ starting at 1 ends at $n \left(\frac{l}{m}\right)^{i-1}$, for $i \geq 1$.

The path $t^{-i}$ starting at $m \left(\frac{l}{n}\right)^{i-1}$ ends at $n \left(\frac{l}{m}\right)^{i-1}$ and similarly for the reverse path $t^i$ starting at $n \left(\frac{l}{m}\right)^{i-1}$, so the distance from one side of $\Omega$ to the other at level $i$ is $i$. $\qquad \square$

**Lemma 6.8.** *Let $w \in \mathrm{BS}(m, n)$ be freely reduced and contain no pinches. Let $p(w)$ be the word in the free monoid generated by $t$ and $t^{-1}$ obtained from $w$ by removing all $a^{\pm 1}$ letters. Consider the path starting at the node 1 in $\Omega$ and following the sequence of edges labeled by $p(w)$. Let $r$ be the label of the node at the end of this path. Then $w^{-1}\langle a \rangle w \cap \langle a \rangle = \langle a^r \rangle$.*

**Proof.** The result follows from Lemma 6.1 and induction on the number of $t^{\pm 1}$ letters in $w$. If $w = a^{\eta_0} t a^{\eta_1}$, $\eta_0, \eta_1 \in \mathbb{Z}$ then $w^{-1}\langle a \rangle w \cap \langle a \rangle = a^{-\eta_1} t^{-1} \langle a \rangle t a^{\eta_1} \cap \langle a \rangle = \langle a^m \rangle$ and the path from 1 labeled $t$ ends at $m$, with a similar result for $w = a^{\eta_0} t^{-1} a^{\eta_1}$.

If $w = u t a^{\eta}$, $\eta \in \mathbb{Z}$, say the path $p(u)$ ends at the node $r$. Then $w^{-1}\langle a \rangle w \cap \langle a \rangle = a^{-\eta} t^{-1} u^{-1} \langle a \rangle u t a^{\eta} \cap \langle a \rangle = a^{-\eta} t^{-1} \langle a^r \rangle t a^{\eta} \cap \langle a \rangle = \langle a^s \rangle$ where $(r, s)$ is an edge labeled $t$ in $\Omega$. A similar argument applies for $w = u t^{-1} a^{\eta}$. $\qquad \square$

**Lemma 6.9.** *Let $w \in \{a^{\pm 1}, t^{\pm 1}\}^*$ with $t$-exponent sum $\rho(w)$, and let $t_{\max}(w)$ be the maximum $t$-exponent sum of any prefix of $w$. Write $w = w_1 w_2$ where $w_1$ has $t$-exponent sum $t_{\max}(w)$, and $w_2$ has $t$-exponent sum $\mu(w) = \rho(w) - t_{\max}(w) \leq 0$. In other words, $u_1$ is the longest prefix of $u$ that has maximum $t$-exponent sum, and the $t$-exponent sum of $u_2$ is nonpositive. Let $r$ be the number of $t^{-1}$ letters in $w$. Then if $R > r$, the path $p(t^R w)$ starting at 1 ends at distance $|\mu(w)|$ from the left side of $\Omega$, and at level $R + t_{\max}(w)$.*

**Proof.** If $w$ has length 0 then the statement is true. Suppose for induction the statement is true for all words of length $k$ and let $w$ have length $k + 1$.

If $w = u a^{\pm 1}$ then $t_{\max}(w) = t_{\max}(u)$, $\mu(w) = \mu(u)$ and $p(t^R w) = p(t^R u)$ so the statement is true.

If $w = ut^{-1}$ then $t_{\max}(w) = t_{\max}(u), \mu(w) = \mu(u) - 1$. The path $p(t^R u)$ ends at level $R + t_{\max}(w)$ distance $|\mu(u)|$ from the left side of $\Omega$ by inductive assumption. By Corollary 6.7 the length of level $R + t_{\max}(u)$ is $R + t_{\max}(u)$, so since $r < R$, the path does not reach the right side of $\Omega$, so appending $t^{-1}$ to the path does not change level, and the distance from the left side is increased by 1, so the statement is true for $w$ since $|\mu(w)| = |\mu(u) - 1| = |\mu(u)| + 1$.

If $w = ut$, then we have two cases. If $t_{\max}(w) = t_{\max}(u) + 1$, then $\mu(w) = 0 = \mu(u)$, so the statement is true since $p(t^R u)$ ends on the left side of $\Omega$ and appending a $t$ edge moves it down one level along a vertical edge.

Otherwise $t_{\max}(w) = t_{\max}(u)$. In this case $\mu(u) < 0$, so by inductive assumption $p(t^R u)$ ends at level $R + t_{\max}(u)$ and positive distance $|\mu(u)|$ from the left side of $\Omega$, so appending a $t$ edge does not change the level, and decreases the distance from the left by 1, proving the claim. $\qquad\square$

**Proposition 6.10.** *Assume neither $m$ or $n$ is a divisor of the other. If $x \in \mathrm{BS}(m, n)$ has $t$-exponent sum $\rho$, then $s(x) = \left(\frac{\mathrm{lcm}(m,n)}{|n|}\right)^{\rho}$ if $\rho \geq 0$, and*
$$s(x) = \left(\frac{\mathrm{lcm}(m,n)}{|m|}\right)^{|\rho|} \text{ if } \rho < 0.$$

**Proof.** We first consider the case $\rho \geq 0$. Take $w$ conjugate to $x$ such that $w^k$ is freely reduced and has no pinches, given by Corollary 6.5. If $w$ has no $t^{\pm 1}$ letters then clearly the closure $\langle a \rangle$ is minimising for $w$ are we are done. If $w$ contains no $t^{-1}$ letters, then $p(w^k)$ is a path in $\Omega$ from 1 to $m \left(\frac{\mathrm{lcm}(m,n)}{n}\right)^{\rho^k - 1}$ down the left side $\rho^k$ edges, where $\rho$ is the number of $t$ letters in $w$. It follows (from Lemma 6.8) that $w^{-k}\langle a \rangle w^k \cap \langle a \rangle = \langle a^{m\left(\frac{\mathrm{lcm}(m,n)}{n}\right)^{\rho^k - 1}} \rangle$, so by Theorem 2.3
$$\begin{aligned} s(w) &= \lim_{k \to \infty} \left[V : V \cap w^{-k} V w^k\right]^{\frac{1}{k}} \\ &= \lim_{k \to \infty} \left(|m| \left(\frac{\mathrm{lcm}(m,n)}{|n|}\right)^{\rho^k - 1}\right)^{\frac{1}{k}} \\ &= \left(\frac{l}{|n|}\right)^{\rho} \end{aligned}$$
where $V$ is the closure of $\langle a \rangle$.

Now suppose $w$ contains both $t^{-1}$ and $t$ letters. Then $w$ is conjugate to a word $u = tvt^{-1}a^{\eta}, \eta \in \mathbb{Z}$. Let $N$ be the number of $t^{-1}$ letters in $u$. We will compute the scale of the words $t^{2N} u^k t^{-2N}$ which is conjugate to $x$, is freely reduced and contains no pinches.

Let $t_{\max}$ be the maximum $t$-exponent sum of any prefix of $u$, and write $u = u_1 u_2$ where $u_1$ has $t$-exponent sum $t_{\max}$, and $w_2$ has $t$-exponent sum $\mu = \rho - t_{\max} \leq 0$ as in Lemma 6.9. Then by Lemma 6.9, the path labeled $p(t^{2N}u)$ starting at $1$ ends at distance $\mu$ from the left side of $\Omega$, and at level $2N + t_{\max}$. Put $L = 2N + t_{\max}$ and call the endpoint of $p(t^{2N}u)$ the point $P_1$.

Now $p(t^{2N}u^2)$ travels along $p(t^{2N}u)$ to $P_1$, then along a path labeled $p(u)$. Since $u$ has $N$ $t^{-1}$ letters, $p(t^{2N}u^2)$ also does not reach the right side of $\Omega$.

Since $t_{\max} = \rho + |\mu|$, the path $p(u_1)$ starting at $P$ must reach the left side and either travel along it for at least one $t$ edge (if $\rho > 0$), or if $\rho = 0$ it will stay on level $L$ and end on the left side.

In the case that $\rho = 0$, $p(t^{2n}u^k)$ will end at the point $P_1$ for all $k \geq 1$. Thus $p(t^{2n}u^k t^{-2N})$ will end at the same point in $\Omega$ for all $k \geq 1$, so the index of $u^{-k}\langle a\rangle u^k \cap \langle a\rangle$ is constant as $k$ increases, and by Theorem 2.3 the scale of $u$ is $1$.

In the case that $\rho > 0$ and $p(t^{2N}uu_1)$ travels down a level from the point $P_1$, it must end on the left side since $u_1$ has maximal $t$-exponent sum. Since the $t$-exponent sum of $u_2$ is nonpositive, the path $p(u_2)$ from the left side stays at this level, and ends distance $|\mu|$ from the left side once again. Let $P_2$ be the endpoint of $p(t^{2N}u^2)$. Then the geodesic path from $P_1$ to $P_2$ in $\Omega$ is $t^\mu t^d t^{-\mu}$ (since the left side is closer than the right at level $2N + t_{\max}$), where $d$ is the difference in levels, and since $u$ is also a path from $P_1$ to $P_2$ and has $t$-exponent sum $\rho$, we must have $d = \rho$.

Inductively we have that $p(t^{2N}u^k)$ ends at a point $P_k$ at level $L + \rho^{k-1}$ at distance $|\mu|$ from the left side. For $k$ sufficiently large, the point $P_k$ lies further than $2N$ from the right side of $\Omega$, so $p(t^{2N}u^k t^{-2N})$ ends at a point labeled

$$l \left(\frac{l}{n}\right)^{L + \rho^{k-1} - |\mu| - 2N} \left(\frac{l}{m}\right)^{|\mu| - 1 + 2N}$$

$$= l \left(\frac{l}{n}\right)^{t_{\max} + \rho^{k-1} - |\mu|} \left(\frac{l}{m}\right)^{2N + |\mu| - 1}.$$

Applying Theorem 2.3 we have

$$s(x) = \lim_{k \to \infty} \left[ V : V \cap (t^{2N}u^k t^{-2N})^{-1} V (t^{2N}u^k t^{-2N}) \right]^{\frac{1}{k}}$$

$$= \lim_{k \to \infty} \left( l \left(\tfrac{l}{|n|}\right)^{t_{\max} + \rho^{k-1} - |\mu|} \left(\tfrac{l}{|m|}\right)^{2N + |\mu| - 1} \right)^{\frac{1}{k}}$$

$$= \left(\tfrac{l}{|n|}\right)^{\rho}$$

where $V$ is the closure of $\langle a\rangle$.

The case that $\rho < 0$ is treated with the same argument as above, exchanging the left side of $\Omega$ for the right, and considering $t^{-2N}u^k t^{2N}$. $\square$

Putting the results of this section together, and noting that scales of elements of $G_{m,\pm m}$ are 1, we have

**Corollary 6.11.** *If $x \in BS(m, n)$ for $m, n \neq 0$ has t-exponent sum $\rho$ then*
$$s(x) = \left(\frac{\operatorname{lcm}(m,n)}{|n|}\right)^\rho \text{ if } \rho \geq 0 \text{ and } \left(\frac{\operatorname{lcm}(m,n)}{|m|}\right)^{|\rho|} \text{ if } \rho < 0.$$

And so

**Theorem 6.12.** *The set of scales of elements in $G_{m,n}$ for $m, n \neq 0$ is*
$$\left\{ \left(\frac{\operatorname{lcm}(m,n)}{|n|}\right)^\rho, \left(\frac{\operatorname{lcm}(m,n)}{|m|}\right)^\rho \;\middle|\; \rho \in \mathbb{N} \right\}.$$

## 7. The Modular Function, Flat Rank, and Scale-Multiplicative Subsemigroups

The *modular function* of a locally compact group $G$ is the homomorphism $\Delta : G \to \mathbb{R}^+$ which measures how far a left-invariant Haar integral on $G$ is from being right-invariant (see for example [12]).

In [23, 24] the second author proves that in a totally disconnected locally compact group, $\Delta(x) = s(x)/s(x^{-1})$. It follows from Corollary 6.11 that the modular function for elements of $BS(m, n)$ of $t$-exponent sum $\rho$ is $\left|\frac{m}{n}\right|^\rho$.

**Lemma 7.1.** *If $x \in G_{m,n}$ then there is some $w \in BS(m, n)$ such that $s(x) = s(w)$ and $s(x^{-1}) = s(w^{-1})$.*

**Proof.** Let $s(x) = c$ and $s(x^{-1}) = d$. Let $\iota$ be the anti-automorphism of $G_{m,n}$ which sends each group element to its inverse. Since $s$ and $\iota$ are continuous, the sets $s^{-1}(c)$ and $\iota(s^{-1}(d))$ are open, so their intersection $U$ is open and contains $x$. Since $BS(m, n)$ is dense in $G_{m,n}$, $U$ contains an element $w \in BS(m, n)$, so $s(w) = c$, and $s(w^{-1}) = d$. $\square$

It follows that $\Delta(x) = \left|\frac{m}{n}\right|^\rho$ for some $\rho \in \mathbb{Z}$ for all $x \in G_{m,n}$.

**Corollary 7.2.** *The t-exponent map extends continuously to $G_{m,n}$ and is a group homomorphism.*

**Proof.** The map $\phi : G_{m,n} \to \mathbb{Z}$ defined by $\phi(x) = \log_{m/n}(\Delta(x))$ is a well defined group homomorphism, and $\phi \mid_{\mathrm{BS}(m,n)}$ is the $t$-exponent sum map.

$\square$

Now that each element of $G_{m,n}$ has a well defined notion of $t$-exponent sum, we can sharpen the main theorem to

**Corollary 7.3.** *The scale of $x \in G_{m,n}$ is $\left(\frac{\mathrm{lcm}(m,n)}{c}\right)^{|\rho|}$ where $\rho$ is the $t$-exponent sum of $x$, $c = |m|$ if $\rho \geq 0$ and $c = |n|$ if $\rho < 0$.*

A subgroup $P$ of a totally disconnected locally compact group is said to be *flat* if some compact open subgroup $V$ is minimizing for all $x \in P$. For example, if $x \in G$ is any element then $\langle x \rangle$ is flat by Proposition 2.1($iii$). Let $P$ be flat in $G$, with $V$ minimizing for each $x \in P$, consider the set $P_1 = \{x \in P \mid s(x) = s(x^{-1}) = 1\}$. It follows from Proposition 2.1($v$) that $P_1$ is a subgroup of $P$, and by Proposition 2.1($iv$) that $P_1$ is normal. This subgroup is called the *uniscalar subgroup* of $P$. The second author showed that $P$ modulo its uniscalar subgroup is free abelian of some rank $r \in \mathbb{N} \cup \{\infty\}$ [25]. The number $r$ is called the *flat rank* of $P$. Define the *flat rank* of a totally disconnected locally compact group $G$ to be the supremum of the flat rank over all flat subgroups of $G$.

**Proposition 7.4.** *For $m, n \neq 0$, the flat rank of $G_{m,n}$ is 1 for $|m| \neq |n|$, and 0 for $|m| = |n|$.*

**Proof.** If $|m| = |n|$, then $G_{m,n}$ has the discrete topology and all elements have scale 1, and so the flat rank of any flat subgroup is 0.

Suppose that $|m| \neq |n|$ and let $P$ be a flat subgroup of $G_{m,n}$. Let $E$ be the kernel of the $t$-exponent map defined in Corollary 7.2, which is the set of elements $x$ in $G_{m,n}$ with $s(x) = s(x^{-1}) = 1$. Then $P/P_1$ embeds into $G_{m,n}/E$. Since the $t$-exponent sum map is surjective, $G_{m,n}/E$ is isomorphic to $\mathbb{Z}$. Hence $P/P_1$ embeds into $\mathbb{Z}$ and the rank of $P/P_1$ is at most 1. Since $\langle t \rangle$ is a flat subgroup of flat rank 1, the flat rank of $G_{m,n}$ is equal to 1. $\square$

In [2] the following notion is introduced.

**Definition 7.5.** Let $G$ be a totally disconnected locally compact group with scale function $s : G \to \mathbb{N}$. A subsemigroup $S$ of $G$ is called *scale-multiplicative* if $s(xy) = s(x)s(y)$ for all $x, y \in S$, and is *maximal* if it is not contained in any other scale-multiplicative subsemigroup of $G$.

It follows immediately from Corollaries 6.11 and 7.2 that $G_{m,n}$ has exactly two maximal scale-multiplicative subsemigroups for all nonzero $m, n$.

**Corollary 7.6.** *The maximal scale-multiplicative subsemigroups of $G_{m,n}$ are*

$$S_+ = \{g \in G_{m,n} \mid s(g) \geq 0\}, \quad S_- = \{g \in G_{m,n} \mid s(g) \leq 0\}.$$

## 8. The Local Structure of $G_{m,n}$

In this section we give a more detailed description of the closure of $\langle a \rangle$ in $G_{m,n}$. Using this description and the machinery developed by the second author in [23–25] the computation of scale becomes much faster. Since we wish to reprove results from the previous two sections, we only assume results from Sections 2 to 5. We do however continue to abuse notation and identify elements and subsets of $\mathrm{BS}(m,n)$ with their images under the embedding $\pi$.

Recall that an inverse (or projective) system of groups and homomorphisms is a family of groups $(A_i)_{i \in I}$ indexed by a directed poset $(I, \leq)$, and a family of homomorphisms $f_{ij} : A_j \to A_i$ for all $i \leq j$ with the following properties:

(1) $f_{ii}$ is the identity on $A_i$,
(2) $f_{ik} = f_{ij} \circ f_{jk}$ for all $i \leq j \leq k$.

The *inverse limit* of the inverse system $((A_i)_{i \in I}, (f_{ij})_{i \leq j \in I})$ is the subgroup

$$\varprojlim_{i \in I} A_i = \left\{ (a_i)_{i \in I} \in \prod_{i \in I} A_i \mid a_i = f_{ij}(a_j) \text{ for all } i \leq j \text{ in } I \right\}$$

of the direct product $\prod_{i \in I} A_i$.

An example of an inverse limit is the additive group of $p$-adic integers — take $A_i = \mathbb{Z}/p^i\mathbb{Z}$ and $f_{ij}$ the remainder map modulo $p^i$.

If each $A_i$ is finite, the inverse limit is called a *profinite group*. It is a consequence of Tychonov's theorem that profinite groups are compact. Conversely, every compact totally disconnected group is isomorphic to a profinite group, [17, Theorem 1.1.12]. If a topological group contains a dense cyclic subgroup, it is said to be *monothetic* [12, Definition 9.2]. The closure of $\langle a \rangle$ in $G_{m,n}$ is compact, totally disconnected and monothetic. Hence, by [12, Theorem 25.16],

$$\overline{\langle a \rangle} \cong \prod_{p \in \Gamma} Z_p, \tag{4}$$

where $\Gamma$ is the set of prime numbers and $Z_p$ is isomorphic to the $p$-adic integers, $\mathbb{Z}_p$, or a (finite) quotient of this group.

The first aim of this section is determine the groups $Z_p$ by examining the details of the construction of $G/\!/H$ given in Section 5 for this particular case. As seen in Equation (3), that construction identifies $\overline{\langle a \rangle}$ with a subgroup of a product of finite permutation groups which, since $\langle a \rangle$ is cyclic, act by cyclicly permuting the cosets in $\langle a \rangle.g\langle a \rangle$ for each $g$ in $\mathrm{BS}(m,n)$. Therefore Equation 3 implies that

$$\overline{\langle a \rangle} \leq \prod \mathbb{Z}/d\mathbb{Z},$$

where the product is indexed by the cosets $g \langle a \rangle$ in $\mathrm{BS}(m,n)/\langle a \rangle$, and $d$ is the order of the corresponding orbit $\langle a \rangle.g\langle a \rangle$.

**Proposition 8.1.** *The compact open subgroup $\overline{\langle a \rangle}$ has an open subgroup $V$ such that*

$$V \cong \prod \{\mathbb{Z}_p \mid p \text{ is a prime divisor of } \mathrm{lcm}(m,n)/m \text{ or } \mathrm{lcm}(m,n)/n\}.$$

*The quotient $\overline{\langle a \rangle}/V$ is a finite cyclic group with order dividing $\gcd(m,n)$.*

**Proof.** Let $w\langle a \rangle \neq \langle a \rangle$ be a coset and write $w = a^{\eta_1}t^{\epsilon_1}a^{\eta_2}t^{\epsilon_2}\dots a^{\eta_k}t^{\epsilon_k}$ with $\epsilon_i = \pm 1$ and $k > 0$. To determine the size of the $\langle a \rangle$-orbit of $w\langle a \rangle$, consider $d$ with $a^d w\langle a \rangle = w\langle a \rangle$. It will be useful to compute $e$ such that $a^d w = wa^e$.

The defining relation for $\mathrm{BS}(m,n)$ implies that if $a^d w = wa^e$ then

$$a^d w = a^{\eta_1}t^{\epsilon_1}a^{\eta_2}a^{d'}t^{\epsilon_2}\dots a^{\eta_k}t^{\epsilon_k},$$

where $d' = d\frac{m}{n}$ or $d\frac{n}{m}$ depending on whether $\epsilon_1$ is equal to 1 or $-1$. Continuing to push $d'$ past $t^{\epsilon_i}$ for $i = 2, \dots, k$, we find that

$$e = d\left(\frac{m}{n}\right)^{\rho} = d\left(\frac{\frac{\mathrm{lcm}(m,n)}{n}}{\frac{\mathrm{lcm}(m,n)}{m}}\right)^{\rho}, \tag{5}$$

where $\rho$ is the $t$-exponent of $w$. Indeed, this calculation goes through provided that the $\gcd(m,n)$ and sufficiently high powers of $\frac{\mathrm{lcm}(m,n)}{n}$ and $\frac{\mathrm{lcm}(m,n)}{m}$ divide $d$. Hence there are $\langle a \rangle$-orbits with orders

$$d = \gcd(m,n)\left(\frac{\mathrm{lcm}(m,n)}{m}\right)^{r}\left(\frac{\mathrm{lcm}(m,n)}{n}\right)^{s} \tag{6}$$

for some $r, s \in \mathbb{N}$. Since $\frac{\mathrm{lcm}(m,n)}{n}$ and $\frac{\mathrm{lcm}(m,n)}{m}$ are relatively prime and the $t$-exponent of $w$ may take any integer value, it follows from (5) that $r$

and $s$ may take arbitrarily high values. Therefore, if $p$ is a prime divisor of $\frac{\text{lcm}(m,n)}{n}$ or $\frac{\text{lcm}(m,n)}{m}$, then $Z_p$ in (4) is isomorphic to $\mathbb{Z}_p$.

To see that all orbits have the orders given in (6), choose a coset representative $w$ that is freely reduced and has no pinches. Then the equation $w^{-1}a^d w = a^e$ implies that $t^{-\epsilon_1}a^d t^{\epsilon_1}$ is a pinch. Hence either $m$ or $n$ divides $d$, depending on whether $\epsilon_1$ equals $-1$ or $1$. Reduction of pinches continues for all $\epsilon_i$ provided that sufficient powers of $\frac{\text{lcm}(m,n)}{m}$ and $\frac{\text{lcm}(m,n)}{n}$ divide $d$, thus showing that all orbits have orders as given in (6), as claimed. This implies that, if $p$ does not divide $\gcd(m,n)$, $\frac{\text{lcm}(m,n)}{n}$ or $\frac{\text{lcm}(m,n)}{m}$, then $Z_p$ in (4) is trivial, and that, if $p$ divides $\gcd(m,n)$ but not $\frac{\text{lcm}(m,n)}{n}$ or $\frac{\text{lcm}(m,n)}{m}$, then $Z_p$ is finite cyclic with order dividing $\gcd(m,n)$. $\qquad\square$

It follows from Proposition 8.1 that $G_{m,n}$ is a closed subgroup of a product of $p$-adic Lie groups with $p$ belonging to

$$\mathfrak{p} := \left\{ p \mid p \text{ is a prime divisor of } \frac{\text{lcm}(m,n)}{n} \text{ or } \frac{\text{lcm}(m,n)}{m} \right\}.$$

Hence $G_{m,n}$ belongs to the class $\mathbb{A}_{\mathfrak{p}}$ of groups defined in [10]. It is shown in [10] that Hausdorff groups in the variety generated by $p$-adic Lie groups, for $p \in \mathfrak{p}$, may be approximated by groups in $\mathbb{A}_{\mathfrak{p}}$.

When the $t$-exponent of the coset representative $w$ equals 0, the above calculations yield further information.

**Corollary 8.2.** *If the $t$-exponent of $w \in \text{BS}(m,n)$ is equal to 0, then there is $d > 0$ such that $w$ centralizes $\langle a^d \rangle$ and (the image under $\pi$ of) $w$ centralizes $\overline{\langle a^d \rangle}$.*

Using this description we now revisit Corollary 7.2 where the notion of $t$-exponent sum was extended to $G_{m,n}$. Let $x \in G_{m,n}$.

Recall that every locally compact group admits a (left-invariant) Haar measure, which is finitely additive, invariant under left translation by group elements, and unique up to rescaling. Let $\mu$ be such a measure on $G_{m,n}$, scaled so that $\mu(\overline{\langle a \rangle}) = 1$. Then $\mu(\overline{\langle a^j \rangle}) = \frac{1}{j}$ for any positive integer $j$ since the measure is translation invariant (so all cosets have the same measure) and finitely additive. The modular function $\Delta : G \to \mathbb{R}^+$ is defined as follows. For any $g \in G$ and compact set $A \subseteq G$, $\mu(Ag) = \Delta(g)\mu(A)$. In the case that $G$ is totally disconnected, as mentioned in Section 7, the second author showed that the modular function always takes rational values. Since $\Delta$ is a continuous map from $G_{m,n}$ to $\mathbb{Q}^+$, and $\text{BS}(m,n)$ is dense in $G_{m,n}$, there exists some $\tilde{x}$ in $\text{BS}(m,n)$ such that $\Delta(x) = \Delta(\tilde{x})$. Then $\tilde{x} = t^p w$

where $w$ is a word of zero $t$-exponent sum. Since $\Delta$ is a homomorphism, $\Delta(\tilde{x}) = (\Delta(t))^p \Delta(w)$.

By Corollary 8.2 the word $w$ centralizes a compact open subgroup, and so $\Delta(w) = 1$. Since $\mu$ is left invariant,

$$\mu(t^{-1}\overline{\langle a^n \rangle}) = \mu(\overline{\langle a^n \rangle}) = \frac{1}{|n|}.$$

Translating the compact open subgroup $t^{-1}\overline{\langle a^n \rangle}$ by $t$ on the right we have

$$\mu(t^{-1}\overline{\langle a^n \rangle}t) = \mu(\overline{\langle a^m \rangle}) = \frac{1}{|m|}.$$

It follows that

$$\frac{1}{|n|} = \mu(t^{-1}\overline{\langle a^n \rangle}) = \Delta(t)\mu(t^{-1}\overline{\langle a^n \rangle}t) = \Delta(t)\frac{1}{|m|}$$

and so $\Delta(t) = \left|\frac{m}{n}\right|$.

Putting this together we have $\Delta(x) = \Delta(\tilde{x}) = (\Delta(t))^p \Delta(w) = \left|\frac{m}{n}\right|^p$, and so we may define the function $\phi : G_{m,n} \to \mathbb{Z}$ as in the proof of Corollary 7.2 which extends the $t$-exponent sum to all of $G_{m,n}$.

The *quasi-centre*, $\mathrm{QZ}(G)$, of the topological group $G$ is defined in [4] to be set of elements of $G$ that centralize an open subgroup of $G$. Since the intersection of any two open subgroups is an open subgroup and the set of open subgroups is invariant under automorphisms, $\mathrm{QZ}(G)$ is a characteristic subgroup of $G$. It is typically not closed, but when $G$ has an open abelian subgroup it is open.

**Corollary 8.3.** *The quasi-centre of $G_{m,n}$ is equal to $\ker \Delta$.*

**Proof.** Since $\overline{\langle a \rangle}$ is abelian, it is contained in $\mathrm{QZ}(G_{m,n})$. The smallest subgroup of $G_{m,n}$ containing $\overline{\langle a \rangle}$ and the image under $\pi$ of all elements of $\mathrm{BS}(m, n)$ having $t$-exponent 0 is $\ker \Delta$. Hence $\mathrm{QZ}(G_{m,n})$ contains this subgroup. The reverse inclusion holds because every element of the quasi-centre is unimodular. $\qquad\square$

Computation of the scale of $x \in G_{m,n}$ may now be carried out more directly with the aid of the relevant structure theory as described in [23]. Recall in particular that the compact, open subgroup, $V$ is *tidy* for $x$ if, setting $V_+ = \bigcap_{k\geq 0} x^k V x^{-k}$ and $V_- = \bigcap_{k\geq 0} x^{-k} V x^k$, we have

$$V = V_+ V_- \quad \text{and} \quad \bigcup_{k\geq 0} x^k V_+ x^{-k} \text{ is closed.}$$

When $V$ is tidy for $x$ we have $s(x) = [xV_+x^{-1} : V_+]$ (and so tidy subgroups are minimising).

**Proposition 8.4.** *Let* $x \in G_{m,n}$ *and denote the $t$-exponent of $x$ by $\rho$. Then there is a compact, open subgroup, $V$, tidy for $x$ with $V = V_+V_-$ where*

$$V_+ \cong \prod \{\mathbb{Z}_p \mid p \text{ is a prime divisor of } \operatorname{lcm}(m, n)/n\}$$

$$and \quad V_- \cong \prod \{\mathbb{Z}_p \mid p \text{ is a prime divisor of } \operatorname{lcm}(m, n)/m\}$$

*if $\rho$ is positive and vice versa if $\rho$ is negative. The scale of $x$ is*

$$s(x) = \begin{cases} \left(\dfrac{\operatorname{lcm}(m, n)}{|n|}\right)^{\rho}, & \text{if } \rho \geq 0 \\[2ex] \left(\dfrac{\operatorname{lcm}(m, n)}{|m|}\right)^{-\rho}, & \text{if } \rho \leq 0 \end{cases}.$$

**Proof.** Write $x = t^{\rho}(t^{-\rho}x) = t^{\rho}w$ where $w \in G_{m,n}$ is in $\ker \Delta$. By Corollary 8.3, there is an open subgroup $V \leq \overline{\langle a \rangle}$ that is centralised by $w$. It follows from Proposition 8.1 and [23, Lemma 1] that we may, by passing to a subgroup if necessary, suppose that

$$V \cong \prod \{\mathbb{Z}_p \mid p \text{ is a prime divisor of } \operatorname{lcm}(m, n)/m \text{ or } \operatorname{lcm}(m, n)/n\} \tag{7}$$

and $V = V_+V_-$. Note that, since $V$ is an open subgroup of $\overline{\langle a \rangle}$, there is $d > 0$ such that $V = \overline{\langle a^d \rangle}$ and it may also be supposed that $m$, $n$ and $(n/m)^{\rho}$ divide $d$.

Since $w$ centralises $V$, we have $xVx^{-1} = t^{\rho}Vt^{-\rho}$. Hence

$$xVx^{-1} = \overline{\langle t^{\rho}a^dt^{-\rho} \rangle} = \overline{\langle a^e \rangle} \text{ where } e = d\left(\frac{n}{m}\right)^{\rho}.$$

Then under the isomorphism (7) of $V$ with the product of groups of $p$-adic integers, which converts multiplicative notation to additive, conjugation by $x$ multiplies each factor $\mathbb{Z}_p$ by $(\frac{m}{n})^{\rho}$. It follows that, if $\rho$ is positive, conjugation by $x$ expands the factor $\mathbb{Z}_p$ if $p$ divides $\frac{\operatorname{lcm}(m,n)}{n}$ and contracts it if $p$ divides $\frac{\operatorname{lcm}(m,n)}{m}$, and *vice versa* if $\rho$ is negative. Therefore $V_+$ and $V_-$ are as claimed. It further follows that $\bigcap_{k \in \mathbb{Z}} x^k V x^{-k}$ is trivial whence, by [3, Lemma 3.31(3)], $V$ is tidy for $x$. Therefore $s(x) = [xV_+x^{-1} : V_+]$ and has the claimed values. $\qquad \square$

# References

[1]    Hyman Bass. Covering theory for graphs of groups. *J. Pure Appl. Algebra*, 89(1-2):3–47, 1993.

[2]    Udo Baumgartner, Jacqui Ramagge, and George A. Willis. Scale-multiplicative semigroups and geometry: automorphism groups of trees. *Groups Geom. Dyn.*, 10(3):1051–1075, 2016.

[3]    Udo Baumgartner and George A. Willis. Contraction groups and scales of automorphisms of totally disconnected locally compact groups. *Israel J. Math.*, 142:221–248, 2004.

[4]    Marc Burger and Shahar Mozes. Groups acting on trees: From local to global structure. *Inst. Hautes Études Sci. Publ. Math.*, (92):113–150 (2001), 2000.

[5]    Murray Elder. A context-free and a 1-counter geodesic language for a Baumslag-Solitar group. *Theoret. Comput. Sci.*, 339(2-3):344–371, 2005.

[6]    Murray Elder. A linear-time algorithm to compute geodesics in solvable Baumslag-solitar groups. *Illinois J. Math.*, 54(1):109–128, 2010.

[7]    Murray Elder, Gillian Elston and Gretchen Ostheimer. On groups that have normal forms computable in logspace. *J. Algebra*, 381:260–281, 2013.

[8]    Murray Elder and Susan Hermiller. Minimal almost convexity. *J. Group Theory*, 8(2):239–266, 2005.

[9]    Świtosław R. Gal and Tadeusz Januszkiewicz. New a-T-menable HNN-extensions. *J. Lie Theory*, 13(2):383–385, 2003.

[10]   Helge Glöckner. Locally compact groups built up from *p*-adic Lie groups, for *p* in a given set of primes. *J. Group Theory*, 9(4):427–454, 2006.

[11]   Helge Glöckner and George A. Willis. Topologization of Hecke pairs and Hecke $C^*$-algebras. In *Proceedings of the 16th Summer Conference on General Topology and its Applications (New York)*, volume 26, pages 565–591, 2001/02.

[12]   Edwin Hewitt and Kenneth A. Ross. *Abstract harmonic analysis. Vol. I*, volume 115 of *Grundlehren der Mathematischen Wissenschaften [Fundamental Principles of Mathematical Sciences]*. Springer-Verlag, Berlin-New York, second edition, 1979. Structure of topological groups, integration theory, group representations.

[13]   Roger C. Lyndon and Paul E. Schupp. *Combinatorial group theory*. Springer-Verlag, Berlin-New York, 1977. Ergebnisse der Mathematik und ihrer Grenzgebiete, Band 89.

[14]   Rögnvaldur G. Möller. Structure theory of totally disconnected locally compact groups via graphs and permutations. *Canad. J. Math.*, 54(4):795–827, 2002.

[15]   C. Wayne Patty. *Foundations of topology*. The Prindle, Weber & Schmidt Series in Advanced Mathematics. PWS-KENT Publishing Co., Boston, MA, 1993.

[16]   C. D. Reid and P. R. Wesolek. Homomorphisms into totally disconnected, locally compact groups with dense image. *ArXiv e-prints*, September 2015.

[17]   Luis Ribes and Pavel Zalesskii. *Profinite groups*, volume 40 of *Ergebnisse der Mathematik und ihrer Grenzgebiete. 3. Folge. A Series of Modern Surveys in Mathematics [Results in Mathematics and Related Areas. 3rd Series. A Series of Modern Surveys in Mathematics]*. Springer-Verlag, Berlin, second edition, 2010.

[18]   G. Schlichting. Operationen mit periodischen Stabilisatoren. *Arch. Math. (Basel)*, 34(2):97–99, 1980.

[19]   Jean-Pierre Serre. *Trees*. Springer Monographs in Mathematics. Springer-Verlag, Berlin, 2003. Translated from the French original by John Stillwell, Corrected 2nd printing of the 1980 English translation.

[20]   Yehuda Shalom and George A. Willis. Commensurated subgroups of arithmetic groups, totally disconnected groups and adelic rigidity. *Geom. Funct. Anal.*, 23(5):1631–1683, 2013.

[21]   Kroum Tzanev. Hecke $C^*$-algebras and amenability. *J. Operator Theory*, 50(1):169–178, 2003.

[22]   D. Van Dantzig. Zur topologischen Algebra. III. Brouwersche und Cantorsche Gruppen. *Compositio Math.*, 3:408–426, 1936.

[23]   G. Willis. The structure of totally disconnected, locally compact groups. *Math. Ann.*, 300(2):341–363, 1994.

[24]   George A. Willis. Further properties of the scale function on a totally disconnected group. *J. Algebra*, 237(1):142–164, 2001.

[25]   George A. Willis. Tidy subgroups for commuting automorphisms of totally disconnected groups: An analogue of simultaneous triangularisation of matrices. *New York J. Math.*, 10:1–35 (electronic), 2004.

Chapter 5

# Elementary and Universal Theories of Nonabelian Commutative Transitive and CSA Groups

B. Fine[1], A.M. Gaglione[2], and D. Spellman[3]

[1] *Department of Mathematics, Fairfield University*
*Fairfield, CT 06430*
*ben1902@aol.com*

[2] *Department of Mathematics, U.S. Naval Academy*
*Annapolis, MD 21402*
*agaglione@aol.com*
*http://www.usna.edu*

[3] *5147 Whitaker Ave., Philadelphia, PA 19124*

*Dedicated to Dennis M. Spellman and Gerhard Rosenberger on the occasion*
*of their seventeeth birthdays*

**ABSTRACT.** Given a nonabelian commutative transitive (CT) group $G$, we show that the elementary theory of $G$ is axiomatizable by the set $H(G)$ of Horn sentences true in $G$ togeher with CT. We also obtain results about the elementary theory of a group $N$ free in the variety of all groups nilpotent of class at most $c$. Finally for a nonabelian CSA group $G$ we show that its universal theory is axiomatizable by the set $Q(G)$ of quasi-identities true in $G$ together with CT and the diagram of $G$. This last result generalizes a result of Myasnikov and Remeslennikov.

## 1. A Generalization of Commutative Transitivity

**Definition 1.1.** A group $G$ is **commutative transitive** or **CT** provided the centralizer in $G$, $C_G(g)$, of every nontrivial element $g \in G \backslash \{1\}$ is abelian.

2000 *Mathematics Subject Classification.* Primary 20E26; Secondary 20E05, 20E06.
*Key words and phrase.* Commutative transitive, commutative transitive of level $n$, reduced product, ultraproduct, elmentary class, $G$-group, Horn sentence, Conjugacy separated abelian, Quasi-identity.

We note that commutative transitivity is captured by the universal sentence

$$\forall x, y, z \left( ((y \neq 1) \wedge ([x, y] = 1) \wedge ([y, z] = 1)) \rightarrow ([x, z] = 1) \right).$$

**Definition 1.2.** ([FGMS]) A group $G$ is commutative transitive of level $n$ or $\mathrm{CT}(n)$ provided the centralizer in $G$, $C_G(g)$, of every element, $g \notin Z_n(G)$, omitted by the $n$-th term of the upper central series of $G$ is abelian.

We note that commutative transitivity of level $n$ is captured by the universal sentence

$$\forall x, y, z, w_1, \ldots, w_n \left( (([w_1, \ldots, w_n, y] \neq 1) \right.$$
$$\left. \wedge ([x, y] = 1) \wedge ([y, z] = 1)) \rightarrow ([x, z]) = 1) \right).$$

If $c \geq 2$ is an integer and $N$ is a group free in the variety $\mathcal{N}_c$ of all groups nilpotent of class at most $c$, then $N$ satisfies $\mathrm{CT}(c-1)$

**Definition 1.3.** A group $G$ is **conjugately separated abelian** or **CSA** provided every maximal abelian subgroup $M$ in $G$ is **malnormal** in $G$ in the sense that $g^{-1} M g \cap M \neq \{1\}$ implies $g \in M$.

It is not hard to show that any CSA group is CT (but not conversely). We will show this later on in this paper (see Section 5). Thus CSA is captured by the universal sentences

$$\forall x, y, z \left( ((y \neq 1) \wedge ([x, y] = 1) \wedge ([y, z] = 1)) \rightarrow ([x, z] = 1) \right)$$

and

$$\forall x, y, z \left( ((x \neq 1) \wedge (y \neq 1) \wedge ([x, y] = 1) \right.$$
$$\left. \wedge ([x, z^{-1} y z] = 1) \rightarrow ([z, y] = 1) \right).$$

## 2. Reduced Products

Fix a nonempty set $I$. If $A \subseteq I$ let $A'$ be its complement $I \backslash A$ in the Boolean algebra $\mathcal{P}(I) = \{J : J \subseteq I\}$ of all subsets of $I$. By a **filter** on $I$ is meant the dual of a proper ideal in $\mathcal{P}(I)$. Specifically, $D \subseteq \mathcal{P}(I)$ is a filter on $I$ provided the following four properties are satisfied:

(1) $I \in D$
(2) $\phi \notin D$

(3) $A \cap B \in D$ whenever $A, B \in D$

(4) $B \in D$ whenever $A \subseteq B \subseteq I$ and $A \in D$

A maximal (with respect to set inclusion) filter $D$ on $I$ is an **ultrafilter** on $I$. It is easy to see that the filter $D$ on $I$ is an ultrafilter on $I$ if and only if for every $A \subseteq I$, exactly one of $A$ or $A'$ lies in $D$. The singleton $\{I\}$ is called the **trivial filter** on $I$.

Suppose that $(\Gamma_i)_{i \in I}$ is a family of (not necessarily distinct) groups indexed by $I$. The direct product $P$ of the family is, as usual, the set of all choice functions $\gamma : I \to \cup_{i \in I} \Gamma_i$, $i \xrightarrow{\gamma} \gamma_i \in \Gamma_i$ for all $i \in I$. Here multiplication is defined componentwise. Fix a group $G$. If $\Gamma$ is any group containing a distinguished copy of $G$ as a subgroup, then $\Gamma$ is a $G$-**group**. If each $\Gamma_i$ is a $G$-group, the map $\varphi : G \to P$, $\varphi(g)(i) = g$ for all $i \in I, g \in G$, embeds $G$ into $P$ and thus $P$ is a $G$-group also in this way.

Now let $D$ be a filter on $I$. Define a binary relation on $P$ by $\alpha = \beta$ a.e. if and only if $\{i \in I : \alpha(i) = \beta(i)\} \in D$. It is easy to see that this is a congruence on $P$ and so $K_D = \{\alpha \in P : \alpha = 1 \text{ a.e.}\}$ is a subgroup normal in $P$. The quotient group

$$R_D = P/K_D$$

is the **reduced product** of the family $(\Gamma_i)_{i \in I}$ modulo the filter $D$ on $I$. We write $[\gamma]_D$ for the coset $\gamma K_D$. If each $\Gamma_i$ is a $G$-group, then we get a well-defined embedding $\varphi_D : G \to R_D$ via $g \mapsto [\varphi(g)]_D$ where $\varphi : G \to P$ is the embedding $\varphi(g)(i) = g$ for all $i \in I, g \in G$. Thus $R_D$ may be viewed as a $G$-group in this way. If $D$ is the trivial filter $\{I\}$ on $I$, then $R_D$ is isomorphic to $P$. Hence, direct products may be viewed as special cases of reduced products. In the opposite extreme, when $D$ is an ultrafilter on $I$, $R_D$ is called the **untraproduct** of the family $(\Gamma_i)_{i \in I}$ modulo the ultrafilter $D$ on $I$. In the case that each $\Gamma_i$ is the same group $\Gamma$ the ultraproduct is called an **ultrapower** of $\Gamma$.

## 3. Preliminaries from Logic

Let $L_0 = L_0[\{1\}]$ be the first-order language with equality containing a binary operation symbol $\bullet$, a unary operation symbol $^{-1}$ and a constant symbol 1. If $G$ is a nontrivial group then $L_0[G]$ shall be the extension of $L_0$ obtained by adjoining the elements of $G \backslash \{1\}$ as new constant symbols. Thus, an $L_0$-structure is a set $\Gamma$ provided with a binary operation $\Gamma^2 \to \Gamma$, $(\gamma_1, \gamma_2) \mapsto \gamma_1 \bullet \gamma_2$ which we write as $\gamma_1 \gamma_2$, a unary operation $\Gamma \to \Gamma$,

$\gamma \mapsto \gamma^{-1}$ and a distinguished constant $1 \in \Gamma$. An $L_0[G]$–structure is a set $\Gamma$ provided with a binary operation $\Gamma^2 \to \Gamma$, $(\gamma_1, \gamma_2) \mapsto \gamma_1 \gamma_2$, a unary operation $\Gamma \to \Gamma$, $\gamma \mapsto \gamma^{-1}$ and one distinguished element $g_\Gamma$ for each element $g \in G$. We remark that there is no *a priori* prohibition against $g_\Gamma = h_\Gamma$ when $g$ and $h$ are distinct elements of $G$.

A **sentence** of $L_0[G]$ is a formula of $L_0[G]$ containing no unquantified occurrence of any variable. Quantifier free sentences are possible. For example, the sentence $1^{-1} = 1$ of $L_0$ is true in every group but can be false in an $L_0$–structure which is not a group. A class $\mathcal{X}$ of $L_0[G]$–structures is **elementary** provided there is at least one set $S$ of sentences of $L_0[G]$ such that $\mathcal{X}$ is precisely the class of all models of $S$. That is, $\Gamma \in \mathcal{X}$ if and only if $\sigma$ is true in $\Gamma$ for every $\sigma \in S$. Every sentence $\sigma$ of $L_0[G]$ is logically equivalent to at least one sentence $\pi_\sigma$ of $L_0[G]$ in so-called **prenex normal form**. That is $\pi_\sigma$ has the form

$$Q_1 x_1 \cdots Q_n x_n \varphi(x_1, \ldots, x_n)$$

where $(x_1, \ldots, x_n)$ is a tuple of distinct variables, each $Q_i$, $1 \leq i \leq n$, is a quantifier ($\forall$ or $\exists$) and $\varphi(x_1, \ldots, x_n)$ is a formula of $L_0[G]$ containing no quantifiers and containing occurrences of at most the variables $x_1, \ldots, x_n$. In that event $\varphi(x_1, \ldots, x_n)$ is called the **matrix** of the sentence $\pi_\sigma$. If all the $Q_i$ are universal quantifiers, then $\pi_\sigma$ is called a **universal sentence** of $L_0[G]$. Examples of universal sentences are the group axioms, namely:

- $\forall x_1 x_2 x_3 \left( (x_1 \bullet x_2) \bullet x_3 = x_1 \bullet (x_2 \bullet x_3) \right)$
- $\forall x \left( x \bullet 1 = x \right)$
- $\forall x \left( x \bullet x^{-1} = 1 \right)$.

Modulo the group axioms every **atomic sentence** of $L_0[G]$ may be written in the form $w(g_1, \ldots, g_n) = 1$ where $w(x_1, \ldots, x_n)$ is a word on the distinct variables $x_1, \ldots, x_n$ and their formal inverses and $(g_1, \ldots, g_n)$ is a tuple of (not necessarily distinct) elements of $G$. The **diagram of $G$**, henceforth diag($G$), is the set of all atomic and negated atomic sentences of $L_0[G]$ true in $G$. Any model of the group axioms and diag($G$) must contain a distinguished copy of $G$ as a subgroup. Indeed, if $\Gamma$ is a model of the group axioms and diag($G$) we cannot have $g_\Gamma = h_\Gamma$ when $g \neq h$ as the negated atomic sentence $gh^{-1} \neq 1$ lies in diag($G$). Thus, a model of the group axioms and diag($G$) is a $G$–group (and conversely).

Every quantifier-free formula $\varphi(x_1, \ldots, x_n)$ of $L_0[G]$ is equivalent to at least one quantifier-free formula $c_\varphi(x_1, \ldots, x_n)$ of $L_0[G]$ in so called

**conjunctive normal form.** That is, $c_\varphi(x_1, \ldots, x_n)$ is a conjunction of disjunctions of atomic and negated atomic formulas of $L_0[G]$. We note that $\varphi(x_1, \ldots, x_n)$ is also equivalent to at least one quantifier-free formula $d_\varphi(x_1, \ldots, x_n)$ of $L_0[G]$ in **disjunctive normal form** - with obvious modifications. (Modulo the group axioms an atomic formula of $L_0[G]$ is one of the form $w(g_1, \ldots, g_m, x_1, \ldots, x_n) = 1$ where $w(y_1, \ldots, y_m, x_1, \ldots, x_n)$ is a word on the distinct variables $y_1, \ldots, y_m, x_1, \ldots, x_n$ and their formal inverses and $(g_1, .., g_m)$ is a tuple from $G$.)

Now suppose that the prenex normal form sentence $\sigma$ of $L_0[G]$ has matrix written in conjunctive normal form. $\sigma$ is a **Horn sentence** of $L_0[G]$ provided in each conjunct at most one disjunct is atomic. One verifies that Horn sentences are preserved in reduced products. More significantly, there is a converse. See, e.g. [CK].

**Theorem 3.1.** *Let $\mathcal{X}$ be an elementary class. Then $\mathcal{X}$ is closed under taking reduced products if and only if it has at least one set of Horn axioms.*

Theorem 3.1 should be compared to the classical theorem of Los asserting that any elementary class whatsoever is closed under taking ultraproducts.

**Remark 3.2.** There is a discrepancy between the definitions of Horn sentence in Grätzer [G] and Chang and Keisler [CK]. A Horn sentence according to Grätzer coincides with our definition while a Horn sentence according to Chang and Keisler is a conjunction of Horn sentences according to Grätzer. Since the model classes with regard to either definition coincide it makes no difference for our purposes which definition of Horn sentence we adopt.

Let us fix a group $G$. Let $\Gamma$ be an $L_0[G]$-structure. Then the **theory** of $\Gamma$, denoted $\text{Th}(\Gamma)$, is the set of all sentences of $L_0[G]$ true in $\Gamma$. Note that $\text{diag}(G) \subseteq \text{Th}(G)$. Moreover, $\text{diag}(G)$ is also contained in the set $H(G)$ of all Horn sentences of $L_0[G]$ true in $G$. Given two (not necessarily distinct) $L_0[G]$-structures $\Gamma_1$ and $\Gamma_2$ we write $\Gamma_1 \equiv \Gamma_2$ and say $\Gamma_1$ and $\Gamma_2$ are **elementarily equivalent** provided $\text{Th}(\Gamma_1) = \text{Th}(\Gamma_2)$.

**Remark 3.3.** We have suppressed the dependence on $G$ in the notations $\text{Th}$, $H$ and $\equiv$. Generally $G$ is understood from the context.

We conclude this section with a characterization of elementary classes that may be found, for example, in Chang and Keisler [CK].

**Theorem 3.4.** *A class of $L_0[G]$-structures is elementary if and only if it is closed under elementary equivalence and ultraproducts.*

**Remark 3.5.** By Los' classical theorem first-order sentences are preserved in ultraproducts.

## 4. The Elementary Theory Case

**Lemma 4.1.** (*The Principal Lemma*) *Let $I$ be a nonempty set and $D$ be a filter on $I$ which is not an ultrafilter. Fix $A \subseteq I$ such that neither $A$ nor $A'$ lies in $D$.*

(1) *Let $(\Gamma_i)_{i \in I}$ be a family of nonabelian groups indexed by $I$. Let $\Gamma$ be the reduced product of the family modulo $D$. Then $\Gamma$ is not CT.*

(2) *Fix integers $c$ and $r$ each at least 2, Let $N$ be a group free of rank $r$ in the variety $\mathbf{N}_c$ of all groups nilpotent of class at most $c$. Let $(\Gamma_i)_{i \in I}$ be a family of $N$-groups lying in $\mathcal{N}_c$ indexed by $I$. Let $\Gamma$ be the reduced product of the family modulo $D$. Then $\Gamma$ does not satisfy $CT(c-1)$.*

**Proof.** (1) For each $i \in I$ choose $u_i, v_i \in \Gamma_i$ such that $[u_i, v_i] \neq 1$. Let

$$s_i = \begin{cases} u_i & \text{if } i \in A \\ 1 & \text{if } i \in A' \end{cases}$$

and

$$t_i = \begin{cases} 1 & \text{if } i \in A \\ v_i & \text{if } i \in A' \end{cases}$$

and let $s = [(s_i)_{i \in I}]_D$, $t = [(t_i)_{i \in I}]_D$, and $u = [(u_i)_{i \in I}]_D$. Then $s \neq 1$ since $\{i \in I : s_i = 1\} = A' \notin D$. Now $[u, s] = [s, t] = 1$ since $\{i \in I : [u_i, s_i] = 1\} = \{i \in I : [s_i, t_i] = 1\} = I \in D$, but $[u, t] \neq 1$ since $\{i \in I : [u_i, t_i] = 1\} = A \notin D$. Hence $P$ is not CT.

(2) Let $N$ be freely generated relative to $\mathcal{N}_c$ by $\{a_1, \ldots, a_r\}$ and assume the generators have been ordered according to their given subscripts. Let $e$ be the weight $c - 1$ Engel element $[a_2, a_1, \ldots, a_1]$ so that the element $[e, a_{r-1}]$ is a basic commutator of weight $c$ - hence nontrivial. Let

$$\delta : N \to \prod_{i \in I} \Gamma_i$$

be the diagonal embedding given by $\delta(x)(i) = x$ for all $i \in I$, $x \in N$ and let $d : N \to \Gamma$ be the embedding given by $d(x) = [\delta(x)]_D$ for all $x \in N$. Let

$$s_i = \begin{cases} a_{r-1} & \text{if } i \in A \\ 1 & \text{if } i \in A' \end{cases}$$

and

$$t_i = \begin{cases} 1 & \text{if } i \in A \\ a_r & \text{if } i \in A' \end{cases}$$

and set $s = [(s_i)_{i \in I}]_D$ and $t = [(t_i)_{i \in I}]_D$. Then $[d(a_2), d(a_1), \ldots, d(a_1), s] = [d(e), s] \neq 1$ since $\{i \in I : [e, s_i] = 1\} = A' \notin D$. Now $[d(a_{r-1}), s] = [s, t] = 1$ since $\{i \in I : [a_{r-1}, s_i] = 1\} = \{i \in I : [s_i, t_i] = 1\} = I \in D$ but $[d(a_{r-1}), t] \neq 1$ since $\{i \in I : d(a_{r-1}), t] = 1\} = A \notin D$. Therefore, $\Gamma$ violates $\text{CT}(c-1)$. ∎

**Theorem 4.2.** (1) *Let $G$ be a nonabelian CT-group. Then $Th(G)$ with respect to the language $L_0[G]$ is axiomatizable by the set $H(G)$ of all Horn sentences of $L_0[G]$ true in $G$ together with CT.*

(2) *Let $c$ and $r$ be integers each at least 2. Let $N$ be a group free of rank $r$ in the variety $\mathcal{N}_c$ of all groups nilpotent of class at most $c$. Then $Th(N)$ with respect to the language $L_0[N]$ is axiomatizable by the set $H(N)$ of all Horn sentences of $L_0[N]$ true in $N$ together with $CT(c-1)$.*

**Proof.** (1) We define a family of classes of $G$-groups by recursion on the ordinals as follows.

Let $\mathcal{X}_0$ be the isomorphism class of $G$. If $\alpha = \beta+1$ is a successor ordinal, then $\mathcal{X}_\alpha$ is the class of all $G$-groups elementarily equivalent to a reduced product of (not necessarily distinct) groups in $\mathcal{X}_\beta$. If, however, $\alpha$ is a limit ordinal, then $\mathcal{X}_\alpha = \cup_{\beta<\alpha}\mathcal{X}_\beta$. Observe that the family $\mathcal{X}_\alpha$ is nested according to subscripts. If we let $\mathcal{X}$ be the union of the $\mathcal{X}_\alpha$ over all ordinals $\alpha$, then $\mathcal{X}$ is clearly the least class of $G$-groups containing $G$ and closed under elementary equivalence and reduced products. Thus, since ultraproducts are special cases of reduced products it is the least Horn axiomatizable class containing $G$ by Theorems 3.1 and 3.4. That is $\mathcal{X}$ is the model class of $H(G)$. We then argue by transfinite induction using Lemma 4.1 that CT members of $\mathcal{X}_\alpha$ are those elementarily equivalent to $G$. The induction proceeds as follows. We show that every CT member of $\mathcal{X}_\alpha$ is elementarily equivalent to $G$. This obviously holds for $\mathcal{X}_0$. Suppose it holds for all ordinals $\beta$, $0 \leq \beta < \alpha$. Let $\Gamma$ be a CT member of $\mathcal{X}_\alpha$. Then, for some ordinal $\beta$, $0 \leq \beta < \alpha$, $\Gamma$ is elementarily equivalent to a reduced product of a family of $G$-groups (hence nonabelian) in $\mathcal{X}_\beta$. By Lemma 4.1, the reduced product must be an ultraproduct so the groups in the family are almost everywhere CT by Los' Theorem. By inductive hypothesis, they are almost everywhere elementarily equivalent to $G$. So again, by Los' Theorem, $\Gamma \equiv G$. That completes the induction and proves the theorem in this case.

(2) is proven similarly.   ∎

## 5.  Algebraic Geometry over Groups and a Theorem of Myasnikov and Remeslennikov

Let $R$ be a Noetherian integral domain, $A$ be a commutative $R$-algebra, and $n$ be a positive integer. Let $S$ be a subset of the polynomial ring $R[x_1, \ldots, x_n]$, $V(S) = \{(a_1, \ldots, a_n) \in A^n : f(a_1, \ldots, a_n) = 0 \; \forall f \in S\}$ is an **affine algebraic subset** of $A^n$. These form a closed subbase for the **Zariski topology** on $A^n$. The fact that $R$ is Noetherian guarantees that this topology is a Noetherian topology. That is, every descending chain

$$F_1 \supseteq F_2 \supseteq \cdots \supseteq F_k \supseteq \cdots$$

of closed subsets of $A^n$ stabilizes after finitely many steps: $F_N = F_{N+1} = F_{N+2} = \cdots$ for some $N$. Consequently, every affine algebraic subset of $A^n$ decomposes as a finite union of irreducible ones.

In analogy with the above classical situation, G. Baumslag, A.G. Myasnikov, and V.N. Remeslennikov introduced in [BMR] the theory of algebraic geometry over groups. The analogy of an $R$-algebra $A$ is a $G$-group $H$ as defined previously. To get at the analog of integral domain one needs a good definition of zero divisor. Of course, the identity element of the group plays the role of zero. A definition of zero divisor which works in this context is the following: $g$ is a **zero divisor** provided there is an $h \neq 1$ such that $[g, h^k] = 1$ for every $k$. Here $h^k$ denotes the conjugate $k^{-1}hk$. Thus, the group $G$ is a **domain** provided $G \neq \{1\}$ and 1 is the only zero divisor. If $G$ is a domain, then an easy induction shows that if $(g_1, \ldots, g_n) \in (G \backslash \{1\})^n$, then there is an $(h_1, \ldots, h_{n-1}) \in G^{n-1}$ such that the left-normed commutator $[g_1, g_2^{h_1}, g_3^{h_2}, \ldots, g_n^{h_{n-1}}]$ is nontrivial. If $n = 1$ we interpret the commutator as $g_1$. Examples of domains are nonabelian

CSA groups as previously defined. First of all, also as previously stated, we have

**Proposition 1.** *Every CSA group is CT.*

**Proof.** For suppose the group $G$ is CSA, $M_1$ and $M_2$ are maximal abelian subgroups in $G$, and $1 \neq m \in M_1 \cap M_2$. Given $m_1 \in M_1$ one has $1 \neq m \in M_2 \cap M_2^{m_1}$. So $m_1 \in M_2$ since $G$ is CSA. Since $m_1$ was arbitrary, $M_1 \leq M_2$. By maximality, $M_1 = M_2$ and so, as claimed, $G$ is CT. ∎

Now suppose that the group $G$ is nonabelian and CSA. Let $(g, h) \in (G \backslash \{1\})^2$. If $[g, h^k] = 1$ for every $k \in G$, then $1 \neq g \in C_G(h) \cap G_G h)^k$ for every $k$. So $G = C_G(h)$ is abelian (as $G$ is CT). This is a contradiction. Hence, $G$ is a domain. Let $H$ be a $G$-group. For each positive integer $n$, let $X_n = \langle x_1, \ldots, x_n; \rangle$ be a group free of rank $n$ with basis $\{x_1, \ldots, x_n\}$. For each subset $S$ of the free product $G * X_n$ the solution set

$$V(S) = \{(h_1, \ldots, h_n) \in H^n : w(h_1, \ldots, h_n) = 1 \ \forall w \in S\}$$

is an **affine algebraic subset** of $H^n$. These form a closed subbase for the **Zariski topology** on $H^n$. $H$ is $G$-**equationally Noetherian** provided, for every positive intger $n$ and every subset $S \subseteq G * X_n$, there is a finite subset $S_0 \subseteq S$ such that $V(S) = V(S_0)$. This is the analog of a Noetherian base ring $R$ and is sufficient (and necessary) to guarantee that, for every $n$, the Zariski topology on $H^n$ be Noetherian. In particular, if $H$ is $G$-equationally Noetherian, then one has, for every $n$, a decomposition of affine algebraic subsets into finite unions of irreducible ones. We are almost prepared to state a theorem of A.G. Myasnikov and V.N. Remeslennikov. We need some more definitions and notation.

Modulo the group axioms, a **quasi-identity** of $L_0[G]$ is a universal sentence of the form

$$\forall \overline{x} \left( \bigwedge_i (u_i(\overline{g}, \overline{x}) = 1) \rightarrow (w(\overline{g}, \overline{x}) = 1) \right).$$

Here $\overline{x}$ abbreviates a tuple of variables and $\overline{g}$ abbreviates a tuple from $G$. Note that every quasi-identity of $L_0[G]$ is equivalent to a universal Horn sentence of $L_0[G]$ (see below). We let $Q(G)$ be the set of all quasi-identities of $L_0[G]$ true in $G$ and $Th_\forall(G)$ be the set of all universal sentences of $L_0[G]$ true in $G$. This is called the **universal theory of** $G$ with respect to $L_0[G]$.

**Remark 5.1.** (1) The above quasi-identity is equivalent to the universal Horn sentence

$$\forall \overline{x} \left( \bigvee_i (u_i(\overline{g}, \overline{x}) \neq 1) \vee (w(\overline{g}, \overline{x}) = 1) \right).$$

(2) *An identity* $\forall \overline{x} \, (w(\overline{x}) = 1)$ *is equivalent to the quasi-identity*

$$\forall \overline{x} \, ((1 = 1) \rightarrow (w(\overline{x}) = 1)).$$

(3) *The group axioms themselves are considered (up to equivalence) special cases of quasi-identities.*

**Theorem 5.2.** (*A.G. Myasnikov and V.N. Remeslennikov* [MR]) *Let the group* $G$ *be nonabelian, CSA, and* $G$-*equationally Noetherian. Then the universal theory of* $G$ *with respect to* $L_0[G]$, $Th_\forall(G)$, *is axiomatizable by* $Q(G) \cup diag(G) \cup \{CT\}$.

We shall show that the hypothesis that $G$ be $G$-equationally Noetherian is not needed. Thus

**Theorem 5.3.** *Let the group* $G$ *be nonabelian and CSA. Then the universal theory of* $G$ *with respect to* $L_0[G]$, $Th_\forall(G)$, *is axiomatizable by* $Q(G) \cup diag(G) \cup \{CT\}$.

**Proof.** Let $G$ be nonabelian and CSA. Suppose that $Th_\forall(G)$ is not axiomatizable by $Q(G) \cup diag(G) \cup \{CT\}$. Then there would be a model $\Gamma$ of $Q(G) \cup diag(G) \cup \{CT\}$ which violates some universal sentence of $L_0[G]$ true in $G$. Equivalently, $\Gamma$ satisfies some existential senence, $\varphi$, of $L_0[G]$ false in $G$. Writing the matrix of $\varphi$ in disjunctive normal form, we may assume $\varphi$ is

$$\exists \overline{x} \left( \bigvee_i \left( \bigwedge_{p=1}^{m_i} (u_{i,p}(\overline{g}, \overline{x}) = 1) \wedge \bigwedge_{q=1}^{n_i} (w_{i,q}(\overline{g}, \overline{x}) \neq 1) \right) \right).$$

Thus, $\varphi$ is equivalent to the disjunction $\bigvee_i \varphi_i$ where $\varphi_i$ is

$$\exists \overline{x} \left( \bigwedge_{p=1}^{m_i} (u_{i,p}(\overline{g}, \overline{x}) = 1) \wedge \bigwedge_{q=1}^{n_i} (w_{i,q}(\overline{g}, \overline{x}) \neq 1) \right).$$

Each $\varphi_i$ must be false in $G$; otherwise, $\bigvee_i \varphi_i$ (and so also $\varphi$) would be true in $G$-contrary to assumption. The only way $\bigvee_i \varphi_i$ could hold in $\Gamma$ is if at least one disjunct $\varphi_i$ held in $\Gamma$. Say $\varphi_{i_0}$ holds in $\Gamma$ and suppress the dependence on $i_0$ notationally to write $\varphi_{i_0}$ as

$$\exists \overline{x} \left( \bigwedge_{p=1}^{m} (u_p(\overline{g}, \overline{x}) = 1) \wedge \bigwedge_{q=1}^{n} (w_q(\overline{g}, \overline{x}) \neq 1) \right).$$

We first claim that $n > 0$. Suppose not. Then $\forall \overline{x} \left( \bigvee_{p=1}^{m} (u_p(\overline{g}, \overline{x}) \neq 1) \right)$ would hold in $G$. Since $G$ is nonabelian it is nontrivial. Fix $g_0 \in G \backslash \{1\}$. Then the quasi-identity

$$\forall \overline{x} \left( \bigwedge_{p=1}^{m} (u_p(\overline{g}, \overline{x}) = 1) \rightarrow (g_0 = 1) \right)$$

and the negated atomic sentence $g_0 \neq 1$ would simultaneously hold in $G$. Thus the quasi-identity lies in $Q(G)$ and the negated atomic sentence lies in $diag(G)$. But $\Gamma$ is a model of $Q(G) \cup diag(G)$; so, $\Gamma$ would be a model of

$$\forall \overline{x} \left( \bigwedge_{p=1}^{m} (u_p(\overline{g}, \overline{x}) = 1) \rightarrow (g_0 = 1) \right) \quad \text{and} \quad g_0 \neq 1$$

and thus a model of $\forall \overline{x} \left( \bigvee_{p=1}^{m} (u_p(\overline{g}, \overline{x}) \neq 1) \right)$ contradicting $\Gamma$ being a model of $\varphi_{i_o}$. The contradiction shows that, as claimed, $n > 0$.

Since $\Gamma$ is a model of $diag(G)$ it is a $G$-group and thus nonabelian. We claim $\Gamma$ is CSA, and hence a domain. Suppose that $\Gamma$ is not CSA. Then, since $\Gamma$ is CT, the following existential sentence would hold in $\Gamma$.

$$\exists x, y, z \ ((x \neq 1) \wedge (y \neq 1) \wedge ([x, y] = 1) \wedge ([x, y^z] = 1) \wedge ([y, z] \neq 1)).$$

Let $x = \xi, y = \eta$, and $z = \varrho$ verify the above sentence in $\Gamma$. From $[\xi, \eta] = 1$, we get $[\varrho^{-1}\xi\varrho, \varrho^{-1}\eta\varrho] = 1$. Since $\eta \neq 1$, $\eta^\varrho$ is also nontrivial. Now $\xi$ commutes with $\eta^\varrho$ which commutes with $\xi^\varrho$; so by CT since $\varrho^{-1}\eta\varrho \neq 1$, we get $[\xi, \varrho^{-1}\xi\varrho] = 1$. If $[\xi, \varrho] = 1$, then, since $\eta$ commutes wth $\xi \neq 1$, we would have $\eta$ commuting with $\varrho$ contradicting $[\eta, \varrho] \neq 1$. Hence,

$$\exists x, z \ (([x, z] \neq 1) \wedge ([x, x^z] = 1))$$

is true in $\Gamma$. But the above existential sentence is equivalent to the negation of the quasi-identity

$$\forall x \ (([x, x^z] = 1) \rightarrow ([x, z] = 1)).$$

We claim the above quasi-identity holds in $G$. For if $x = 1$, then certainly $[x, z] = 1$ and, if $x \neq 1$, then $1 \neq x \in C_G(x) \cap C_G(x)^z$ and so since $G$ is CSA, $[x, z] = 1$ in that case also.

Since $\Gamma$ is a model of $Q(G)$, that is a contradiction. The contradiction shows that, as claimed $\Gamma$ is a domain. From that fact and the result that $\varphi_{i_0}$ holds in $\Gamma$, we get that the following existential sentence $\psi$ holds in $\Gamma$.

$$\exists \overline{x}, \overline{y} \left( \bigwedge_{p=1}^{m} (u_p(\overline{g}, \overline{x}) = 1) \wedge ([w_1(\overline{g}, \overline{x}), \ w_2(\overline{g}, \overline{x})^{y_1}, \ldots, w_n(\overline{g}, \overline{x})^{y_{n-1}}] \neq 1) \right).$$

Here $\overline{y}$ abbreviates the tuple $(y_1, \ldots, y_{n-1})$. (We interpret $\psi$ as $\varphi_{i_0}$ if $n = 1$.) But $\psi$ is equivalent to the negation of the quasi-identity

$$\forall \overline{x}, \overline{y} \left( \bigwedge_{p=1}^{m} (u_p(\overline{g}, \overline{x}) = 1) \rightarrow ([w_1(\overline{g}, \overline{x}), \ w_2(\overline{g}, \overline{x})^{y_1}, \ldots, w_n(\overline{g}, \overline{x})^{y_{n-1}}] = 1) \right).$$

We claim the above quasi-identity holds in $G$. If it did not $\psi$ would hold in $G$ and there would be $\overline{x}, \overline{y}$ in $G$ such that all the $u_p(\overline{g}, \overline{x}) = 1$ while

$$[w_1(\overline{g}, \overline{x}), \ w_2(\overline{g}, \overline{x})^{y_1}, \ldots, w_n(\overline{g}, \overline{x})^{y_{n-1}}] \neq 1$$

and so all the $w_q(\overline{g}, \overline{x}) \neq 1$ in $G$. Thus, $\varphi_{i_0}$ holds in $G$. That is a contradiction. The contradiction shows that the quasi-identity

$$\forall \overline{x}, \overline{y} \left( \bigwedge_{p=1}^{m} (u_p(\overline{g}, \overline{x}) = 1) \rightarrow ([w_1(\overline{g}, \overline{x}), \ w_2(\overline{g}, \overline{x})^{y_1}, \ldots, w_n(\overline{g}, \overline{x})^{y_{n-1}}] = 1) \right)$$

must hold in $G$. Since $\Gamma$ is a model of $Q(G)$, the above quasi-identity must hold in $\Gamma$. But that contradicts its negation (up to logical equivalence) $\psi$ holding in $\Gamma$. The contradiction shows that $Q(G) \cup diag(G) \cup \{CT\}$ is a set of axioms for $Th_\forall(G)$. ∎

We conclude this section with a family of examples to show that CSA cannot be weakened to CT in the hypotheses of Theorem 5.3. For each odd integer $n \geq 3$, let $G_n = \langle a, b; a^2 = b^n = 1, ba = ab^{n-1} \rangle$ be the dihedral group of order $2n$. Let $\delta : G_n \rightarrow G_n^2$ be the diagonal embedding $g \mapsto (g, g)$. $G_n^2$ becomes a $G_n$-group under this embedding. $G_n$ is CT but not CSA. Let $B = \langle b \rangle$ and let $\Gamma$ be the subgroup of $G_n^2$ generated by $\delta(a)$ and $B^2$. Observe that $\delta(G_n) \leq \Gamma$ so that $\Gamma$ is a $G_n$-subgroup of $G_n^2$. Thus $\Gamma$ is a model of $diag(G_n)$. Since $Q(G_n)$ is preserved in direct products of $G_n$-groups and in $G_n$-subgroups, $\Gamma$ is a model of $Q(G_n)$. We claim that $\Gamma$ is CT. First of all every element of $\Gamma$ is uniquely of the form $(a^m b^p, a^m b^q)$ where $0 \leq m \leq 1$

and $0 \leq p, q \leq n - 1$ (so that $\Gamma$ has exactly $2n^2$ elements). Suppose $\gamma = (a^m b^p, a^m b^q) \in \Gamma \backslash \{1\}$. Assume first that $m = 0$. Then $\gamma = (b^p, b^q)$ where at least one of $p$ or $q$ is nonzero. In that event, $C_\Gamma(\gamma) = B^2$ is abelian. Now suppose $m = 1$. Then $\gamma = (ab^p, ab^q)$. Since $C_{G_n}(ab^k) = \langle ab^k \rangle$ for all $k$, $0 \leq k \leq n - 1$, and since neither component of  nontrivial element of $C_\Gamma(\gamma)$ can be 1, we have $C_\Gamma(\gamma) = \Gamma \cap (\langle ab^p \rangle \times \langle ab^q \rangle)$ which is again abelian. Thus $\Gamma$ satisfies CT and is a model of $Q(G_n) \cup diag(G_n) \cup \{CT\}$. If $\Gamma$ were a model of $Th_\forall(G_n)$ it would satisfy

$$\forall x \left( \bigvee_{g \in G_n} (x = g) \right)$$

and have exactly $2n$ elements. That contradicts $|\Gamma| = 2n^2$. Therefore, CSA cannot be weakened to CT in the hypotheses of the Theorem 5.3.

## 6.  Question

(A.G. Myasnikov) Let $c$ and $r$ be integers with $c \geq 2$ and $r \geq \max\{2, c-1\}$. Let $N$ be a group free of rank $r$ relative to the variety $\mathcal{N}_c$ of all groups nilpotent of class at most $c$. Is $Th_\forall(N)$ axiomatizable by $Q(N) \cup diag(N) \cup \{CT(c-1)\}$?

## References

[BMR]     G. Baumslag, A.G. Myasnikov, and V.N. Remeslennikov, "Algebraic geometry over groups," J. Algebra 219, (1999), 16–79.
[CK]      C.C. Chang and H.J. Keisler, **Model Theory**, North-Holland, Amsterdam, 1992.
[FGMS]    B. Fine, A.M. Gaglione, A.G. Myasnikov, and D. Spellman, "Discriminating groups," J. Group Theory 4, (2001), 463–474.
[G]       G. Grätzer, **Universal Algebra**, Van Nostrand, Princeton, 1968.
[MR]      A.G. Myasnikov and V.N. Remeslennikov, "Algebraic geometry over group II: logical foundations," J. Algebra, 234, (2000), 225–276.

# Chapter 6

# Commutative Transitivity and the CSA Property

Benjamin Fine[1], Anthony Gaglione[2], Gerhard Rosenberger[3],
and Dennis Spellman[4]

[1] *Department of Mathematics, Fairfield University*
*Fairfield, Connecticut 06430, United States*

[2] *Department of Mathematics, United States Naval Academy*
*Annapolis, Maryland 21402, United States*

[3] *Fachbereich Mathematik, University of Hamburg Bundestrasse 55*
*20146 Hamburg, Germany*

[4] *Department of Statistics, Temple University*
*Philadelphia, Pennsylvania 19122, United States*

*Dedicated to Gerhard Rosenberger and Dennis Spellman*
*on their Seventieth Birthdays*

**ABSTRACT.** A group is $G$ *commutative transitive* or CT if commuting is
transitive on nontrivial elements. A group $G$ is CSA or *conjugately separated
abelian* if maximal abelian subgroups are malnormal. These concepts have
played a prominent role in the studies of fully residually free groups, limit
groups and discriminating groups. They also play a role in the solution to the
Tarski problems. CSA always implies CT however the class of CSA groups
is a proper subclass of the class of CT groups. For limit groups and finitely
generated elementary free groups they are equivalent. In this paper we survey
the relationship between the two concepts and also examine a series of related
results. In [FGRS 2] it was proved that a finite CSA group must be abelian and
if $G$ is CT then $G$ is not CSA if and only if $G$ contains a nonabelian subgroup
$G_0$ which contains a nontrivial abelian subgroup $H$ that is normal in $G_0$. In
[FGRS] it was shown that the groups $PSL(2, K)$ for $K$ a field are crucial in
studying the relationship between CT and CSA. $PSK(2, K)$ is never CSA and
here we show when it is CT.

*AMS Subject Classification*: Primary 20F67; Secondary 20F65, 20E06, 20E07
*Keywords*: Commutative tranistive, CSA group, Tarski problems, monolithic group.

## 1. Introduction

A group $G$ is *commutative transitive*, which we will abbreviate by CT, if commutativity is transitive on nonidentity elements. Commutative transitivity is a simple idea that suprisingly has had a wide-ranging impact on many areas of algebra in general and group theory in particular. The paper [FR] contains a great deal of information about CT groups in general.

A group $G$ is CSA or *conjugately separated abelian* if maximal abelian subgroups are malnormal (see Section 2). CSA implies CT (see Section 2) but the class of CSA groups is a proper subclass of the class of CT groups. These two concepts and their relationship played a major role in the proof of the Tarski problems (see [FGMRS]). A result of Gaglione and Spellman [GS] and independently Remeslennikov [Re] showed that for nonabelian residually free groups, being CT is equivalent to having the same universal theory as a nonabelian free group (see Section 2). This result was one of the initial important steps in the solution of the Tarski conjectures (see Section 2).

The term *commutative transitive* was coined in [F] relative to free groups and Fuchsian groups yet the concept appeared in the literature substantially earlier. In some papers a CT group is referred to as *centralizer abelian* or CA-group since being CT is easily shown to be equivalent to having all centralizers of nontrivial elements abelian.

Finite CT groups were studied originally by Weisner [W] in 1925. He proved that finite CT groups are either solvable or simple. However there was a mistake in his proof. Yu-Fen Wu in 1997 [Wu] corrected the mistake and reproved Weisner's result. She also proved that a finite solvable CT group is the semidirect product of its Fitting subgroup $F$, which must be abelian by a fixed point free group of automorphisms of $F$. Earlier Suzuki [Su], in 1957, using character theory proved that every finite nonabelian simple CT group is isomorphic to some $PSL(2, 2^f), f \geq 2$.

In [FGRS 2] the relationship between CT and CSA was examined. In that paper it was proved that a finite CSA group must be abelian and hence it can be shown (see Section 4) that a finite CT group that is not simple and not CSA must have a nontrivial abelian normal subgroup. For infinite groups it was proved that a group $G$ that has a composition series and is CT but not CSA either contains a nontrivial normal abelian subgroup or is monolithic. Further if $G$ is monolithic with monolith isomorphic to $PSL(2, K)$ for a field of characteristic 2, then $G$ is isomorphic to $PSL(2, K)$ for a field of characteristic 2.

In this paper, we examine and survey more fully the relationship between the two concepts, CT and CSA. The CSA property implies CT however there do exist groups, both finite and infinite which are CT but not CSA. For limit groups, however, as well as elementary free groups and some related groups, the two concepts are equivalent. We provide a quick proof of this for limit groups in Section 3. The equivalence of CT and CSA carries over to the class of $B\mathcal{X}$-groups introduced by Ciobanu, Fine and Rosenberger (see [CFR]). We discuss this in Section 4. In this paper we retrace many of the proofs that can be found in [FGRS 2] and [FGMRS]. We feel that this material is interesting and important and is appropriate for a conference proceedings.

## 2. Basic Material on CT and CSA Groups

A group G is *commutative transitive* or CT if commutativity is transitive on nontrivial elements. That is

$$[x, y] = 1 \quad \text{and} \quad [y, z] = 1 \implies [x, z] = 1$$

provided $x, y, z$ are nontrivial.

It is straightforward that being commutative transitive is equivalent to the property that the centralizer of every nontrivial element is abelian. For this reason CT groups are sometimes called CA groups or centralizer-abelian groups.

G is CT if and only if it satisfies the universal sentence

$$\forall x, y, z(((y \neq 1) \wedge (xy = yx) \wedge (yz = zy)) \rightarrow (xz = zx)).$$

It is also clear that if $Z(G) \neq \{1\}$ and $G$ is CT then $G$ is abelian.

It is clear (and follows directly from the fact that CT is captured by a universal sentence) that subgroups of CT groups are CT.

There are many examples of classes of CT groups including free groups, torsion-free hyperbolic groups and free solvable groups. The paper [FR] and the book [FGMRS] contains many more examples.

Myasnikov and Remeslennikov in their study of fully residually free groups introduced the concept of a CSA group (conjugately separated abelian group). Recall that if $G$ is a group and $H$ a subgroup of $G$ then $H$ is *malnormal* in $G$ or *conjugately separated* in $G$ provided $g^{-1}Hg \cap H = 1$ unless $g \in H$. A group $G$ is a *CSA-group* or *conjugately separated abelian group* provided the maximal abelian subgroups are malnormal.

Each CSA group must be CT. The converse however is not true in general.

**Lemma 2.1.** *The class of CSA groups is a proper subclass of the class of CT groups.*

**Proof.** We first show that every CSA-group is commutative transitive. Let $G$ be a group in which maximal abelian subgroups are malnormal and suppose that $M_1$ and $M_2$ are maximal abelian subgroups in $G$ with $z \neq 1$ lying in $M_1 \cap M_2$. Could we have $M_1 \neq M_2$? Suppose that $w \in M_1 \setminus M_2$. Then $w^{-1}zw = z$ is a non-trivial element of $w^{-1}M_2w \cap M_2$ so that $w \in M_2$. This is impossible and therefore $M_1 \subset M_2$. By maximality we then get $M_1 = M_2$. Hence, $G$ is commutative transitive whenever all maximal abelian subgroups are malnormal.

We now show that there do exist CT groups that are not CSA. In any non-abelian CSA-group the only abelian normal subgroup is the trivial subgroup 1. To see this suppose that $N$ is any normal abelian subgroup of the non-abelian CSA-group $G$. Then $N$ is contained in a maximal abelian subgroup $M$. Let $g \notin M$. Then

$$N = g^{-1}Ng \cap N \subset g^{-1}Mg \cap M.$$

The fact $N \neq 1$ would imply that $g \in M$ which is a contradiction.

Now let $p$ and $q$ be distinct primes with $p$ a divisor of $q - 1$. Let $G$ be the non-abelian group of order $pq$. Then it is not difficult to prove that the centralizer of every non-trivial element of $G$ is cyclic of order either $p$ or $q$. Thus $G$ is commutative transitive. However, the (necessarily unique) Sylow q-subgroup of $G$ is normal in $G$. Hence from the argument above $G$ cannot be CSA. □

In the next section we give many more examples of CT but non CSA groups.

Although the class of CSA groups is a proper subclass of the CT groups, in the presence of full residual freeness (in fact even in the presence of just residual freeness) they are equivalent. Fully residually free groups play a prominent role in the solution of the Tarski problems (see [FGMRS]). Finitely generated fully residually free groups are also known as *limit groups* since they arise (as initially observed by Sela [Se 1-5]) as limits of homomorphisms into free groups.

Recall that a group $G$ is *residually free* if for each non-trivial $g \in G$ there is a free group $F_g$ and an epimorphism $h_g : G \to F_g$ such that $h_g(g) \neq 1$.

Equivalently for each $g \in G$ there is a normal subgroup $N_g$ such that $G/N_g$ is free and $g \notin N_g$. The group $G$ is *fully residually free* provided to every finite set $S \subset G \setminus \{1\}$ of non-trivial elements of $G$ there is a free group $F_S$ and an epimorphism $h_S : G \to F_S$ such that $h_S(g) \neq 1$ for all $g \in S$.

A beautiful theorem due independently to Gaglione and Spellman [GS] and Remeslennikov [Re] ties together full residual freeness, CT and the property of being universally free. A group is universally free if it has exactly the same universal theory as a nonabelian free group (see [FGMRS]). It is straightforward that all nonabelian free groups have the same universal theory. This result in some sense is the beginning of the solution of the Tarski problems.

**Theorem 2.1 (GS).** [Re] *If a nonabelian group $G$ is residually free then the following are equivalent*:

(1) *$G$ is fully residually free*
(2) *$G$ is CT*
(3) *$G$ is CSA*
(4) *$G$ is universally free.*

## 3. Equivalences Between CT and CSA and the Groups $PSL(2, K)$

As mentioned, CSA always implies CT but the class of CSA groups is a proper subclass of the class of CT groups. In this section we prove that $PSL(2, K)$ is never CSA. However if $K$ has characteristic 2 then $PSL(2, K)$ is always CT while $PSL(2, \mathbb{R})$ and $PSL(\mathbb{Q}_p)$ are also CT. The groups $PSL(2, K)$ for characteristic an odd prime $p$ are never CT. $PSL(2, \mathbb{C})$ and more generally $PSL(2, K)$ where $K$ is an algebraically closed field of characteristic 0 is never CT. Thus several of these types of groups provide an infinite number of examples, both finite and infinite of CT non CSA groups. We also prove that a finite CSA group must be abelian. Wu [Wu] proves that there exist finite solvable CT groups for every solvability class. Hence the nonabelian ones provide more examples of CT non CSA groups.

We first consider some cases where CT and CSA are equivalent.

**Lemma 3.1.** *If $G$ is residually free then $CT \equiv CSA$.*

**Proof.** We assume a theorem of Benjamin Baumslag that CT and residually free imply fully residually free (see [BB]). CSA always imples CT so we assume that $G$ is CT and show that it must be CSA. From Baumslag's theorem since $G$ is residually free and CT it must be fully residually free so that we can assume that $G$ is a fully residually free group with more than one element. Let $u \in G \setminus \{1\}$ and let $M$ be its centralizer which we will denote by $C_G(u)$. Then $M$ is maximal abelian in $G$. We claim that $M$ is malnormal in $G$. If $G$ is abelian, then $M = G$ and the conclusion follows trivially. Suppose that $G$ is non-abelian. Suppose that $w = g^{-1}zg \neq 1$ lies in $g^{-1}Mg \cap M$. If $g \notin M$ then $[g, u] \neq 1$. Thus, there is a free group $F$ and an epimorphism $h : G \to F, x \to \overline{x}$, such that $\overline{w} \neq 1$ and $[\overline{g}, \overline{u}] \neq 1$. Let $C = C_F(\overline{u})$ . Then $\overline{w} \in \overline{g}^{-1}C\overline{g} \cap C$. However the maximal abelian subgroups in a free group are malnormal. This implies $\overline{g} \in C$, contradicting $[\overline{g}, \overline{u}] \neq 1$. This contradiction shows that $g^{-1}Mg \cap M \neq 1$ implies $g \in M$ and hence the maximal abelian subgroups in $G$ are malnormal.    □

Ciobanu, Fine and Rosenberger [CFR] generalized Benjamin Baumslag's theorem to what are called the class of $B\mathcal{X}$-groups. A class of groups $\mathcal{X}$ satisfies the property $B\mathcal{X}$ if a group $G$ is fully residually $\mathcal{X}$ if and only if $G$ is residually $\mathcal{X}$ and CT.

With this definition B. Baumslag's original theorem says that the class of free groups $\mathcal{F}$ satisfies $B\mathcal{F}$.

In [CFR] it was proved that a class of groups $\mathcal{X}$ satisfies $B\mathcal{X}$ under very mild conditions and hence the classes of groups for which this is true is quite extensive. In any class of groups satisfying $B\mathcal{X}$ the properties CT and CSA are equivalent.

**Theorem 3.1.** (see [CFR]) Let $\mathcal{X}$ be a class of groups such that each nonabelian $H \in \mathcal{X}$ is CSA. Let $G$ be a nonabelian and residually $\mathcal{X}$ group. Then the following are equivalent

(1) $G$ is fully residually $\mathcal{X}$
(2) $G$ is CSA
(3) $G$ is CT.

Therefore the class $\mathcal{X}$ has the property $B\mathcal{X}$.

It follows that a class of groups $\mathcal{X}$ satisfies $B\mathcal{X}$ if each nonabelian $H \in \mathcal{X}$ is CSA. Examples of $B\mathcal{X}$ classes abound. In particular in [CFR] the following are listed.

**Theorem 3.2.** Each of the following classes satisfies $B\mathcal{X}$:

(1) The class of nonabelian free groups.
(2) The class of limit groups.

(3) *The class of noncyclic torsion-free hyperbolic groups (see* [GKM]).
(4) *The class of noncyclic one-relator groups with only odd torsion (see* [GKM]).
(5) *The class of cocompact Fuchsian groups with only odd torsion.*
(6) *The class of noncyclic groups acting freely on $\Lambda$-trees where $\Lambda$ is an ordered abelian group (see* [CFR]).
(7) *The class of noncyclic free products of cyclics with only odd torsion (see* [GKM]).
(8) *The class of noncyclic torsion-free RG-groups (see* [FMgrRR] *and* [CFR]).
(9) *The class of conjugacy pinched one-relator groups of the following form*

$$G = \langle F, t; tut^{-1} = v \rangle$$

*where $F$ is a free group of rank $n \geq 1$ and $u, v$ are nontrivial elements of $F$ that are not proper powers in $F$ and for which $\langle u \rangle \cap x \langle v \rangle x^{-1} = \{1\}$ for all $x \in F$.*

For the rest of this section we will concentrate on the situations where CT and CSA are not equivalent. That is we will examine CT non-CSA groups. We saw that CT is given by a universal sentence and hence is captured by subgroups. The same is true for CSA.

**Lemma 3.2.** *If $G$ is a CSA group and $H \subset G$ then $H$ is CSA.*

**Proof.** We first give a direct proof (see [GKM]). Let $G$ be a CSA group and let $H$ be a subgroup of $G$. Let $A_H$ be a maximal abelian subgroup of $H$. We must show that $A_H$ is malnormal in $H$. Let $x \in H$ with $xA_Hx^{-1} \cap A_H \neq \{1\}$. $A_H$ is contained in a maximal abelian subgroup $A_G$ of $G$. Since $G$ is CSA it follows that $A_G$ is malnormal in $G$ and so $x \in A_G$. Then $x \in (A_G \cap H) \subset A_H$ and hence $A_H$ is malnormal in $H$.

The result also follows from the fact that CSA can also be described in terms of universal sentences. In particular the CSA property is described by the following pair of universal sentences.

$$(CT :) \forall x, y, z(((y \neq 1) \wedge ([x, y] = 1) \wedge ([y, z] = 1)) \rightarrow ([x, z] = 1))$$

$$(MAL :) \forall x, y, z((x \neq 1) \wedge (y \neq 1) \wedge ([x, y] = 1) \wedge ([x, z^{-1}yz] = 1))$$
$$\rightarrow ([y, z] = 1)) \qquad \square$$

Recall that the infinite dihedral group is the free product $D = \mathbb{Z}_2 \star \mathbb{Z}_2$. The group $D$ then has the presentation $D = \langle x, y; x^2 = y^2 = 1 \rangle$. Then $xxyx^{-1} = yxyy^{-1} = yx = (xy)^{-1}$ and hence $D$ is not CSA.

**Lemma 3.3.** *If the group $G$ contains a subgroup isomorphic to the infinite dihedral group then $G$ is not CSA.*

**Corollary 3.1.** *The modular group $M = PSL(2, \mathbb{Z})$ is not CSA.*

**Proof.** The modular group $M$ is isomorphic to the free product $\mathbb{Z}_2 \star \mathbb{Z}_3$ of a cyclic group of order 2 and a cyclic group of order 3. Such a free product contains as a subgroup the free product $\mathbb{Z}_2 \star \mathbb{Z}_2$, a subgroup isomorphic to the infinite dihedral group. Therefore by Lemma 3.4 $M$ cannot be CSA. □

**Corollary 3.2.** *The group $PSL(2, \mathbb{Q})$ where $\mathbb{Q}$ is the field of rational numbers is not CSA.*

**Proof.** $PSL(2, \mathbb{Q})$ contains $M$ as a subgroup and hence cannot be CSA. □

Because of Wu Fen's work on finite CT groups (see [Wu]) the groups $PSL(2, K)$, where $K$ is a field, figure prominently in the analysis of CT and CSA groups. In [Wu] she also proves that there are finite solvable CT groups for every solvability class.

We prove the following two results. The first is that $PSL(2, K)$ for any field is never CSA.

**Theorem 3.3.** *Suppose that $K$ is a field. Then the group $PSL(2, K)$ is not CSA.*

**Proof.** We consider the characteristic of $K$ and handle each characteristic separately. If $K$ is a field of characteristic $p \neq 2$ then the group $PSL(2, K)$ is not CT and hence it cannot be CSA.

Now let $K$ be a field of characteristic 0. Then $K$ contains a subfield isomorphic to $\mathbb{Q}$. Hence $PSL(2, K)$ contains a subgroup isomorphic to $PSL(2, \mathbb{Q})$. From Lemma 3.3 $PSL(2, \mathbb{Q})$ is not CSA and therefore $PSL(2, K)$ cannot be CSA.

Finally let $K$ be a field of characteristic 2. Let $F = \mathbb{Z}_2$ be the two-element field. Then $K$ contains a subfield isomorphic to $F$ and hence $PSL(2, F) = PSL(2, \mathbb{Z}_2)$ is a subgroup of $PSL(2, K)$. However $PSL(2, \mathbb{Z}_2)$ is nonabelian of order 6 and hence is isomorphic to $S_3$ the symmetric group on 3 symbols. This group has an abelian normal subgroup or order 3 and hence is not CSA. It follows that $PSL(2, K)$ cannot be CSA. □

The next theorem handles the CT property for $PSL(2, K)$. It is more complex than for CSA.

**Theorem 3.4.** *Suppose that $K$ is a field.*

(1) *If* $\operatorname{char}(K) = 2$ *then the group $PSL(2, K)$ is CT.*
(2) *If* $\operatorname{char}(K) = p$ *where $p$ is an odd prime the group $PSL(2, K)$ is not CT.*
(3) *If* $\operatorname{char}(K) = 0$ *then the group $PSL(2, K)$ is CT if $-1$ is not a sum of two squares in $K$ and not CT if $-1$ is a sum of two squares in $K$. In particular if $K = \mathbb{R}$, the real numbers or $K = \mathbb{Q}_p$ the p-adic numbers or any subfield of these then $PSL(2, K)$ is CT. On the other hand $PSL(2, \mathbb{C})$ is not CT and more generally $PSL(2, K)$ is not CT for any algebraically closed field of characteristic 0.*

It follows from these two theorems that in the class of groups $PSL(2, K)$ there are infinitely many examples, both of finite order and infinite order of CT non CSA groups.

**Proof.** (Theorem 3.4) We do each characteristic separately with characteristic $p$ the simplest.

**Lemma 3.4.** *If $K$ is a field of characteristic $p \neq 2$ then the group $PSL(2, K)$ is not CT.*

**Proof.** From Wu's result a finite CT group must either be solvable or isomorphic to $PSL(2, K)$ where $K$ is a field of characteristic 2. Hence if $p \neq 2$ we must have that $PSL(2, \mathbb{Z}_p)$ is not CT for the finite field $K = \mathbb{Z}_p$. If $K$ is a field of characteristic $p \neq 2$ then $PSL(2, K)$ will contain $PSL(2, \mathbb{Z}_p)$ as a subgroup. Since the CT property is captured by subgroups it follows that $PSL(2, K)$ cannot be CT. $\qquad\square$

**Lemma 3.5.** *If $\operatorname{char}(K) = 0$ then the group $PSL(2, K)$ is CT if $-1$ is not a sum of two squarees in $K$ and not CT if $-1$ is a sum of two squares in $K$. In particular if $K = \mathbb{R}$, the real numbers or $K = \mathbb{Q}_p$ the p-adic numbers or any subfield of these then $PSL(2, K)$ is CT. Further $PSL(2, K)$ is CT for any real field. On the other hand $PSL(2, \mathbb{C})$ is not CT and more generally $PSL(2, K)$ is not CT for any algebraically closed field of characterisitc 0.*

**Proof.** Suppose that $K$ is a field with $\operatorname{char}(K) = 0$ and suppose that there do not exist elements $x, y \in K$ such that $x^2 + y^2 = -1$. Let $A, B, C$ be nontrivial elements of $PSL(2, K)$ wtih $AB = BA$ and $BC = CB$. Since $K$ can be embedded in an algebraic closure $k$ we may assume that each of $A, B, C$ has one or two eigenvalues within $k$.

Case 1: $B$ has one eigenvalue in $k$. Then this eigenvalue is already in $K$. After a suitable conjugation in $PSL(2, K)$ we may assume that

$$B = \pm \begin{pmatrix} 1 & 1 \\ 0 & 1 \end{pmatrix}.$$

Let

$$A = \pm \begin{pmatrix} a & b \\ c & d \end{pmatrix} \text{ in } PSL(2, K).$$

From $AB = BA$ we get that either $c = 0$ or $a = d = -b, -c = 2a$. We must have $c = 0$ for otherwise this implies that

$$1 = ad - bc = a^2 - 2a^2 = -a^2 - 0^2$$

and hence $-1$ is a sum of two squares in $K$ contrary to assumption. Further $AB = BA$ then leads to $a = d = \pm 1$. Hence $A$ has the form

$$A = \pm \begin{pmatrix} 1 & \alpha \\ 0 & 1 \end{pmatrix} \text{ in } PSL(2, K).$$

An analogous statement holds for $C$ since $BC = CB$. Therefore $C$ has this form also and hence $AC = CA$ in Case 1.

Case 2: $B$ has two eigenvalues in $k$. After a suitable conjugation in $PSL(2, k)$ we may assume that

$$B = \pm \begin{pmatrix} \alpha & 0 \\ 0 & \alpha^{-1} \end{pmatrix} \in PSL(2, k).$$

Since $B$ is nontrivial we have $\alpha \neq \pm 1$.
    Let

$$A = \pm \begin{pmatrix} a & b \\ c & d \end{pmatrix} \text{ in } PSL(2, k).$$

Then from $AB = BA$ we get that either $b = c = 0$ or $c \neq 0, c\alpha = -c\alpha^{-1}$ or $b \neq 0, b\alpha = -\alpha^{-1}b$.
    If $c \neq 0$ or $b \neq 0$ then $\alpha = -\alpha^{-1}$ and hence the trace, $\operatorname{tr}(B) = 0$.
    Analogously for $C$. Let

$$C = \pm \begin{pmatrix} e & f \\ g & h \end{pmatrix} \text{ in } PSL(2, k).$$

Then from $BC = CB$ we get that either $f = g = 0$ or $\operatorname{tr}(B) = 0$.

If $b = c = f = g = 0$ then $AC = CA$ and hence here in case 2 we may assume that $\mathrm{tr}(B) = 0$.

We now consider $A, B, C \in PSL(2, K)$ with $\mathrm{tr}(B) = 0$. Let

$$B = \pm \begin{pmatrix} \alpha & \beta \\ \gamma & -\alpha \end{pmatrix}.$$

We have that $\gamma \neq 0$ for if $\gamma = 0$ then $-\alpha^2 = -\alpha^2 + 0^2 = 1$ contrary to assumption that $-1$ is not a sum of squares.

Hence by conjugation we may assume that $B$ has the form

$$B = \pm \begin{pmatrix} 0 & 1 \\ -1 & 0 \end{pmatrix}.$$

Let

$$A = \pm \begin{pmatrix} a & b \\ c & d \end{pmatrix}.$$

Then from $AB = BA$ we get that either

$$A = \pm \begin{pmatrix} x & -y \\ y & x \end{pmatrix} \quad \text{or} \quad A = \pm \begin{pmatrix} -x & y \\ y & x \end{pmatrix}.$$

If

$$A = \pm \begin{pmatrix} -x & y \\ y & x \end{pmatrix}$$

then

$$-x^2 - y^2 = 1$$

contradicting the assumption on $-1$ not being a sum of squares. Therefore

$$A = \pm \begin{pmatrix} x & -y \\ y & x \end{pmatrix}.$$

Analogously let

$$C = \pm \begin{pmatrix} e & f \\ g & h \end{pmatrix}.$$

Then from $BC = CB$ we get that

$$C = \pm \begin{pmatrix} u & -v \\ v & u \end{pmatrix}.$$

But in this case $AC = CA$.

Therefore altogether $PSL(2, K)$ is CT if $-1$ is not a sum of two squares in $K$.

Now suppose that $-1 = x^2 + y^2$ in $K$. Since $K$ has characteristic 0 we have $\mathbb{Q} \subset K$. Let $\alpha, \beta$ be nonzero elements of $\mathbb{Q}$ such that $\alpha^2 + \beta^2 = 1$. For example let $\alpha = \frac{3}{5}, \beta = \frac{4}{5}$. Now let

$$A = \pm \begin{pmatrix} -x & y \\ y & x \end{pmatrix}$$

$$B = \pm \begin{pmatrix} 0 & 1 \\ -1 & 0 \end{pmatrix}$$

and

$$C = \pm \begin{pmatrix} \dfrac{3}{5} & \dfrac{4}{5} \\ -\dfrac{4}{5} & \dfrac{3}{5} \end{pmatrix}.$$

Then $A, B, C$ are three nontrivial elements of $PSL(2, K)$ with $AB = BA$, $BC = CB$ but $AC \neq CA$ so the group is not CT.

Notice that if $-1$ is itself a square in $K$ the group $PSL(2, K)$ then cannot be CT. In particular this is true for the complex numbers $\mathbb{C}$ and more generally for any algebraically closed field $K$. We give an example in $PSL(2, \mathbb{C})$ to clarify this.

In $PSL(2, \mathbb{C})$ we have the projective matrices

$$T = \pm \begin{pmatrix} 0 & 1 \\ -1 & 0 \end{pmatrix}, \quad U = \pm \begin{pmatrix} i & 0 \\ 0 & -i \end{pmatrix}, \quad V = \pm \begin{pmatrix} \alpha & 0 \\ 0 & \dfrac{1}{\alpha} \end{pmatrix}, \quad \alpha \neq \pm 1.$$

As linear fractional transformations these are

$$T : z' = -\frac{1}{z}, \quad U : z' = -z, \quad V : z' = \alpha^2 z.$$

By a direct computation $U$ commutes with $T$ and $V$ but $T$ and $V$ do not commute. Therefore $PSL(2, \mathbb{C})$ is not CT.

Exactly the analogous example works in any field $K$ of characteristic 0 where $-1$ is a square. Therefore the example holds in $PSL(2, K)$ for any algebraically closed field of characteristic 0.

Notice further that the lemma applies to $PSL(2, \mathbb{R})$ for the real numbers $\mathbb{R}$ and for any subfield of $\mathbb{R}$, in particular any algebraic number field, and for any subgroup of $PSL(2, \mathbb{R})$. Hence any Fuchsian group is CT.

In general, fields where $-1$ is not a sum of squares are called real fields and have been extensively studied (see [La]). If $-1$ is not a sum of squares then it is not a sum of two squares and hence $PSL(2, K)$ is CT for any real field. □

**Lemma 3.6.** *Let $K$ be a field of characteristic 2. Then $PSL(2, K)$ is CT.*

**Proof.** Let $K$ be a field of characteristic 2 and let $F = \mathbb{Z}_2$ be the two element field. Clearly $F$ is a subfield of $K$. For a field of characteristic 2 we have $PSL(2, K) = SL(2, K)$ so we show that $SL(2, K)$ is CT. Now $PSL(2, F) \cong S_3$, the symmetric group on three symbols. This group is CT so we now may assume that $|K| \geq 4$.

Let $\overline{K}$ be an algebraic closure of $K$ and let $k$ be the algebraic closure of $F$ in $\overline{K}$. Then we have the tower of fields.

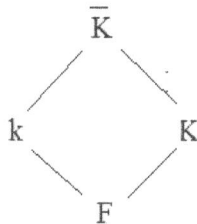

$$
\begin{array}{ccc}
 & \overline{K} & \\
\diagup & & \diagdown \\
k & & K \\
\diagdown & & \diagup \\
 & F &
\end{array}
$$

Tower of Fields 1

From [Hod, p. 47] we have that one form of Hilbert's Nullstellensatz says that if $A$ is an algebraically closed field and $E$ is a finite systems of equations and inequations with coefficients from $A$, such that some field extending $A$ contains a solution of $E$, then $A$ already contains a solution of $E$ (see also Jacobson [Ja], p. 425). It follows from this that an existentially closed field (see [Ja] for a definition) is the same thing as an algebraically closed field. We now use a bit of model theory. We refer the reader to [BeS] or [FGMRS] for a discussion of ultrapowers.

Since $k$ and $\overline{K}$ are algebraically closed they are existentially closed and hence, since $k \subset \overline{K}$ we must have the universal equivalence $k \equiv_\forall \overline{K}$.

By Lemma 3.8 in [BeS] (p. 187) the field $\overline{K}$ embeds in an ultrapower $^\star k = k^I / D$ of $k$. We now have the diagram;

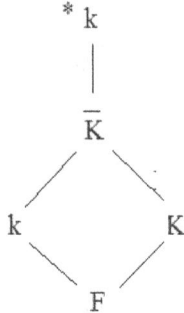

$$
\begin{array}{c}
{}^\star k \\
| \\
\overline{K} \\
\diagup \quad \diagdown \\
k \qquad\qquad K \\
\diagdown \quad \diagup \\
F
\end{array}
$$

Tower of Fields 2

For a field $E$ making the obvious identification of $^\star SL(2, E) = SL(2, E)^I / D$ with $SL(2, ^\star E) = SL(2, E^I / D)$ we have the diagram of group inclusions.

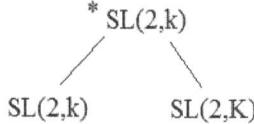

$$
\begin{array}{c}
{}^\star SL(2,k) \\
\diagup \qquad \diagdown \\
SL(2,k) \qquad\qquad SL(2,K)
\end{array}
$$

Group Inclusions

From the result of Wu we have that $SL(2, k)$ is CT since $k$ is locally finite. Further $^\star SL(2, k) \equiv_\forall SL(2, k)$ and hence $^\star SL(2, k)$ is CT since CT is expressed as a universal sentence in the language $L_0$ of group theory. But the CT property is inherited by subgroups and therefore it follows that $SL(2, K)$ is CT.    □

These three lemmas complete the proof of Theorem 3.1.    □

We now prove:

**Theorem 3.5.** *Let $G$ be a finite CSA group. Then $G$ is abelian.*

**Proof.** Let $G$ be a finite CSA group. Since CSA implies CT we then have $G$ is a finite CT group. From Wu's theorem $G$ is then either solvable or isomorphic to $PSL(2, K)$ for a finite field of characteristic 2. If $G \cong PSL(2, K)$ then from Theorem 3.2 $G$ cannot be CSA. Hence $G$ must be solvable. If the solvability class is $d > 1$ then the element of the derived series $G^{d-1}$ is a nontrivial abelian normal subgroup and hence $G$ cannot be CSA in this case. It follows that the solvability class must be $d = 1$ and therefore $G$ is abelian. □

## 4. Infinite CT non-CSA Groups

For finite groups a CT group is either solvable or simple. From Theorem 3.3 and Wu's analysis we have that finite CSA groups are abelian. Wu proves that there exist finite CT groups of every solvability class and hence the nonabelian ones (those of solvability class $d > 1$ provide examples of finite CT non CSA groups. The situation for infinite CT groups is more complicated. First notice that the infinite Tarksi simple groups (see [FGRS 1] are both CT and CSA while the free product of two finite CT non CSA groups such as $S_3 \star S_3$ is an infinite CT but non-CSA group.

**Lemma 4.1.** *Let $G, H$ be any two CT non-CSA groups. Then the free product $G \star H$ is also CT non CSA.*

**Proof.** The free product of CT groups is again CT so $G \star H$ is CT. However $G$ can be considered as a subgroup of the free product so $G \star H$ cannot be CSA. □

We now give several results characterizing infinite CT non CSA groups. Notice that if a group $G$ contains a nontrivial abelian normal subgroup then it cannot be CSA. This is almost enough to characterize when a CT group is not CSA.

**Theorem 4.1.** *A CT group $G$ is not CSA if and only if $G$ contains a nonabelian subgroup $G_0$ which itself contains a nontrivial abelian subgroup $H$ which is normal in $G_0$*

**Proof.** Suppose that $G$ contains a nonabelian subgroup $G_0$ which itself contains a nontrivial abelian subgroup $H$ which is normal in $G_0$. Then $G_0$ cannot be CSA and hence $G$ cannot be CSA.

Conversely suppose that $G$ is CT but not CSA. Recall that in the presence of the group axioms the CSA property is captured by the pair of universal sentences

$$(CT :)\forall x, y, z(((y \neq 1) \wedge ([x, y] = 1) \wedge ([y, z] = 1)) \to ([x, z] = 1))$$

$$(MAL :)\forall x, y, z((x \neq 1) \wedge (y \neq 1) \wedge ([x, y] = 1) \wedge ([x, z^{-1}yz] = 1)$$
$$\to ([y, z] = 1))$$

It follows that being CT but not CSA is captured (in the presence of the group axioms) by

$$(CT :)\forall x, y, z(((y \neq 1) \wedge ([x, y] = 1) \wedge ([y, z] = 1) \to ([x, z] = 1))$$

$$(NOTMAL :)\exists x, y, z((x \neq 1)$$
$$\wedge (y \neq 1) \wedge ([x, y] = 1) \wedge ([x, z^{-1}yz] = 1) \wedge ([y, z] \neq 1)).$$

Now suppose that $G$ is CT but not CSA. Let $g, h, k \in G$ such that if $g = x, h = y, z = k$ then these three elements verify NOTMAL in $G$. Consider the subgroup $G_0 = \langle g, h, k \rangle$ of $G$. This is nonabelian since $[h, k] \neq 1$.

Now consider the subgroup $A = \langle h \rangle^{G_0}$, the normal closure of $\langle h \rangle$ in $G_0$. Since $h \neq 1$ the subgroup $A$ is nontrivial. We claim that $A$ is abelian which will complete the proof.

Now $g$ and $h$ commute with $h$ and further $g \neq 1$ commutes with $k^{-1}hk$ and commutes with $h$ so by CT, $k^{-1}hk$ commutes with $h$. From the fact that $k^{-1}hk$ commutes with $h$ we get that $k(k^{-1}hk)k^{-1}$ commutes with $khk^{-1}$ and so $khk^{-1}$ commutes with $h$. Hence if $u \in \{g, g^{-1}, h, h^{-1}, k, k^{-1}\}$ then $u h^{\epsilon} u^{-1}$ commutes with $h$ where $\epsilon = \pm 1$. It follows that these conjugates commute with each other.

If $A$ were not abelian there would be a word $w(x, y, z)$ of shortest length such that

$$w(g, h, k)^{-1}hw(g, h, k)$$

did not commute with $h$. Choose such a $w$ with $w = v(x, y, z)u$ and $u \in \{g, g^{-1}, h, h^{-1}, k, k^{-1}\}$. Then by minimality $v(g, h, k)^{-1}hv(g, h, k)$ commutes with $h$. Let $\bar{u}$ be the value of $u$ in $G_0$ so that $\bar{u} \in \{g, g^{-1}, h, h^{-1}, k, k^{-1}\}$. From $v(g, h, k)^{-1}hv(g, h, k)$ commuting with $h$ we get that $\bar{u}^{-1}v(g, h, k)^{-1}hv(g, h, k)\bar{u}$ commutes with $\bar{u}^{-1}h\bar{u}$. But $\bar{u}^{-1}h\bar{u}$ commutes with $h$ so by CT

$$w(g, h, k)^{-1}hw(g, h, k) = \bar{u}^{-1}v(g, h, k)^{-1}hv(g, h, k)\bar{u}$$

commutes with $h$ contradicting our choice of $w(x, y, z)$. Therefore this contradiction shows that $A$ must be abelian.    □

To proceed further we need the following concept. A group $G$ is *monolithic* if $G$ contains a unique nontrivial minimal normal subgroup $N$ (see [N]). This subgroup is then called the *monolith*. Our first result is the following:

**Theorem 4.2.** *Let $G$ be a nontrivial CT group which contains no nontrivial abelian normal subgroup. If $G$ has a composition series then $G$ is monolithic whose monolith $N$ is a simple nonabelian CT group.*

**Proof.** Notice that if the monolith $N \cong PSL(2, K)$ for $K$ a field of characteristic 2, which is the situation for finite CT groups with no abelian normal subgroups then $G$ would not be CSA. However as pointed out above there do exist simple nonabelian CT groups that are CSA.

If $H$ is a group then a descending chain of subgroups

$$H = H_0 \supset H_1 \supset \cdots H_n$$

is a *chief series* (see [Hal] p. 124) from $H$ to $H_n$ provided $H_i$ is normal in $H$ for all $i = 0, 1, ..., n$ and for all $i = 0, ..., n$, $H_i$ is maximal normal in $H_{i-1}$.

Let $G$ be a CT group with a composition series and no abelian normal subgroup. Since $G$ is assumed to have a composition series it follows from Theorem 8.6.1 ([Ha], p. 131) that $G$ has a chief series

$$G = G_0 \supset G_1 \supset \cdots \supset G_{n-1} \supset G_n = \{1\}.$$

Let $M = G_{n-1}$. Then $G_n = \{1\}$ is maximal normal in $M$ and hence there is no subgroup $N$ normal in $G$ such that $M \supset N \supset \{1\}$. It follows that $M$ is a minimal normal subgroup in $G$.

We claim that $M$ is unique. Suppose that $M_1$ and $M_2$ are minimal normal subgroups of $G$ with $M_1 \neq M_2$. By assumption neither is abelian. By minimality $M_1 \cap M_2 = \{1\}$. It follows that the subgroup $H = \langle M_1, M_2 \rangle$ generated by $M_1, M_2$ is their direct product. That is $H = M_1 \times M_2$. However a direct product of nonabelian groups is not CT a contradiction since $G$ is CT and CT is inherited by subgroups. Therefore $M$ is a unique minimal normal subgroup and hence $G$ is monolithic with monolith $M$.

Again from Theorem 8.6.1 in [Ha] $M$ is a direct power $A^m$ of a simple group $A$. Since $G$ contains no normal abelian subgroup it follows that $A^m$ is not abelian and hence $A$ is not abelian. If $m > 1$ the monolith $A^m$ is not CT, again a contradiction and therefore $m = 1$ and the monolith is a nonabelian simple CT group. $\qquad\square$

We next prove the following result stated in the introduction.

**Theorem 4.3.** *A finite CT group that is not simple and not CSA must have a nontrivial abelian normal subgroup.*

**Proof.** Let $G$ be a finite CT group that is not simple and not CSA. Since $G$ is finite it must have a composition series. Suppose further to deduce a contradiction that the $G$ does not contain any nontrivial abelian normal subgroup. By Theorem 4.2, a finite CT, non-CSA group must be monolithic with monolith a finite simple CT group. Since the group does not contain any nontrivial abelian normal subgroup, then the monolith must be a finite nonabelian simple group and hence isomorphic to $PSL(2, 2^f)$ for $f \geq 2$. Then by Theorem 4.3 the group itself coincides with $PSL(2, 2^f)$. This contradicts that $G$ is not simple completing the proof. $\qquad\square$

We now consider monolithic groups with monolith isomorphic to $PSL(2, K)$ for a field $K$ of characteristic 2.

**Theorem 4.4.** *Let $G$ be a monolithic group with monolith isomorphic to $PSL(2, K)$ where $K$ is a field of characteristic 2 with $|K| \geq 4$. Then if $G$ is CT we must have $G \cong PSL(2, K)$ and hence $G$ is non CSA.*

We need two preliminary results before we prove this theorem.

**Lemma 4.2.** *Let $G$ be a monolithic group with monolith $M$ isomorphic to $PSL(2, K)$ where $K$ is a field of characteristic 2. If $G$ is CT, $G$ then embeds into $\mathrm{Aut}(M)$.*

**Proof.** Since $M$ is normal in $G$ we get a map $\phi$ from $G$ to $\mathrm{Aut}(M)$ by mapping $g \in G$ to conjugation on $M$ by $g$. Now $M = SL(2, K)$ is nonabelian. Choose $a, b \in M$ such that $ab \neq ba$ and suppose that $z \in \ker(\phi)$. Then $zaz^{-1} = a$ and $zbz^{-1} = b$. Now $z$ commutes with both $a$ and $b$. If $z \neq 1$ and $ab \neq ba$ this contradicts the assumption that $G$ is CT. Hence $\ker(\phi) = \{1\}$ and hence $\phi$ is an embedding. $\qquad\square$

Recall that an algebraic structure is *rigid* if it admits only the identity automorphism.

**Theorem 4.5.** *Let $K$ be a field of characteristic 2 with $|K| \geq 4$. Then $K$ is not rigid.*

**Proof.** Let $K$ be a field of characteristic 2 with $|K| \geq 4$. Then the map $\sigma : K \to K$ given by $\sigma(x) = x^2$ for all $x \in K$ is an injective homomorphism.

If it is surjective we are done since the only roots of $x^2 - x$ over $K$ are 0 or 1 and we thus get a nontrivial automorphism.

Assume now that $K$ contains an element $\theta$ which is not a square in $K$ and assume that $K$ is rigid. Now consider the simple group $M = SL(2, K)$. Recall that $SL(2, K) = PSL(2, K)$. By Theorem 15.3.2 in [Sc] we have that $\mathrm{Aut}(M)$ is complete. Since we assumed that $K$ is rigid $\mathrm{Aut}(M)$ consists solely of inner automorphisms and hence $M$ is isomorphic to $\mathrm{Aut}(M)$ and hence $M$ is complete. By Theorem 15.3.1 in [Sc] if $H$ is any overgroup of $M$ in which $M$ is normal then $H$ has the internal direct product representation $M \times C_H(M)$ where $C_H(M)$ is the centralizer of $H$ in $M$.

Thus since $M = SL(2, K)$ is normal in $H = GL(2, K)$, $H$ has the internal direct product representation $H = M \times C_H(M)$. We claim that the centralizer of $M$ in $H$ consists of the nonzero scalar matrices. It is straightforward that any matrix in $H$ which commutes with both

$$X = \begin{pmatrix} 1 & 1 \\ 0 & 1 \end{pmatrix} \quad \text{and} \quad Y = \begin{pmatrix} 0 & 1 \\ 1 & 0 \end{pmatrix}$$

must be scalar.

Now fix an element $\theta \in K$ which is not a square in $K$ and let

$$C = \begin{pmatrix} \theta & 0 \\ 0 & 1 \end{pmatrix}.$$

Then $C$ has a unique representation of the form $AB$ where $A \in M$ and $B \in C_H(M)$.

Since $B \in C_H(M)$ the matrix for $B$ is a scalar matrix $\begin{pmatrix} \beta & 0 \\ 0 & \beta \end{pmatrix}$ for some nonzero $\beta \in K$. But then

$$\theta = \det \begin{pmatrix} \theta & 0 \\ 0 & 1 \end{pmatrix} = \det(A)\det(B) = \beta^2.$$

Then $\theta$ is a square in $K$ contrary to assumption. It follows that $K$ cannot be rigid. $\qquad\square$

We now give the proof of Theorem 4.4.

**Proof.** Let $G$ be a monolithic group with monolith isomorphic to $PSL(2, K)$ where $K$ is a field of characteristic 2 and $|K| \geq 4$ and suppose that $G$ is CT.

From Theorem 4.3 we have that $PSL(2, K)$ is not rigid. From [Wan] we have that an automorphism of $SL(2, K)$ is of the form $A \mapsto PA^\sigma P^{-1}$

or $A \mapsto P(A^\iota)^t P^{-1}$ where $\sigma$ is an automorphism of $K$, $\iota$ is an antiautomorphism of $K$ and $A^t$ is the transpose of $A$. Since $K$ is commutative being a field any anti-automorphism is already an automorphism. Further the transpose operator is an anti-automorphism but not an automorphism of $SL(2, K)$. It follows then that the maps of the form $A \mapsto P(A^\iota)^t P^{-1}$ do not occur here.

Since $K$ is not rigid there is a nonidentity automorphism of $K$. Let $\sigma$ be such a nonidentity automorphism and let $\rho$ be the nontrivial automorphism of $PSL(2, K)$ given by

$$\rho \begin{pmatrix} a & b \\ c & d \end{pmatrix} \mapsto \begin{pmatrix} \sigma(a) & \sigma(b) \\ \sigma(c) & \sigma(d) \end{pmatrix}.$$

Now let $A, B$ be the elements of $SL(2, K)$ given by

$$A = \begin{pmatrix} 1 & 1 \\ 0 & 1 \end{pmatrix} \quad \text{and} \quad B = \begin{pmatrix} 1 & 0 \\ 1 & 1 \end{pmatrix}.$$

Let $\alpha$ denote conjugation in $SL(2, K)$ by $A$ and $\beta$ denote conjugation in $SL(2, K)$ by $B$.

By direct computation we have that in $\text{Aut}(SL(2, K)$ the automorphisms $\alpha, \beta$ both commute with $\tau$, that is $\alpha\tau = \tau\alpha$ and $\tau\beta = \beta\tau$. However again by direct computation we have $\alpha\beta \neq \beta\alpha$.

By Lemma 4.1, $G$ embeds in $\text{Aut}(M) = \text{Aut}(SL(2, K))$ and it follows that the image of $G$ in the automorphism group must also be CT. This combined with the computations above imply that no transformation of the form

$$\rho \begin{pmatrix} a & b \\ c & d \end{pmatrix} \mapsto \begin{pmatrix} \sigma(a) & \sigma(b) \\ \sigma(c) & \sigma(d) \end{pmatrix}$$

for a nontrivial automorphism $\sigma$ of $K$ can occur in the image of $G$ in $\text{Aut}(M)$. Thus the image of $G$ in $\text{Aut}(M)$ must consist solely of inner automorphisms of $M$. Hence for every $g \in G$ there exists an $h \in M$ such that $gxg^{-1} = hxh^{-1}$ for all $x \in M$. Therefore

$$g^{-1}h \in C_G(M) = \ker(G \to \text{Aut}(M)) = \{1\}.$$

Thus $g = h \in M$ and since $g, h$ were arbitrary $G = M$ completing the proof.  □

Finally recall that a class of groups $\mathcal{X}$ is *axiomatic* if this class is defined in terms of a set of first order sentences (see [FGMRS]) or axioms. We have seen that the CT property is given by the group axioms together with

$$(CT :) \forall x, y, z(((y \neq 1) \wedge ([x, y] = 1) \wedge ([y, z] = 1) \rightarrow ([x, z] = 1))$$

while the class of CSA groups is captured by

$$(CT :) \forall x, y, z(((y \neq 1) \wedge ([x, y] = 1) \wedge ([y, z] = 1) \rightarrow ([x, z] = 1))$$

$$(MAL :) \forall x, y, z((x \neq 1) \wedge (y \neq 1) \wedge ([x, y] = 1) \wedge ([x, z^{-1}yz] = 1)$$
$$\rightarrow ([y, z] = 1)).$$

Finally being CT but not CSA is captured (in the presence of the group axioms) by

$$(CT :) \forall x, y, z(((y \neq 1) \wedge ([x, y] = 1) \wedge ([y, z] = 1) \rightarrow ([x, z] = 1))$$

$$(NOTMAL :) \exists x, y, z((x \neq 1)$$
$$\wedge (y \neq 1) \wedge ([x, y] = 1) \wedge ([x, z^{-1}yz] = 1) \wedge ([y, z] \neq 1)).$$

If $\mathcal{G}$ represents the class of CT groups, $\mathcal{H}$ the class of CSA groups and $\mathcal{M} = \mathcal{G} \cap (\mathcal{H})^{\lrcorner}$ the class of CT but not CSA groups, then all three classes are axiomatic.

**Theorem 4.6.** *The class $\mathcal{G}$ of CT groups, the class $\mathcal{H}$ of CSA groups and the class $\mathcal{M} = \mathcal{G} \cap (\mathcal{H})^{\lrcorner}$ of CT non CSA groups are all axiomatic.*

## References

[BB]    B.Baumslag, Residually free groups, **Proc. London Math. Soc.** (3), 17, 1967, 635–645.

[GB]    G.Baumslag, On generalised free products, **Math. Z.**, 78, 1962, 423–438.

[BeS]   J.L. Bell and A.B. Slomson, Models and Ultraproducts: An Introduction, Second Revised Printing, North-Holland, Amsterdam, 1971.

[Bo]    O. Bogopolski, A surface analogue of a theorem of Magnus, **Cont. Math**, vol. 352, 2005, 55–89.

[BoS]   O. Bogopolski and K. Sviridov, A Magnus theorem for some one-relator groups, in **The Zieschang Gedenkschrift**, 2008, 63–73.

[CK]    C.C. Chang and H.J. Keisler, Model Theory, Second Edition, North-Holland, Amsterdam, 1977.

[CFR]   L. Ciobanu, B. Fine and G. Rosenberger, Classes of Groups Generalizing a Theorem of Benjamin Baumslag, to appear.

[F]     B. Fine, Power Conjugacy and SQ-universality in Fuchsian and Kleinian Groups, in: **Modular Functions in Analysis and Number Theory**, University of Pittsburgh Press, 1983, 41–55.

[FMgrRR]  B. Fine, A. Myasnikov, V. gr. Rebel and G. Rosenberger, A Classification of Conjugately Separated Abelian, Commutative Transitive and Restricted Gromov One-Relator Groups, **Result. Math.**, 50, 2007, 183–193.

[FGMRS]  B. Fine, A. Gaglione, A. Myasnikov, G. Rosenberger and D. Spellman, **The Elementary Theory of Groups**, DeGruyter, Berlin 2015.

[FGRS 1]  B. Fine, A. Gaglione, G. Rosenberger and D. Spellman, Something for Nothing: Some Consequences of the Solution to the Tarski Problems, **Groups St. Andrews 2013**, London Mathematical Society Lecture Notes 422, 2015, 242–270.

[FGRS 2]  B. Fine, A. Gaglione, G. Rosenberger and D. Spellman, On Elementary Free groups, **Contemporary Mathematics**, 633, 2015, 41–58.

[FR]  B. Fine, G. Rosenberger, Reflections on Commutative Transitivity, in **Aspects of Infinite Groups**, World Scientific Press, 2009, 112–130.

[GS]  A. Gaglione and D. Spellman, Even More Model Theory of Free Groups, in **Infinite Groups and Group Rings** edited by J. Corson, M. Dixon, M. Evans, F. Rohl, World Scientific Press, 1993, 37–40.

[GLS]  A. Gaglione, S. Lipschutz and D. Spellman, Almost Locally Free Groups and a Theorem of Magnus, **J. Groups, Complexity and Cryptology**, 1, 2009, 181–198..

[GKM]  D. Gildenhuys, O. Kharlampovich and A. Myasnikov, CSA Groups and Separated Free Constructions, **Bull. Austral. Math. Soc.**, 52, 1995, 63–84.

[Hal]  M. Hall, **Group Theory**, Macmillan 1965.

[Ha]  N. Harrison, Real Length Functions in Groups, **Trans. Amer. Math. Soc.**, 174, 1972, 77–106.

[Ho]  J. Howie, Some Results on One-Relator Surface Groups, **Boletin de la Sociedad Matematica Mexicana**, 10, 2004, 255–262.

[Hod]  W. Hodges, **Building Models by Games** Cambridge University Press, 1985.

[Ja]  N. Jacobson, **Basic Algebra**, Dover Mathematics.

[KhM 1]  O. Kharlamapovich and A. Myasnikov, *Irreducible affine varieties over a free group: I. Irreducibility of quadratic equations and Nullstellensatz*, **J. of Algebra**, 200, 1998, 472–516.

[KhM 2]  O. Kharlamapovich and A. Myasnikov, *Affine varieties over a free group: II. Systems in triangular quasi-quadratic form and a description of residually free groups*, **J. of Algebra**, 200, 1998, 517–569.

[KhM 3]  O. Kharlamapovich and A. Myasnikov, *The Implicit Function Theorem over Free groups*, **J. Alg.**, 290, 2005, 1–203.

[KhM 4]  O. Kharlamapovich and A. Myasnikov *Effective JSJ Decompositions* **Cont. Math.**, 378, 2005, 87–211.

[KhM 5]  O. Kharlamapovich and A. Myasnikov, *Elementary Theory of Free Nonabelian Groups* **J. Alg.**, 302, 2006, 451–552.

[La]  S. Lang, **Algebra**, Addison-Wessley, 1984.

[LR]    F. Levin and G. Rosenberger, On Power Commutative and Commutation Transitive Groups, **Proc. Groups St Andrews 1985**, Cambridge University Press, 1986, 249–253.

[LS]    R.C. Lyndon and P.E. Schupp, **Combinatorial Group Theory**, Springer-Verlag 1977.

[MKS]   W. Magnus, A. Karrass and D. Solitar, **Combinatorial Group Theory**, Wiley-Interscience 1966.

[MR]    A. Myasnikov and V. Remeslennikov, Length functions on free exponential groups, **Proc. of Intern. Conference in Analysis and Geometry**, Omsk, 1995, 59–61.

[N]     H. Neumann, **Varieties of Groups**, Springer-Verlag, 1968.

[O]     A. Olshanskii, On Relatively Hyperbolic and G-subgroups of Hyperbolic Groups **Int. J. Alg. and Comput.** 3, 1993, 365–409.

[Re]    V.N. Remeslennikov, ∃-free groups **Siberian Mat. J.**, 30, 1989, 998–1001.

[Sc]    W.R. Scott, **Group Theory**, Dover Reprints.

[Se 1]  Z. Sela, Diophantine Geometry over Groups I: Makanin-Razborov Diagrams, **Publ. Math. de IHES** 93, 2001, 31–105.

[Se 2]  Z. Sela, Diophantine Geometry over Groups II: Completions, Closures and Fromal Solutions, **Israel Jour. of Math.**, 104, 2003, 173–254.

[Se 3]  Z. Sela, Diophantine Geometry over Groups III: Rigid and Solid Solutions, **Israel Jour. of Math.**, 147, 2005, 1–73.

[Se 4]  Z. Sela, Diophantine Geometry over Groups IV: An Itertaive Procedure for Validation of a Sentence, **Israel Jour. of Math.**, 143, 2004, 1–71.

[Se 5]  Z. Sela, Diophantine Geometry over Groups V: Quantifier Elimination, **Israel Jour. of Math.**, 150, 2005, 1–9.

[Su]    M. Suzuki, The Nonexistence of a Certain Type of Simple Groups of Odd Order, **Proc. Amer. Math. Soc.**, 8, 1957, 686–695.

[W]     L. Weisner, Groups in which the normalizer of every element but the identity is abelian, **Bull. Amer. Math. Soc.**, 31, 1925, 413–416.

[Wu]    Y.F. Wu, Groups in which Commutativity is a Transitive Relation, **J. of Algebra**, 207, 1998, 165–181.

[Wan]   Z.K. Wan, A proof of the automorphisms of linear groups over a field of characteristic 2, **Sci. Sin.**, 11, 1962, 1183–1194.

# Chapter 7

# The Universal Theory of Free Burnside Groups of Large Prime Exponent

Anthony M. Gaglione[1], Seymour Lipschutz[2], and Dennis Spellman[3]

[1] *Department of Mathematics, U.S. Naval Academy*
*Annapolis, MD 21402*
*agaglione@aol.com*
*http://www.usna.edu*

[2] *Department of Mathematics, Temple University*
*Philadelphia, PA 19122*
*seymour@temple.edu*

[3] *5147 Whitaker Avenue, Philadelphia, PA 19124*
*dennisspellman1@aol.com*

*Dedicated to Gerhard Rosenberger and Dennis Spellman on their 70th birthdays*

**ABSTRACT.** We characterize the universal theory of the free Burnside groups of large prime exponent and briefly show how to extend and modify our results to arbitrary sufficiently large odd exponent.

## 1. Introduction

In this paper, we continue our investigations begun in [GLS]. Definitions and notation will be that of [GLS] which we encourage the reader to consult. None the less, in an effort to make this paper relatively self-contained, we repeat most of those here. The paper will be divided into sections as follows:

(1) Introduction
(2) Preliminaries from logic
(3) Preliminaries from algebraic geometry over groups

2000 *Mathematics Subject Classification.* Primary 20E26; Secondary 20E05, 20E06
*Keywords and Phrases.* Free Burnside group, universal theory, quasi-identity, quasivariety, commutative transitive, Adian-Sirvanjan number

(4) The case of prime exponent

    (a) 4.1 The universal theory with respect to the base language $L_0$

    (b) 4.2 The universal theory with respect to the extended language $L_0[B]$

(5) The case of general odd exponent

(6) Independence of the axioms.

## 2. Preliminaries from Logic

Let $L_0 = L_0[\{1\}]$ be the first-order language with equality containing a binary operation symbol $\bullet$, a unary operation symbol $^{-1}$ and a constant symbol 1. If $G$ is a group, then we may extend $L_0$ to $L_0[G]$ by appending the elements of $G \setminus \{1\}$ to be new constant symbols.

**Remark 2.1.** To be pedantic one should introduce names $\widehat{g}$ for the elements $g \in G$. We trust no confusion will arise from this abuse of notation.

A **universal sentence** of $L_0[G]$ is one of the form $\forall \, \overline{x} \, \varphi(\overline{x})$ where $\overline{x}$ is a tuple of distinct variables and $\varphi(\overline{x})$ is a formula of $L_0[G]$ containing no quantifiers and containing occurrences of at most the variables in $\overline{x}$. Vacuous quantifications are permitted; so, every quantifier-free sentence of $L_0[G]$ is considered a special case of a universal sentence of $L_0[G]$. A **quasi-identity** of $L_0[G]$ is a universal sentence of $L_0[G]$ of the form

$$\forall \, \overline{x} \, \left( \bigwedge_i (S_i(\overline{x}) = s_i(\overline{x})) \to (T(\overline{x}) = t(\overline{x})) \right)$$

where $S_i(\overline{x})$, $s_i(\overline{x})$, $T(\overline{x})$ and $t(\overline{x})$ are terms of $L_0[G]$. An **identity** or **law** of $L_0[G]$ is a universal sentence of the form

$$\forall \, \overline{x} \, (T(\overline{x}) = t(\overline{x}))$$

where $T(\overline{x})$ and $t(\overline{x})$ are terms of $L_0[G]$. Since the above law is equivalent to the quasi-identity

$$\forall \, \overline{x} \, ((1 = 1) \to (T(\overline{x}) = t(\overline{x}))),$$

every law may be considered a special case of a quasi-identity.

A **group** shall be an $L_0$-structure satisfying the laws

- $\forall x_1 x_2 x_3 \, ((x_1 \bullet x_2) \bullet x_3 = x_1 \bullet (x_2 \bullet x_3))$
- $\forall x \, (x \bullet 1 = x)$
- $\forall x \, (x \bullet x^{-1} = 1)$

Henceforth we shall suppress the binary operation symbol $\bullet$ in favor of juxtaposition. Furthermore, modulo the group axioms, we may express up to logical equivalence any quasi-identity $\forall\,\overline{x}\ \left(\bigwedge_i\ (S_i(\overline{x}) = s_i(\overline{x})) \rightarrow (T(\overline{x}) = t(\overline{x}))\right)$ and any identity $\forall\,\overline{x}\,(T(\overline{x}) = t(\overline{x}))$ as

$$\forall\,\overline{x}\ \left(\bigwedge_i (u_i(\overline{x}) = 1) \rightarrow (w(\overline{x}) = 1)\right)$$

and

$$\forall\,\overline{x}\ (w(\overline{x}) = 1)$$

respectively where the $u_i(\overline{x})$ and $w(\overline{x})$ are words on the elements of $G$ and the variables in $\overline{x}$ and their formal inverses. If $G$ is any group then the **diagram of** $G$ shall be the set of all atomic and negated atomic sentences of $L_0[G]$ true in $G$. Modulo the group axioms, these may be taken as $u(g_1,\ldots,g_n) = 1$ and $w(g_1,\ldots,g_n) \neq 1$ where $u(x_1,\ldots,x_n)$ and $w(x_1,\ldots,x_n)$ are words on $x_1,\ldots,x_n$ ($n$ arbitrary) and their formal inverses and $(g_1,\ldots,g_n)$ is a tuple of elements from $G$, and the above equations and inequations are true in $G$. It is easy to see that a model of the group axioms and the diagram of $G$ is just a group $H$ containing a distinguished copy of $G$ as a subgroup. In the parlance of [BMR], such $H$ are $G$-groups.

The notions of $G$-**subgroup** and $G$-**homomorphism** are now readily apparent. In particular, a $G$-homomorphism from a $G$-group $H_1$ to a $G$-group $H_2$ is a group homomorphism from $H_1$ to $H_2$ which is the identity on $G$ (see [BMR]). Two $G$-groups $H_1$ and $H_2$ are $G$-**universally equivalent** provided they satisfy exactly the same universal sentences of $L_0[G]$. In the case $G = \{1\}$ we may abbreviate $G$-universally equivalent to **universally equivalent**. An **existential sentence** of $L_0[G]$ is one of the form $\exists\,\overline{x}\ \varphi(\overline{x})$ where $\overline{x}$ is a tuple of distinct variables and $\varphi(\overline{x})$ is a formula of $L_0[G]$ containing no quantifiers but containing occurrences of at most the variables in $\overline{x}$. Quantifier-free sentences of $L_0[G]$ are considered special cases of existential sentences of $L_0[G]$. An existential sentence of $L_0[G]$ of the form

$$\exists\,\overline{x}\ \left(\bigwedge_j (S_j(\overline{x}) = s_j(\overline{x})) \wedge \bigwedge_k (T_k(\overline{x}) \neq t_k(\overline{x}))\right)$$

is a **primitive sentence** of $L_0[G]$. Modulo the group axioms the above may be rewritten (up to logical equivalence) in the form

$$\exists \, \overline{x} \left( \bigwedge_j (u_j(\overline{x}) = 1) \wedge \bigwedge_k (w_k(\overline{x}) \neq 1) \right)$$

where the $u_j(\overline{x})$ and $w_k(\overline{x})$ are words on the elements of $G$ and the variables in $\overline{x}$ and their formal inverses.

The above sentence asserts that a finite system of equations and inequations

$$\begin{cases} u_j(x_1, \ldots, x_n) = 1, & 1 \leq j \leq J \\ w_k(x_1, \ldots, x_n) \neq 1 & 1 \leq k \leq K \end{cases}$$

on finitely many unknowns $x_1, \ldots, x_n$ has a solution. It is well-known and easy to prove that if $H_0$ is a $G$-subgroup of a $G$-group $H$, then $H_0$ and $H$ are $G$-universally equivalent if and only if every finite system (in finitely many unknowns) of equations and inequations having a solution in $H$ already has a solution in $H_0$. Equivalently, such $H_0$ and $H$ are $G$-universally equivalent if and only if every primitive sentence of $L_0[G]$ true in $H$ must also be true in $H_0$. Thus, if $H_0$ is a $G$-subgroup of the $G$-group $H$, then a sufficient condition for $H_0$ and $H$ to be $G$-universally equivalent is for $H_0$ to $G$-discriminate $H$ according to Definition 2.3 below.

**Definition 2.2.** Let $H_0$ and $H$ be $G$-groups. $H_0$ **$G$-separates** $H$ provided to every nontrivial element $h \in H \backslash \{1\}$ there is a $G$-homomorphism $H \to H_0$ which does not annihilate $h$. In the case $G = \{1\}$ we write $H_0$ **separates** $H$ for $H_0$ $G$-separates $H$.

**Definition 2.3.** Let $H_0$ and $H$ be $G$-groups. $H_0$ **$G$-discriminates** $H$ provided to every finite nonempty set $S \subseteq H \backslash \{1\}$ there is a $G$-homomorphism $H \to H_0$ which does not annihilate any element of $S$. In the case $G = \{1\}$ we write $H_0$ **discriminates** $H$ for $H_0$ $G$-discriminates $H$.

A **quasivariety** of groups is the model class of the group axioms together with a set of quasi-identities of $L_0$. A **variety** of groups is the model class of the group axioms together with a set of laws of $L_0$. The **trivial variety E** is the model class of the group axioms and the law $\forall x \, (x = 1)$. All other varieties of groups are **nontrivial**. As is well-known, every nontrivial variety **V** of groups admits free objects of every rank $r \geq 1$. (Convention: $\{1\}$ is free of rank 0 in **V**.) We use the notation $F_r(\mathbf{V})$ for a group free of rank $r$ in **V**. In this paper, we shall encounter the varieties:

- **O** of all groups.
- $\mathbf{B}_n$ of all groups satisfying the law $\forall x \ (x^n = 1)$. Here $n > 1$ is an integer and $\mathbf{B}_n$ is called the **Burnside variety of exponent** $n$.
- $\mathbf{K}_n$ of all locally finite groups satisfying the law $\forall x \ (x^n = 1)$. Here $n > 1$ is an integer and $\mathbf{K}_n$ is called the **Kostrikin variety of exponent** $n$.

That $\mathbf{K}_n$ is a variety was first established by Kostrikin [K] for prime $n$ and subsequently by Zelmanov [Z] for every positive integer $n$. Moreover, the fact that $\mathbf{K}_n$ is a variety is equivalent to the positive solution of the Restricted Burnside Problem.

**Definition 2.4.** Let **V** be a nontrivial variety of groups and let $G$ be a group in **V**. $G$ is *V*-**freely separated** provided it is separated by a group $F_\omega(\mathbf{V})$ free of countably infinite rank in **V**.

**Definition 2.5.** Let **V** be a nontrivial variety of groups and let $G$ be a group in **V**. $G$ is *V*-**freely discriminated** provided it is discriminated by a group $F_\omega(\mathbf{V})$ free of countably infinite rank in **V**.

**Remark 2.6.** We view the set $\omega = \{0, 1, 2, \ldots\}$ of nonnegative integers (provided with its natural order) as the first limit ordinal which we identify with the first infinite cardinal.

## 3. Algebraic Geometry Over Groups

Let $G$ be a group, $H$ be a $G$-group and $n$ be a positive integer. A formula of $L_0[G]$ of the form $w(x_1, \ldots, x_n) = 1$, where $w(x_1, \ldots, x_n)$ is a word on the elements of $G$ and the variables $x_1, \ldots, x_n$ and their inverses $x_1^{-1}, \ldots, x_n^{-1}$, is an **equation over** $G$ in $x_1, \ldots, x_n$. If $S$ is a set of equations over $G$ in $x_1, \ldots, x_n$ then a **solution** $(h_1, \ldots, h_n) \in H^n$ of the system $S$ in $H$ satisfies $w(h_1, \ldots, h_n) = 1$ for all equations $w(x_1, \ldots, x_n) = 1$ lying in $S$. The **solution set** of the system $S$ in $H$ is the set $V_H(S)$ of all solutions $\overline{h} \in H^n$ of $S$ in $H$. A subset $A \subseteq H^n$ is an **affine algebraic subset** of $H^n$ provided there is a set $S$ of equations over $G$ such that $A = V_H(S)$. The **Zariski topology** on $H^n$ is determined by taking the affine algebraic subsets of $H^n$ as a closed subbase. $H$ is said to be $G$-equationally **Noetherian** provided, for every integer $n > 0$, the Zariski topology on $H^n$ is a **Noetherian topology** in the sense that every descending chain of closed subsets

$$F_1 \supseteq F_2 \supseteq \cdots \supseteq F_k \supseteq$$

stabilizes $F_N = F_{N+1} = F_{N+2} = \cdots$ after finitely many steps. It is shown in [BMR] that $H$ $G$-equationally Noetherian is equivalent to the condition that for every integer $n > 0$ and every system $S$ of equations in the variables $x_1, \ldots, x_n$, there is a finite subset $S_0 \subseteq S$ such that $V_H(S) = V_H(S_0)$. It will be convenient for us to make the following definition.

**Definition 3.1.** Let $G$ be a group, $H$ be a $G$-group and $n$ be a positive integer. $H$ is $n$-partially $G$-**equationally Noetherian** provided to every system $S$ of equations over $G$ in the variables $x_1, \ldots, x_n$ there is a finite subset $S_0 \subseteq S$ such that $V_H(S) = V_H(S_0)$.

Thus, $H$ is $G$-equationally Noetherian if and only if it is $n$-partially $G$-equationally Noetherian for every positive integer $n$.

**Definition 3.2.** A group is **commutative transitive** or **CT** provided it satisfies the universal sentence

$$\forall x, y, z(((y \neq 1) \wedge (xy = yx) \wedge (yz = zy)) \rightarrow (xz = zx)).$$

The following is due to Harrison [H].

**Proposition 1.** *Let $G$ be a group. The following are equivalent in pairs.*

(1) *$G$ is CT.*
(2) *The centralizer $C_G(g)$ in $G$ of every element $g \in G \backslash \{1\}$ is abelian.*
(3) *$M_1 \cap M_2 = \{1\}$ whenever $M_1$ and $M_2$ are distinct maximal abelian subgroups of $G$.*

**Definition 3.3.** A group $G$ is **conjugately separated abelian** or **CSA** provided every maximal abelian subgroup $M$ in $G$ is **malnormal** in $G$ in the sense that $g^{-1}Mg \cap M \neq \{1\}$ implies $g \in M$.

**Proposition 2.** *Let $G$ be a group. If $G$ is CSA, then it is CT.*

**Proof.** Let $M_1$ and $M_2$ be maximal abelian subgroups of the CSA-group $G$. Suppose $g \neq 1$ lies in $M_1 \cap M_2$. Let $m_1 \in M_1$. Then $g = m_1^{-1}gm_1$ is a nontrivial element of $m_1^{-1}M_2m_1 \cap M_2$. So $m_1 \in M_2$ since $G$ is CSA. But $m_1 \in M_1$ was arbitrary, thus $M_1 \leq M_2$. By maximality $M_1 = M_2$. $\square$

Thus, a group is CSA provided it satisfies the universal sentences

$$\forall x, y, z \, (((y \neq 1) \wedge (xy = yx) \wedge (yz = zy)) \rightarrow (xz = zx))$$

and

$$\forall x, y, z \big( ((x \neq 1) \wedge (y \neq 1) \wedge (xy = yx) \wedge (xz^{-1}yz = z^{-1}yzx)) \\ \rightarrow (yz = zy) \big).$$

No nonabelian CSA-group can contain a nontrivial abelian normal subgroup. Thus an example of a CT-group which is not CSA is the infinite dihedral group $D_\infty = \langle a, b; a^2 = 1, a^{-1}ba = b^{-1} \rangle$ since $\langle b \rangle$ is nontrivial and normal in $D_\infty$. To see that $D_\infty$ is CT, just recall that $D_\infty$ is the free product $C_2 * C_2$ of two copies of the cyclic group $C_2$ of order 2 and that free products of CT groups are CT. Another example of a CT-group which is not CSA is the alternating group $A_5$. That $A_5$ is CT can be seen by obersving $A_5$ is isomorphic to $\mathrm{PSL}(2,4)$ as there is only one simple group of order 60; moreover, Suzuki [Su] proved that a nonabelian finite simple group is CT if and only if it is isomorphic to $\mathrm{PSL}(2, 2^f)$ for some integer $f \geq 2$. Also $A_5$ is not CSA since CSA being a universal property is inherited by subgroups and $\langle (125) \rangle$ is a nontrivial abelian subgroup of $\langle (12)(34), (125) \rangle$ normal in the nonabelian group $\langle (12)(34), (125) \rangle$.

It was shown in [FGRS] that a CT-group $G$ violates CSA if and only if it contains a nonabelian subgroup $H$ admitting a nontrivial abelian subgroup $A \leq H$ normal in $H$.

The following theorem may be found in [MR].

**Theorem 3.4.** *Let $G$ be a nonabelian CSA-group. If $G$ is $G$-equationally Noetherian, then the universal theory of $G$ with respect to $L_0[G]$ is axiomatizable by the set $Q$ of quasi-identities of $L_0[G]$ true in $G$ together with CT when the models are restricted to $G$-groups — i.e., $\mathrm{diag}(G)$ is tacitly included in the axiomatization.*

Here by the universal theory of $G$ with respect to $L_0[G]$ is meant the set of all universal sentences of $L_0[G]$ true in $G$.

## 4. The Case of Prime Exponent

### 4.1. *The universal theory with respect to the base language $L_0$*

As in [GLS], we call an integer $n > 1$ an **Adian-Sirvanjan-Atabekyan number** provided the following four conditions hold.

(B1) $F_r(\mathbf{B}_n)$ is infinite for all $r \geq 2$; moreover, every finite subgroup of $F_r(\mathbf{B}_n)$ is cyclic.

(B2) The centralizer of every nontrivial element in $F_r(\mathbf{B}_n)$ is cyclic for all $r \geq 2$.

(B3) $F_\omega(\mathbf{B}_n)$ embeds in $F_2(\mathbf{B}_n)$ .

(B4) For all integers $s \geq 3$, $F_s(\mathbf{B}_n)$ is discriminated by the family of retractions $F_s(\mathbf{B}_n) \to F_2(\mathbf{B}_n)$.

(We note that in [GLS] we did not assume (B4) and also the name Atabekyan was omitted.)

Here $F_\omega(\mathbf{B}_n)$ is a group free of countably infinite rank in $\mathbf{B}_n$. We note Adian [A] proved that (B1) and (B2) hold for all odd integers $n \geq 665$ while Sirvanjan [Si] proved that (B3) holds for all odd integers $n \geq 665$. A prime $p$ shall be called an **Adian-Sirvanjan-Atabekyan prime** provided it is an Adian-Sirvanjan-Atabekyan number. Throughout the remainder of Section 4, $p$ shall be a fixed but arbitrary Adian-Sirvanjan-Atabekyan prime.

If $G$ is any group, we let $\mathrm{Th}_\forall(G)$ be the set of all universal sentences of $L_0$ true in $G$ and call $\mathrm{Th}_\forall(G)$ **the universal theory of** $G$. Suppose $B$ is free of rank 2 relative to $\mathbf{B}_p$. From (B3) it easily follows that $\mathrm{Th}_\forall(F_r(\mathbf{B}_p)) = \mathrm{Th}_\forall(B)$ for all $r$, $2 \leq r \leq \omega$.

We say that the nonabelian groups free in $\mathbf{B}_p$ are **universally equivalent with respect to** $L_0$. Now let $p - CyC$ be the following universal sentence of $L_0$:

$$\forall x, y \left( ((x \neq 1) \wedge (xy = yx)) \to \left( \bigvee_{k=0}^{p-1} (y = x^k) \right) \right).$$

In view of (B1) and (B2), $B$ satisfies $p - CyC$. Let $Q_p$ be the set of all quasi-identities of $L_0$ true in $B$. In particular (up to logical equivalence) $Q_p$ contains the group axioms and the law $\forall x(x^p = 1)$. The following is essentially a restatement of Lemma 5.1 of [GLS]. We repeat its proof here.

**Lemma 4.1.** *Every model of* $Q_p \cup \{p - CyC\}$ *is CSA.*

**Proof.** Let $G$ be a model of $Q_p \cup \{p - CyC\}$. Let $a \in G \backslash \{1\}$. Then (letting for each integer $n > 0$, $C_n$ denote a group cyclic of order $n$) $C_G(a) \cong C_p$. Suppose $g^{-1}bg \in \langle a \rangle$ for some $1 \neq b \in \langle a \rangle$. Since $\langle a \rangle = \langle b \rangle$ we may assume $b = a$. Then $g^{-1}ag = a^m$ and $a = g^{-p}ag^p = a^{m^p}$. But we may compute exponents modulo $p$ and, by Fermat's Little Theorem, $a^{m^p} = a^m$. Therefore, $a^m = a$ and $g^{-1}ag = a$. So $g \in C_G(a) = \langle a \rangle$. Hence, $G$ is CSA. $\square$

**Theorem 4.2.** $\text{Th}_\forall(B)$ *is axiomatizable by* $Q_p \cup \{p - CyC\}$.

**Proof.** Suppose not. Then there is some model $G$ of $Q_p \cup \{p - CyC\}$ which violates some universal sentence $U$ of $L_0$ true in $B$. The negation $\sim U$ of $U$ is equivalent to an existential sentence $E$ of $L_0$ false in $B$. As is well-known and easy to prove, every existential sentence is equivalent to a disjunction of primitive sentences. Moreover, a disjunction holds in a group if and only if at least one of its disjuncts does. Thus, there is a primitive sentence $\pi$ of $L_0$ false in $B$ which holds in $G$. Let $\pi$ be

$$\exists x_1, \ldots, x_n \left( \bigwedge_{j=1}^{J} (u_j(x_1, \ldots, x_n) = 1) \right) \wedge \left( \bigwedge_{k=1}^{K} (w_k(x_1, \ldots, x_n) \neq 1) \right)$$

Choose $(g_1, \ldots, g_n) \in G^n$ such that $(x_1, \ldots, x_n) = (g_1, \ldots, g_n)$ verifies $\pi$ in $G$. Now we claim the model of $G$ of $Q_p \cup \{p - CyC\}$ cannot be abelian; otherwise, it would be cyclic of exponent dividing $p$ hence embed in $C_p \leq B$ and so be a model of $\text{Th}_\forall(B)$. That contradicts the choice of $G$.

Now we need some notational conventions: $x^y$ shall abbreviate the conjugate $y^{-1}xy$; furthermore, the left-normed commutator $[z_1, \ldots, z_N]$ is defined recursively by $[z_1] = z_1$ and, if $m > 1$ and $[z_1, \ldots, z_{m-1}]$ has already been defined, then $[z_1, \ldots, z_m]$ is

$$[z_1, \ldots, z_{m-1}]^{-1} z_m^{-1} [z_1, \ldots, z_{m-1}] z_m.$$

With these conventions, let $w(x_1, \ldots, x_n, y_1, \ldots, y_{k-1})$ be

$$[w_1(x_1, \ldots, x_n), w_2(x_1, \ldots, x_n)^{y_1}, \ldots, w_k(x_1, \ldots, x_n)^{y_{k-1}}].$$

Since $G$ is a model of $Q_p \cup \{p - CyC\}$ it is CSA by Lemma 4.1. Moreover, $G$ is nonabelian. Thus, in the terminology of [BMR] $G$ is a **domain**. From that it follows that there is $(h_1, \ldots, h_{k-1}) \in G^{k-1}$ such that $w(g_1, \ldots, g_n, h_1, \ldots, h_{k-1}) \neq 1$. Then $G$ satisfies the following existential sentence $\psi$ of $L_0$:

$$\exists x_1, \ldots x_n, y_1, \ldots y_{k-1}$$

$$\left( \bigwedge_{j=1}^{J} (u_j(x_1, \ldots, x_n) = 1) \wedge (w(x_1, \ldots, x_n, y_1, \ldots, y_{k-1}) \neq 1) \right).$$

Could $\psi$ be false in $B$? Then $\sim \psi$ would hold in $B$. But $\sim \psi$ is equivalent to the quasi-identity

$$\forall x_1, \ldots, x_n, y_1, \ldots, y_{k-1}$$

$$\left( \bigwedge_{j=1}^{J} (u_j(x_1, \ldots, x_n) = 1) \rightarrow (w(x_1, \ldots, x_n, y_1, \ldots, y_{k-1}) = 1) \right).$$

Thus, the above quasi-identity lies in $Q_p$ and must hold in the model $G$ of $Q_p \cup \{p - CyC\}$. But that contradicts the fact that $\psi$ holds in $G$. The contradiction shows that there can be no model $G$ of $Q_p \cup \{p - CyC\}$ violating any universal sentence of $L_0$ true in $B$. In view of the consequence of the Gödel-Henkin Completeness Theorem asserting that a sentence $\sigma$ is a formal consequence of a set $\Sigma$ of sentences if and only if every model of $\Sigma$ is also a model of $\sigma$, $Q_p \cup \{p - CyC\}$ axiomatizes $\mathrm{Th}_\forall(B)$.    □

**Theorem 4.3.** *Let $G$ be $\mathbf{B}_p$-freely separated. Then $G$ is $\mathbf{B}_p$-freely discriminated if and only if $G$ satisfies $p - CyC$.*

**Proof.** Suppose first that $G$ is $\mathbf{B}_p$-freely discriminated. We first claim that $G$ is CT. Suppose not. Then there are $a, b, c \in G$ with $b \neq 1$ such that $[a, b] = [b, c] = 1$ but $[a, c] \neq 1$, Thus there is a homomorphism $\varphi : G \rightarrow B = F_\omega(\mathbf{B}_p)$ such that simultaneously $\varphi(b) \neq 1$ and $[\varphi(a), \varphi(b)] = \varphi([a, c]) \neq 1$. But that contradicts the fact that $B$ is CT since $[\varphi(a), \varphi(b)] = \varphi([a, b]) = 1$ and $[\varphi(b), \varphi(c)] = \varphi([b, c]) = 1$. The contradiction shows that $G$ is CT. Thus, given $g \in G \backslash \{1\}$, $C_G(g)$ is abelian. Since $G$ satisfies the law $\forall x (x^p = 1)$, $C_G(g)$ is a nontrivial vector space over the $p$ element field. Thus, if $C_G(g)$ is not cyclic, it must contain a copy of $C_p \times C_p$. In particular, it must contain $p$ distinct nontrivial pairwise commuting elements. List those as $g_1, \ldots, g_p$. Now consider the $p + \binom{p}{2} = \binom{p+1}{2}$ (not necessarily distinct) nontrivial elements $g_1, \ldots, g_p, g_1 g_2^{-1}, \ldots, g_i g_j^{-1}, \ldots, g_{p-1} g_p^{-1}$ of $G$. Since $G$ is $\mathbf{B}_p$-freely discriminated, there is a homomorphism $\varphi : G \rightarrow B = F_\omega(\mathbf{B}_p)$ such that simultaneously $\varphi(g_i) \neq 1, i = 1, \ldots, p$, and $\varphi(g_i)\varphi(g_j)^{-1} \neq 1, 1 \leq i < j \leq p$. Then $\varphi(g_1), \ldots, \varphi(g_p)$ would be $p$ distinct pairwise commuting nontrivial elements of $B$. But that fact contradicts the fact that $B$ satisfies $p - CyC$. The contradiction shows that $G$ satisfies $p - CyC$.

Now suppose $G$ is $\mathbf{B}_p$-freely separated and satisfies $p - CyC$. If $G$ is abelian then it must either be trivial or cyclic of order $p$. In either event $G$ embeds in $C_p = F_1(\mathbf{B}_p)$ which in turn embeds in $B = F_\omega(\mathbf{B}_p)$. An embedding of $G$ into $B$ will not annihilate any element of any finite subset

$S \subseteq G \backslash \{1\}$. So suppose $G$ is nonabelian. Now we claim $G$ is a model of $Q_p \cup \{p - CyC\}$. To see that suppose

$$\forall x_1, \ldots x_n \left( \bigwedge_{j=1}^{J} (u_i(x_1, \ldots, x_n) = 1) \rightarrow (w(x_1, \ldots, x_n) = 1) \right)$$

lies in $Q_p$. If $G$ violated the above quasi-identity, then $G$ would satisfy the existential sentence

$$\exists x_1, \ldots x_n \left( \bigwedge_{j=1}^{J} (u_i(x_1, \ldots, x_n) = 1) \wedge (w(x_1, \ldots, x_n) \neq 1) \right).$$

Let $(g_1, \ldots, g_n) \in G^n$ be such that $(x_1, \ldots, x_n) = (g_1, \ldots, g_n)$ verifies the above existential sentence in $G$. Since $G$ is $\mathbf{B}_p$-freely separated, there is a homomorphism $\varphi : G \rightarrow B = F_\omega(\mathbf{B}_p)$ such that $w(\varphi(g_1), \ldots, \varphi(g_n)) = \varphi(w(g_1, \ldots, g_n)) \neq 1$. But that is impossible since $u_j(\varphi(g_1), \ldots, \varphi(g_n)) = \varphi(u_j(g_1, \ldots, g_n)) = 1$ for all $j = 1, \ldots, J$ and $B$ satisfies the quasi-identity

$$\forall x_1, \ldots x_n \left( \bigwedge_{j=1}^{J} (u_i(x_1, \ldots, x_n) = 1) \rightarrow (w(x_1, \ldots, x_n) = 1) \right).$$

The contradiction shows that, as claimed, $G$ is a model of $Q_p \cup \{p - CyC\}$. Thus, by Lemma 4.1, $G$ is CSA; moreover, it is nonabelian (hence a domain). Let finitely many nontrivial elements $g_1, \ldots, g_n$ of $G$ be given. Let $w(x_1, \ldots, x_n, y_1, \ldots, y_{n-1})$ be

$$[x_1, x_2^{y_1}, \ldots, y_n^{y_{n-1}}].$$

Then, since $G$ is a domain, there is $(h_1, \ldots, h_{n-1}) \in G^{n-1}$ such that

$$[g_1, g_2^{h_1}, \ldots, g_n^{h_{n-1}}] \neq 1.$$

Since $G$ is $\mathbf{B}_p$-freely separated there is a homomorphism $\varphi : G \rightarrow B = F_\omega(\mathbf{B}_p)$ such that

$$\varphi(w(g_1, \ldots, g_n, h_1, \ldots, h_{n-1}) = w(\varphi(g_1), \ldots, \varphi(g_n), \varphi(h_1), \ldots, \varphi(h_{n-1})$$
$$= [\varphi(g_1), \varphi(g_2)^{\varphi(h_1)}, \ldots, \varphi(g_n)^{\varphi(h_{n-1})}] \neq 1.$$

Therefore $\varphi(g_i) \neq 1$ for all $i - 1, \ldots, n$. Hence $G$ is $\mathbf{B}_p$-freely discriminated.

$\square$

**Remark 4.4.** An alternative argument for $p - CyC$ being a sufficient condition for a $\mathbf{B}_p$-freely separated group to be $\mathbf{B}_p$-freely discriminated is the proof of Theorem 5.1 of [GLS].

The proof of Theorem 4.3 yields the following immediate corollary.

**Corollary 1.** *Let* $N = \binom{p+1}{2}$. *A group* $G \in \mathbf{B}_p$ *is* $\mathbf{B}_p$-*freely discriminated if and only if for every* $(g_1, \ldots, g_N) \in (G\backslash\{1\})^N$, *there is a homomorphism* $\varphi : G \to F_\omega(\mathbf{B}_p)$ *which does not annihilate* $g_i$ *for all* $i = 1, \ldots, N$.

### 4.2.  The universal theory with respect to the extended language $L_0[G]$

Let $n > 1$ be an integer and let $*_n$ be the coproduct in $\mathbf{B}_n$. That is $*_n$ is the verbal product corresponding to $\mathbf{B}_n$ as explicated in Section 8 of Chapter 1 of [N]. Let $s \geq 3$ be an integer. V. Atabekyan [At1,2] and subsequently A. Yu. Olshanskii [O] (see Theorem 39.2 in the English transaltion of this reference.) proved that, for all sufficiently large odd integers $n$,

$$F_s(\mathbf{B}_n) = F_{s-2}(\mathbf{B}_n) *_n F_2(\mathbf{B}_n)$$

is discriminated by the family of retractions

$$F_s(\mathbf{B}_n) \to F_2(\mathbf{B}_n).$$

Suppse $p$ is an Adian-Sirvanjan-Atabekyan prime and $s \geq 3$ is an integer. Fix an integer $r$, $2 \leq r < s$. We claim that $F_s(\mathbf{B}_p) = F_{s-r}(\mathbf{B}_p) *_p F_r(\mathbf{B}_p)$ is discriminated by the family of retractions $F_s(\mathbf{B}_p) \to F_r(\mathbf{B}_p)$. To see that suppose $\{a_1, \ldots, a_s\}$ freely generates $F_s(\mathbf{B}_p)$ relative to $\mathbf{B}_p$. Given finitely many nontrivial elmements $f_1, \ldots, f_k$ of $F_s(\mathbf{B}_p)$, there is a retraction $\rho$ from $F_s(\mathbf{B}_p)$ onto $\langle a_1, a_2 \rangle$ which does not annihilate any $f_j$, $j = 1, \ldots, k$. Now define $\varphi_r$ from $F_s(\mathbf{B}_p)$ onto $\langle a_1, \ldots, a_r \rangle$ by

$$\varphi_r(a_i) = \begin{cases} a_i & \text{if } 1 \leq i \leq r \\ \rho(a_i) & \text{if } r+1 \leq i \leq s \end{cases}$$

Now, since $\rho$ is a retraction, $\rho^2 = \rho$ and $\rho\varphi_r(a_i) = \rho(a_i)$ for all $i = 1, \ldots, s$. It follows that $\rho\varphi_r = \rho$; hence, $\rho\varphi_r(f_j) = \rho(f_j) \neq 1$ for all $j = 1, \ldots, k$. Therefore $\varphi_r(f_j) \neq 1$ for all $j = 1, \ldots, k$ and the claim is established.

**Remark 4.5.** The claim is Theorem 39.2 of Olshanskii [O].

We immediately deduce

**Theorem 4.6.** *Let $p$ be an Adian-Sirvanjan-Atabekyan prime and $r \geq 2$ be an integer. Let $B = F_r(\mathbf{B}_p)$. Then, for all integers $s \geq r$, $F_s(\mathbf{B}_p)$ is universally equivalent to $B$ with respect to the extended language $L_0[B]$.*

Evidently we have

**Corollary 2.** *Let $p$ be an Adian-Sirvanjan-Atabekyan prime and $r \geq 2$ be an integer. Let $B = F_r(\mathbf{B}_p)$. Then for all integers $r \leq s \leq t$, $F_s(\mathbf{B}_p)$ and $F_t(\mathbf{B}_p)$ are $B$-universally equivalent.*

**Remark 4.7.** The proof of Theorem 4.6 (and hence also Corollary 2) goes through for any odd composite Adian-Sirvananjan-Atabekyan integer.

As of this writing it is not known, with $B$ as in Theorem 4.6 and Corollary 2 whether or not $B$ is $B$-equationally Noetherian. We next introduce a condition which, if satisfied, would guarantee that $B$ is at least 1-partially $B$-equationally Noetherian. Moreover, the satisfaction of this condition would also lead to an independent proof of the theorem of Atabekyan and Olshanskii when the condition (B4) is not postulated.

The condition was posed as a question to the Kourovka Notebook and will be referred to as such hence.

**The Kourovka Question:** *Let $k \geq 2$ be an integer. Let $B$ be free in the Burnside variety on $\{a_1, \ldots, a_k\}$ satisfying the law $x^p = 1$ where $p$ is an Adian-Sirvanjan-Atabekyan prime. Must every nonidentical one variable equation $w(a_1, \ldots, a_k, x) = 1$ over $B$ have at most finitely many solutions in $B$?*

R. C. Lyndon [L] proved that the solution set $V_F(w)$ of any one variable equation over an absolutely free group $F = \langle f_1, \ldots, f_r; \rangle$ is exhausted by the values in $F$ of finitely many parametric words in finitely many integral parameters. K. Appel [Ap] subsequently showed that these can be taken to be of the form $ab^\nu c$ where $a, b, c \in F$ and $\nu$ is an integral parameter. A reviewer found an error in Appel's argument; however, a rigorous proof has been given by Chiswell and Remeslennikov [CR] and also by Bormotov, Gilman, and Myasnikov [BGM].

Let $k \geq 2$ be an integer. Let $B = F_k(\mathbf{B}_p)$ be free in $\mathbf{B}_p$ on the generating set $\{a_1, \ldots, a_k\}$. If an analogue of Lyndon's result would hold for all such $B$, then the Kourovka Question would have a positive answer. This is so since integral exponents are taken modulo $p$. Thus the parametric words could assume only finitely many distinct values in $B$ as the parameters varied over the integers.

A huge reason for the success of algebraic geometry over groups as an ingredient in the attack on the Tarski questions is the fact that absolutely free groups are equationally Noetherian.

We therefore, by way of analogy, pose the following.

**The EN Question:** *Let $p$ be an Adian-Sirvanjan-Atabekyan prime. Let $B$ be free in* $\mathbf{B}_p$. *Is $B$ a $B$-equationally Noetherian group?*

The next theorem asserts that a positive answer to the Kourovka Question would guarantee at least that $B$ is 1-partially $B$-equationally Noetherian.

**Theorem 4.8.** *Let $r \geq 2$ be an integer. Let $B = F_r(\mathbf{B}_p)$ be freely generated relative to $\mathbf{B}_p$ by $\{a_1, \ldots, a_r\}$. If the Kourovka Question has a positive answer, then $B$ is 1-partially $B$-equationally Noetherian.*

**Proof.** Let $S$ be the system $w_n(a_1, \ldots, a_r, x) = 1$, $0 \leq n < \omega$ of one variable equation over $B$. For each $n$, $0 \leq n < \omega$, let $S_n$ be the finite subsystem $w_m(a_1, \ldots, a_r, x) = 1$, $0 \leq m < n$.

Let $V_n = V_B(S_n) = V_B(w_0) \cap \cdots \cap V_B(w_{n-1})$ be the solution set of $S_n$ over $B$ and let $V = V_S(B)$ be the solution set of $S$ over $B$. If, for any fixed $n$, $0 \leq n < \omega$, $w_n(a_1, \ldots, a_r, x) = 1$ is an identical equation over $B$ then $V_B(w_n) = B$ may safely be removed from the intersection and we assume without loss of generality that none of the equations is identical over $B$. Therefore, should the Kourovka Question have a positive answer, each member of the nonascending chain $V_0 \supseteq V_1 \supseteq \cdots \supseteq V_n \supseteq \cdots$ is finite. Hence, there is $N$, $0 \leq N < \omega$ such that

$$V_N = V_{N+1} = \cdots = \bigcap_{0 \leq n < \omega} V_n = V$$

and so $B$ is 1-partially $B$-equationally Noetherian.  □

Here is a proof that a positive answer to the Kourovka Question would lead to another proof of the result of Atabekyan and Olshanskii. Here we do not postulate (B4).

Let $n \geq 2$ be an integer and $B = F_n(\mathbf{B}_p)$ be free on the generating set $\{a_1, \ldots, a_n\}$ and let

$$B^+ = B *_p \langle a_{n+1}; \ a_{n+1}^p = 1\rangle$$
$$= F_{n+1}(\mathbf{B}_p)$$

be free in $\mathbf{B}_p$ on the generating set $\{a_1, \ldots, a_n, a_{n+1}\}$.

Let $b_j = w_j(a_1, \ldots, a_n, a_{n+1})$, $j = 1, \ldots, J$, be finitely many nontrivial elements of $B^+$. Consider the system $S$ of one variable equations over $B^+$:

$$w_j(a_1, \ldots, a_n, x) = 1, \quad j = 1, \ldots, J.$$

If, for any $j_0$, $1 \leq j_0 \leq J$, $w_{j_0}(a_1, \ldots, a_n, x) = 1$ for all $b \in B$, then

$$w_{j_0}(a_1, \ldots, a_n, x) = 1$$

would have infinitely many solutions in $B^+$. Therefore, if the Kourovka Question has a positive answer, then $w_{j_0}(a_1, \ldots, a_n, x) = 1$ is an identical equation over $B^+$ and, in particular,

$$b_{j_0} = w_{j_0}(a_1, \ldots, a_n, a_{n+1}) = 1.$$

This contradicts the choice of the $b_j$.

It follows that, modulo a positive answer to the Kourovka Question, each equation $w_j(a_1, \ldots, a_n, x) = 1$, $j = 1, \ldots, J$, viewed as an equation over $B$, can have at most finitely many solutions. Hence, there is $b \in B$ which is not a solution to any of the $w_j(a_1, \ldots, a_n, x) = 1$, $1 \leq j \leq J$. Then the mapping $B^+ \to B$ determined by $a_i \mapsto a_i, i = 1, \ldots, n$, $a_{n+1} \mapsto b$ does not annihilate any $b_j$, $1 \leq j \leq J$. Hence, $F_{n+1}(\mathbf{B}_p)$ is discriminated by the family of retractions $F_{n+1}(\mathbf{B}_p) \to F_n(\mathbf{B}_p)$.

It is then easy to see, by taking the composition of appropriate retractions at each step that (for $r > 2$ an integer)

$$F_r(\mathbf{B}_p) \to F_{r-1}(\mathbf{B}_p) \to \cdots \to F_2(\mathbf{B}_p),$$

the family of retractions $F_r(\mathbf{B}_p) \to F_2(\mathbf{B}_p)$ discriminates $F_r(\mathbf{B}_p)$.

We next give some small evidence that the Kourovka Question may have a positive answer.

**Definition 4.9.** Let $k, n$ and $d$ be positive integers. Suppose $k \geq 2$. Suppose also that $n$ is odd and Adian-Sirvanjan-Atabekyan. Let $B = F_k(\mathbf{B}_n)$ be free in $\mathbf{B}_n$ on the generating set $\{a_1, \ldots, a_k\}$. Let $u_i$ lie in $B$, $0 \le i \le d$ and let $\varepsilon_i \in \{1, -1\}$, $1 \le i \le d$. We say the equation

$$w(a_1, \ldots, a_k, x) = u_0 x^{\varepsilon_1} u_1 x^{\varepsilon_2} \cdots u_{d-1} x^{\varepsilon_d} u_d = 1$$

has **degree** $d$ over $B$. When $d = 1$ we call the equation **linear** over $B$ and when $d = 2$ we call this equation **quadratic** over $B$. When $d = 3$ we call this equation **cubic** over $B$.

**Theorem 4.10.** *With the conventions and notation of Definition 4.9, if $d \le 2$, then the equation $w(a_1, \ldots, a_k, x) = 1$ has at most finitely many solutions in $B$ provided it is not an identical equation.*

**Proof.** The linear equation $u_0 x^\varepsilon u_1 = 1$ has the unique solution $x = (u_0^{-1} u_1^{-1})^\varepsilon$. Consider the quadratic equation $u_0 x^{\varepsilon_1} u_1 x^{\varepsilon_2} u_2 = 1$. That is equivalent to the transformed equation $x^{\varepsilon_1} u_1 x^{\varepsilon_2} = u_0^{-1} u_2^{-1}$. Suppose first that $\varepsilon_1 = \varepsilon_2$. Then $uxvxw = 1$ is equivalent to the equation $y^2 = u^{-1} w^{-1} v$, where $y = xv$, the square root exists and is unique if all elements have odd orders.

Now suppose that $\varepsilon_1 \ne \varepsilon_2$ so that $\varepsilon_2 = -\varepsilon_1$. The transformed equation may thus be written $x^\varepsilon u_1 x^{-\varepsilon} = u_0^{-1} u_2^{-1}$. If $u_1$ is not conjugate to $u_0^{-1} u_2^{-1}$ in $B$, then the above is inconsistent over $B$ (so has the finite number zero of solutions). Suppose $u_1$ and $u_0^{-1} u_2^{-1}$ are conjugate over $B$. If $u_1 = 1$ and $u_2 = u_0^{-1}$, then $x^\varepsilon u_1 x^{-\varepsilon} = u_0^{-1} u_2^{-1}$ is an identical equation over $B$ so assume also that $u_1$ (and therefore also $u_0^{-1} u_2^{-1}$) is a nontrivial element of $B$. Suppose first $\varepsilon_1 = 1$. All solutions of $xux^{-1} = v$ belong to the same left coset of $C_B(u)$ and the statement follows from (B2). Similarly if $\varepsilon_1 = -1$, all solutions of $xux^{-1} = v$ belong to the same right coset of $C_B(u)$.    $\square$

We do not know if the analogue of Theorem 4.10 holds for cubic equations. None the less, in the case $n$ is a prime power, we can at least say something about the solution set.

**Theorem 4.11.** *Let $p_0 > 3$ be a prime and let $e > 0$ be an integer. Suppose $q = p_0^e$ is a Adian-Sirvanjan-Atabekyan number. With the notation and conventions of Definition 4.9 and $n = q$ assume the cubic equation*

$$u_0 x^{\varepsilon_1} u_1 x^{\varepsilon_2} u_2 x^{\varepsilon_3} u_3 = 1$$

*is consistent over $B$. Fix a solution $\xi_0$ and let $\xi$ be an arbitrary solution. Then $\xi \equiv \xi_0 (\mathrm{mod}\, \kappa_q(B))$ where $\kappa_q$ is the verbal subgroup operator corresponding to the Kostrikin variety $\mathbf{K}_q$.*

**Proof.** Since modulo the verbal subgroup operator, $\kappa_q(B)$, corresponding to the Kostrikin variety, we have a finite group of prime power exponent, then that, of course, implies that mod $\kappa_q(B)$ the group is nilpotent. The result follows from A.L. Shmelkin [Sh] who proved that a cubic equation has at most one solution in any nilpotent group without 3-torsion.  □

## 5.  The Case of General Odd Exponent

**Definition 5.1.** The model class of a set of quasi-identities of $L_0$ together with the group axioms is a **quasivariety** of groups.

**Theorem 5.2.** *Let* **V** *be a variety of groups. Then the class* $\mathcal{X}$ *of all groups* $G$ *such that every finite subgroup of* $G$ *lies in* **V** *is a quasivariety of groups.*

**Proof.** The variety of all groups is determined, modulo the group axioms, by the 0-variable law $1 = 1$. Clearly the class of all groups $G$ all of whose finite subgroups lie in the variety of all groups coincides with the class of all groups. Certainly the class of all groups is a quasivariety of groups. So assume that **V** is a proper subvariety of the variety of all groups. Let $\mathbb{N} = \omega \backslash \{0\}$ be the set of positive integers. For each $n \in \mathbb{N}$ let $[n]$ be the initial segment $\{k \in \mathbb{N} : k \leq n\}$ of $\mathbb{N}$ and let $\mathcal{F}_n$ be the set of all functions

$$f : [n] \times [n] \to [n],$$

$$(i, j) \xrightarrow{f} f(i, j),$$

such that $[n]$ forms a group under $f$. Let $\mathcal{F} = \bigcup_{n \in \mathbb{N}} \mathcal{F}_n$. If $f \in \mathcal{F}$ let its **modulus** be the (necessarily unique) $n \in \mathbb{N}$ such that $f \in \mathcal{F}_n$ and write $|f|$ for the modulus of $f$. Observe that every finite group is isomorphic to $([|f|], f)$ for some $f \in \mathcal{F}$. In particular, every finite group is a homomorphic image of $([|f|], f)$ for some $f \in \mathcal{F}$.

Suppose $G$ is a group, $f \in \mathcal{F}$, and $(g_1, \ldots, g_{|f|}) \in G^{|f|}$ satisfy $g_i g_j = g_{f(i,j)}$ for all $(i, j) \in [|f|] \times [|f|]$, Then clearly the function $[|f|] \to G$, $i \mapsto g_i$ is a homomorphism from $([|f|], f)$ into $G$. Now let $X = \{x_n : n \in \mathbb{N}\}$ be a set of distinct variables, Fix a set $W$ of nonempty words on $X \cup X^{-1}$ such that the laws $w = 1$ as $w$ varies over $W$, determines **V**. For each $w \in W$ let its **modulus** be the least positive integer $n$ such that $w = 1$ is an $n$-variable law. Write $|w|$ for the modulus of $w$. For each $f \in \mathcal{F}$, $w \in W$, and each $(i_1, \ldots, i_{|w|}) \in [|f|]^{|w|}$ let $\sigma(f, w, i_1, \ldots, i_{|w|})$ be the following quasi-identity

$$\forall x_1, \ldots, x_{|f|} \left( \begin{array}{c} \bigwedge\limits_{(i,j)\in[|f|]\times[|f|]} (x_i x_j = x_{f(i,j)}) \rightarrow \\ (w(x_{i_1}, \ldots, x_{i_{|w|}}) = 1) \end{array} \right).$$

We claim the class of all groups $G$ all of whose finite subgroups lie in $\mathbf{V}$ is the model class of the group axioms and the quasi-identities $\sigma(f, w, i_1, \ldots, i_{|w|})$. To see that observe that, if we fix $f \in \mathcal{F}$, then, modulo the group axioms, the quasi-identities $\sigma(f, w, i_1, \ldots, i_{|w|})$ assert that every subgroup of $G$ which is a homomorphic image of $([|f|], f)$ lies in $\mathbf{V}$.    $\square$

Now let $n$ be an odd Adian-Sirvanjan-Atabekyan number. Let $B = F_2(\mathbf{B}_n)$ be free of rank 2 relative to $\mathbf{B}_n$ and let $Q_n$ be the set of all quasi-identities of $L_0$ true in $B$. Up to logical equivalence, $Q_n$ contains the group axioms and the law $\forall x(x^n = 1)$. Moreover, since every finite subgroup of $B$ is abelian $Q_n$ contains the quasi-identities determining the quasivariety $\mathcal{X}$ of Theorem 5.2 where the variety $\mathbf{V}$ is that of abelian groups. Furthermore, $B$ is CT. Moreover, since every abelian subgroup of $B$ is cyclic of order dividing $n$, we must have the $[a, b] = 1$ implies that $\langle a, b \rangle = \langle c \rangle$ where $c = a^{m_1} b^{m_2}$ for some $m_1, m_2$ with $0 \leq m_1, m_2 < n$. Therefore, $B$ satisfies the following universal sentence $\sigma_n$ of $L_0$:

$$\forall x_1, x_2 \left( \begin{array}{c} (x_1 x_2 = x_2 x_1) \rightarrow \\ \bigvee\limits_{0 \leq m_1, m_2, k_1, k_2 < n} ((x_1 = (x_1^{m_1} x_2^{m_2})^{k_1}) \wedge (x_2 = (x_1^{m_1} x_2^{m_2})^{k_2})) \end{array} \right)$$

**Lemma 5.3.** *Let $G$ be a model of $Q_n \cup \{CT, \sigma_n\}$. Then $G$ is CSA.*

**Proof.** Let $g \in G \backslash \{1\}$. By CT, $C_G(g)$ is abelian. By $\sigma_n$, $C_G(g)$ is locally cyclic. But $C_G(g)$ has exponent dividing $n$. No locally cyclic group of finite exponent can fail to be cyclic. Thus, writing $M$ for $C_G(g)$, $M = \langle m \rangle$ is cyclic of order $d(g)$ where $d(g) \mid n$ and and $d(g) > 1$ (as $g \neq 1$ lies in $M$). Suppose $h^{-1} M h \cap M \neq \{1\}$. Since $h^{-1} M h$ is also maximal abelian in $G$, we must have $h^{-1} M h = M$ by CT. So $x \mapsto h^{-1} x h$ restricts to an automorphism $\alpha$ of $M$. $\alpha$ is determined by $m \mapsto m^k$ where $GCD(k, n) = 1$. Consider the subgroup $G_0 = \langle h, m \rangle$ of $G$. Clearly $M$ is normal in $G_0$. Furthermore, $G_0/M$ is clearly cyclic generated by the coset $Mh$. Then (since $m$ and $h$ — and hence also $Mh$ - have finite order $m \in M$, $Mh \in G_0/M$) $|G_0| = |M| [G_0 : M] < \infty$.

Since every finite subgroup of any group $B$ free in $\mathbf{B}_n$ is abelian $Q_n$ contains the quasi-identities

$$\forall x_1, \ldots, x_{|f|} \left( \bigwedge_{(i,j) \in [|f|] \times [|f|]} \begin{matrix} (x_i x_j = x_{f(i,j)}) \\ ([x_{1_1}, x_{i_2}] = 1) \end{matrix} \rightarrow \right)$$

used in the proof of Theorem 5.2. Therefore, $G_0$ must be abelian. Hence, $h$ commutes with $m \neq 1$ which commutes with $g$. By CT, $h$ commutes with $g$ so $h \in C_G(g) = M$. $\qquad\square$

Now exactly as in Theorem 4.2 of Section 4, we can prove

**Theorem 5.4.** $\mathrm{Th}_\forall(B)$ *is axiomatzable by* $Q_n \cup \{CT, \sigma_n\}$.

Exactly as in Theorem 4.3 of Section 4, we can prove

**Theorem 5.5.** *Let $G$ be $\mathbf{B}_n$-freely separated. Then $G$ is $\mathbf{B}_n$-freely discriminated if and only if, for every $g \in G\backslash\{1\}$, $C_G(g)$ is cyclic of order $d(g)$ where $d(g) \mid n$ and $d(g) > 1$.*

Furthermore, we have the analogues of Theorem 4.6, Corollary 2 and Theorem 4.8.

## 6. Independence of the Axioms

In all that follows $m \leq n$ shall be positive integers, $C_m$ shall be a group cyclic of order $m$ and $B = \langle a_1, a_2; \rangle_{\mathbf{B}_n} = F_2(\mathbf{B}_n)$ shall be a group free in $\mathbf{B}_n$ on the generating set $\{a_1, a_2\}$.

Now let $n$ be a fixed but arbitrary odd Adian-Sirvanjan-Atabekyan number.

**Case P:** $n = p$ *is prime.*

We claim

$(\mathbf{P}_1)$ $Q_p$ is not a consequence of $p - CyC$.
$(\mathbf{P}_2)$ $p - CyC$ in not a consequence of $Q_p$.

We take these one at a time.

$(\mathbf{P}_1)$ Without assuming the law $\forall x(x^p = 1)$, $p-CyC$ merely asserts (in the presence of the group axioms) that the centralizer of every nontrivial element is cyclic or order at most $p$. Thus $C_{p-1}$, satisfies $p - CyC$ but violates the law $\forall x(x^p = 1)$.

(**P**$_2$) $C_p = F_1(\mathbf{B}_p) \leq F_2(\mathbf{B}_p) = B$. Thus, $C_p \times C_p \leq B \times B$. Since quasi-identities are preserved in direct products and subgroups, $C_p \times C_p$ is a model of $Q_p$, But $C_p \times C_p$ violates $p - CyC$.

**Case C:** *n is composite.*

We claim

(**C**$_1$) $Q_n$ is not a consequence of $\{CT, \sigma_n\}$.
(**C**$_2$) $\sigma_n$ is not a consequence of $Q_n \cup \{CT\}$.
(**C**$_3$) $CT$ is not a consequence of $Q_n \cup \{\sigma_n\}$.

(**C**$_1$) We begin with a pair of observations.

(**1**) Since $m \leq n$, $0 \leq m_1, m_2, k_1, k_2 < n$ whenever $0 \leq m_1, m_2$, $k_1, k_2 < m$ so, $\sigma_m$ implies $\sigma_n$.

(**2**) Any group in which every pair of commuting elements lies in a cyclic group of order at most $m$ satisfies $\sigma_m$. Thus, the cyclic group $C_2$ is CT and satisfies $\sigma_2$. Hence it is CT and satisfies $\sigma_n$. But it violates the quasi-identity or law $\forall x(x^n = 1)$ (where $n$ is odd).

(**C**$_2$) $C_n \times C_n \leq B \times B$ is abelian, hence CT, and satisfies $Q_n$ but violates $\sigma_n$.

(**C**$_3$) Let $p$ be a prime divisor of $n$, Let $G$ be the subgroup of $B \times B$ generated by $b_1 = (a_1^{n/p}, a_1)$ and $b_2 = (a_2^{n/p}, a_1)$. $G$ is a model of $Q_n$. Now suppose $X = \{x_1, x_2\}$ and $w_i(x_1, x_2)$ are words of $X \cup X^{-1}$, $i = 1, 2$, such that $w_1(b_1, b_2)$ and $w_2(b_1, b_2)$ commute. Now $w_i(b_1, b_2) = (w_i(a_1^{n/p}, a_2^{n/p}), a_1^{\frac{p}{n}(e_1(w_i)+e_2(w_i))})$ where $e_\nu(w_i)$ is the exponent sum of $x_\nu$ in $w_i(x_1^{n/p}, x_2^{n/p})$, $i = 1, 2$, $\nu = 1, 2$. Since $w_1(a_1^{n/p}, a_2^{n/p})$ and $w_2(a_1^{n/p}, a_2^{n/p})$ commute in $\langle a_1^{n/p}, a_2^{n/p} \rangle \leq B$ and every abelian subgroup of $\langle a_1^{n/p}, a_2^{n/p} \rangle$ is cyclic of order dividing $n$, there must be a word $u(x_1, x_2)$ on $X \cup X^{-1}$ and integers $k_1, k_2$, $1 \leq k_1, k_2 < n$, such that, modulo the law $x^n = 1$,

$$w_i(x_1^{n/p}, x_2^{n/p}) = u(x_1^{n/p}, x_2^{n/p})^{k_i}, \quad i = 1, 2.$$

Then $w_i(b_1, b_2) = \left( u(a_1^{n/p}, a_2^{n/p})^{k_i}, a_1^{\frac{k_i n}{p}(e_1(u)+e_2(u))} \right)$ where $e_\nu(u)$ is the exponent sum on $x_\nu$ in $u(x_1^{n/p}, x_2^{n/p})$, $\nu = 1, 2$. That is $w_i(b_1, b_2) = u(b_1, b_2)^{k_i}$, $i = 1, 2$, so $\langle w_1(b_1, b_2), w_2(b_1, b_2) \rangle \leq \langle u(b_1, b_2) \rangle$ is cyclic of order dividing $n$. Hence, $G$ satisfies $\sigma_n$. Now $b_1^p = (1, a_1^p) = b_2^p$ is a nontrivial element of $G$. $b_1$ commutes with $b_1^p = b_2^p$ which commutes with $b_2$. But $b_1$ and $b_2$ do not commute. Hence, $G$ violates CT.

# References

[A]    S.I. Adian, "Classificaton of periodic words and their application in group theory," **Springer Lecture Notes in Mathematics**, 806, **Burnside Groups**, J.L. Mennicke, Editor, Springer-Verlag, Berlin (1980), 1–40.

[Ap]    K.I. Appel. "One-variable equations in free groups," **Proc. Amer. Math. Soc. 19** (1968), 912–918.

[At1]    V. Atabekyan, "Residual properties of subgroups of free periodic groups," deposited **Vinti** (1986), no. 5381-B86.

[At2]    V.V. Atabekyan, "On simple and free periodic groups," **Vestnik. Moakov.** Univ. Mat. Mekh. 76 (1987), 76–78.

[BMR]    G. Baumslag, A.G. Myasnikov, and V.N. Remeslennikov, "Algebraic geometry over groups I. Algebraic sets and ideal theory," **J. Alg.** 219 (1999), 16–79.

[BGM]    D. Bormotov, R. Gilman, and A.G. Myasnikov, "Solving one-variable equations in free groups," **J. Group Theory** 12 (2009), no. 2, 317–330.

[CR]    I.M. Chiswell and V.N. Remeslennikov, "Equations in free groups with one variable: I," **J. Group Theory** (2000), 445–466.

[C]    P.M. Cohn, **Universal Algebra**, D. Reidel, Dordrecht: Holland, 1965.

[FGRS]    B. Fine, A.M. Gaglione, G. Roesenberger, and D. Spellman, "On CT and CSA groups and related ideas," **J. Group Theory**, to appear.

[GLS]    A.M. Gaglione, S. Lipschutz, and D. Spellman, "Groups universally equivalent to free Burnside groups of prime exponent and a question of Philip Hall," **Aspects of Infinite Groups**, Ed.'s: B. Fine, G. Rosenberger, and D. Spellman, World Scientific, New Jersey, 2008, 131–148.

[H]    N. Harrison, "Real length functions in groups," **Trans. Amer. Math. Soc.**, 174, 1972, 77–106.

[K]    A.I. Kostrikin, "On Burnside's problem," (in Russian) **Izvestiya Akad. Nauk SSSR**, Ser. Mat. 23, (1959), 3–34.

[L]    R.C. Lyndon, "Equations in free groups," **Trans. Amer. Math. Soc.**, 96, 1960, 445–457.

[MR]    A.G. Myasnikov and V.N. Remeslennikov, "Algebraic geometry over groups II: logical foundations," **J.Alg.** 234, (2000), 225–276.

[N]    H. Neumann, **Varieties of Groups**, Springer-Verlag, New York, 1967.

[O]    A. Yu. Olshanskii, **Geometry of Defining Relations in Groups**, Moacow, Nauka 1989 (in Russian); English translation by Kluwer Publ., 1991.

[S]    S. Shelah, "Every two elementarily equivalent models have isomorphic ultrapowers," **Israel J. Math.**, 10 (1972), 224–233.

[Sh]    A.L. Shmelkin, "Complete nilpotent groups," **Algebra i Logica**, Sem., 6 (1967), no. 2, 111–114.

[Si]    V.L. Sirvanjan, "Imbedding of the group $B(\infty, n)$ in the group $B(2, n)$," **Math. USSR-Izv. 4** (1977), 181–199.

[Su]    M. Suzuki, "The nonexistence if a certain type of simple group of odd order," **Proc. Amer. Math. Soc.** 8 (1957), 686–695.

[Z]     E. Zelmanov, "The restricted Burnside problem," **Math. USSR Sbornek** 72, (1992), 543–564.

# Chapter 8

# Primitive Curve Lengths on Pairs of Pants

Jane Gilman

*Mathematics Department*
*Rutgers University, Newark, NJ 07102*
*gilman@rutgers.edu*

*Dedicated to Gerhard Rosenberger and Dennis Spellman on their 70th Birthdays*

**ABSTRACT.** The problem of determining whether or not a non-elementary subgroup of $PSL(2, \mathbb{C})$ is discrete is a long standing one. The importance of two generator subgroups comes from Jørgensen's inequality which has as a corollary the fact that a non-elementary subgroup of $PSL(2, \mathbb{C})$ is discrete if and only if every non-elementary two generator subgroup is. A solution even in the two-generator $PSL(2, \mathbb{R})$ case appears to require an algorithm that relies on the concept of *trace minimizing* that was initiated by Rosenberger and Purzitsky in the 1970's. Their work has lead to many discreteness results and algorithms. Here we show how their concept of trace minimizing leads to a theorem that gives bounds on the hyperbolic lengths of curves on the quotient surface that are the images of primitive generators in the case where the group is discrete and the quotient is a pair of pants. The result follows as a consequence of the Non-Euclidean Euclidean algorithm.

## 1. Introduction

The problem of determining whether or not a non-elementary subgroup of $PSL(2, \mathbb{C})$ is discrete is a long standing one. The importance of two generator subgroups comes from Jørgensen's inequality which has as a corollary the fact that a non-elementary subgroup of $PSL(2, \mathbb{C})$ is discrete

*Date*: submitted 5/25/2015; revised 5/3/2016.
1991 *Mathematics Subject Classification*. primary: 30F40, 20H10; secondary: 57M60, 37F30, 20F10.
*Keywords and phrase*. Pair of Pants, Isometry of hyperbolic space, Fuchsian group, Trace Minimizing, Algorithm, Discreteness, Curve Length, Intersections, non-Euclidean Euclidean Algorithm.

if and only if every non-elementary two generator subgroup is. A solution even in the $PSL(2, \mathbb{R})$ case appears to require an algorithm that relies on the concept of *trace minimizing*. This concept was initiated and its use pioneered by Rosenberger and Purzitsky in the 1970's [17–20]. Their work has lead to many discreteness results and algorithms. For a more recent summary see [7]. Here we show how their concept of trace minimizing leads to a theorem that gives bounds on the hyperbolic lengths of primitive curves on the quotient surface in the case where the group is discrete and the quotient is a pair of pants. A primitive curve is, of course, the image of a primitive generator in a rank two free group.

The main result of this paper appear in Section 6 (Theorems 6.1, 6.2, 6.4, 6.5). When the group is discrete and free, the conjugacy class of any primitive curve corresponds to a rational number. The number of so-called *essential self intersections* of the image of the curve on the quotient is given by a formula that depends upon the rational [13]. In particular, if $r = p/q$ is the rational in lowest terms, let $I(r) = I(p/q)$ denote the number of essential self-intersections. The discreteness algorithm finds the length of the three shortest simple closed curve on the pair of pants, as well as the longest *seam length* on any quotient. The length of a curve on the quotient is given by the translation length of the matrix and is found using the trace of the matrix. We are able to bound the length of any primitive curve in terms of $I(p/q)$ and the lengths of the three shortest geodesics, Theorem 6.4. We obtain other variants on the upper and lower bounds for the translation length of any primitive element of the group using the entries in the continued expansion of $r$ and the translation length of the shortest curves, Theorem 6.1, and we express the longest seam length as a limit, Theorem 6.5.

The results stem from casting the discreteness algorithm as a Non-Euclidean Euclidean algorithm [11].

We note that other authors bound the lengths of curves, primitive or not, in terms of the number of self intersections (see [1, 3, 4] and references given there).

The paper builds on a theory of curves on surfaces and algorithms, especially the Non-Euclidean Euclidean Algorithm (the NEE for short) and is organized with a number of preliminary sections which can be viewed as an exposition for the uninitiated or skipped for those familiar with the theory. The new results appear in Section 6. More specifically in Section 2 notation is set, in Section 3 the Gilman-Maskit algorithm is given (the GM algorithm for short), the $F$-sequence is defined and the Non-Euclidean Euclidean Algorithm is given. Section 4 reviews the standard notation for the hyperbolic hexagon determined by a two-generator group and in Section 5 the winding and unwinding process of the algorithm is explained.

## 2. Preliminaries: Notation and Terminology

In this section we set notation and review terminology. In the next sections we summarize results that will be needed.

In what follows we let $G = \langle A, B \rangle$ be a non-elementary subgroup of $PSL(2, \mathbb{C})$ or equivalently the group of isometries of hyperbolic three-space $\mathbb{H}^3$.

We will concentrate on $PSL(2, \mathbb{R})$ which we identify with the group of isometries of hyperbolic two-space. We recall that a matrix $M = \begin{pmatrix} a & b \\ c & d \end{pmatrix}$ with $ad - bc = 1$ in $SL(2, \mathbb{R})$ acts on the upper-half plane $\mathbb{H}^2 = \{z = x + iy \mid x, y \in \mathbb{R}, y > 0\}$ by $z \to \frac{az+b}{cz+d}$. That is, a matrix induces a fractional linear transformation. Since a matrix and its negative yield the same action, we identify the action with that of $PSL(2, \mathbb{R})$. Further if $\mathbb{H}^2$ is endowed with the hyperbolic metric, these transformations are isometries in the metric. We let $\mathrm{tr}(M)$ denote the trace of $M$.

We recall that elements of the isometry group $\mathbb{H}^2$ are classified by their traces and equivalently by their geometric actions. If $|\mathrm{tr}(M)| > 2$, $M$ is hyperbolic and fixes two points on the boundary of $\mathbb{H}^2$ (i.e. the real axis) and a hyperbolic or non-Euclidean straight line. This is the semi-circle in the upper-half plane perpendicular to the real axis intersecting it at the fixed points of $M$. The transformation moves point on the semi-circle towards one fixed point, the attracting fixed point, and away from the other fixed point, the repelling fixed point and moves points a certain fixed distance in the hyperbolic metric. This distance is known as the translation length of $M$ and we denote it by $T_M$. The semi-circle is termed the axis of $M$ and denoted $Ax_M$. The formula relating the translation length to the trace of the matrix is given below.

However, first we note that if $A$ and $B$ are any generators, there are two possible pull-backs from $PSL(2, \mathbb{R})$ to $SL(2, \mathbb{R})$, one with positive trace and one with negative trace. If we choose pull-backs for both $A$ and $B$ with positive traces, then the trace of any word in the pull-back generators is determined and for simplicity of exposition we denote the trace of an element $X \in G$ by Trace $X$. Note that the trace of any commutator is always positive irrespective of pull backs.

For ease of exposition we do not distinguish rotationally between an element of $X \in G = \langle A, B \rangle \subset PSL(2, \mathbb{R})$ and the corresponding matrix or isometry with the trace Trace $X$ as defined above.

An element $X$ in $SL(2, \mathbb{R})$ or its inverse is conjugate to a fractional linear transformation of the form: $z \mapsto \pm Kz$ for some real number $K > 0$, $K \neq 1$. $K$ is called the multiplier of $X$ and is denoted $K_X$ and $|\text{Trace } X| = \sqrt{K} + \sqrt{K}^{-1}$.

$T_X$ and $K_X$ are related by

$$\cosh \frac{T_X}{2} = \frac{1}{2} |\text{Trace } X| = \frac{1}{2} \left( \sqrt{K} + \frac{1}{\sqrt{K}} \right)$$

where cosh is the hyperbolic cosine.

We note for future use,

$$T_X = T_{X^{-1}} = |\log K_X|.$$

Note that if $|\text{tr}(M)| = 2$, $M$ is parabolic and fixes one point on the real axis, known as its axis. It does have a rotation direction about its fixed point even though it does not have a translations length. If $|\text{tr}(M)| < 2$, $M$ is elliptic and is a rotation about its single fixed point in upper-half plane.

If $A$ and $B$ are any hyperbolic elements, it is well known that their axes have a common perpendicular $L$. Orient the axis towards their attractive fixed points. We say that the pair $(A, B)$ is *coherently oriented* if when $L$ is oriented from the axis of $A$ to the axis of $B$ the attracting fixed points of $A$ and $B$ lie to the left of $L$ and Trace $A \geq$ Trace $B \geq 0$.

If one or both of $A$ and/or $B$ is parabolic, the is still a common perpendicular $L$, either between the axis of the hyperbolic and the axis of the parabolic, that is its fixed point of the parabolic or between the two parabolic fixed points, and the definition of coherent orientation still applies provided if when only one generator is hyperbolic it is $A$.

If $X$ is any geodesic in $\mathbb{H}^2$ we let $H_X$ be the half-turn about $X$.

Now the $PSL(2, \mathbb{R})$ discreteness algorithm divides naturally into two cases, the intertwining case known as the GM algorithm [16] and the intersecting axes algorithm [8]. Here we only address the intertwining algorithm, the case where any pair of hyperbolics encountered have disjoint axes.

## 3. Preliminaries: Summary of Some Prior Results

**Theorem 3.1.** [16] The Gilman-Maskit Geometric $PSL(2, \mathbb{R})$ Algorithm (1991).

Let $G = \langle A, B \rangle$ be non-elementary with $A, B \in PSL(2, \mathbb{R})$. Assume that $A$ and $B$ are a coherently ordered pair of generators and that $A$ and $B$ have disjoint axes in the case when both are hyperbolic. There exists integers $[n_1, n_2, \ldots, n_t]$ that determines the sequence of pairs of generators

$$(A, B) \to (B^{-1}, A^{-1}B^{n_1}) \to (B^{-n_1}A, B(A^{-1}B^{n_1})^{n_2}) \to \cdots \to (C, D)$$

which is a sequence of coherently ordered pairs and which stops at a pair $(C, D)$ after a finite number of replacements and says (i) $G$ is discrete (ii) $G$ is not discrete, or (iii) $G$ is not free.

We emphasize that the replace is done in a *trace minimizing manner*. That is, if we set $(A, B) = (A_0, B_0)$ and at step $j \geq 1$ consider the ordered $(A_j, B_j)$ pair. We can then assume by the algorithm assume

$$|\text{Trace } A_j| \geq |\text{Trace } B_j| \text{ and}$$

and at the next pair $(A_{j+1}, B_{j+1})$, which is equal to the ordered pair $(B_j^{-1}, A_j^{-1}B_j^{n_{j+1}})$, we have

$$|\text{Trace } B_j| \geq |\text{Trace } B_j^{-n_{j+1}}A_j|$$

The algorithm stops when it reaches an element $X$ with Trace $X \leq 2$. We refer to this geometric discreteness algorithm as the GM algorithm.

The sequence of integers has been used in a number of settings. For example it was used to calculate the computational complexity of the algorithm in (1997) by Gilman [9] and by Y. C. Jiang in his (2001) thesis [22] and to calculate the essential self-intersection numbers of primitive curves on the quotient when the group is discrete [13]. However, it was not until 2002 in [13] after the sequence has been used in various settings, that it was given a name.

**Definition 3.2.** [13] (2002) The sequence $[n_1, \ldots, n_t]$ is termed the *F*-sequence or the *Fibonacci sequence* of the algorithm.

The existence of such $n_i$ was implicit in the proof of the GM geometric discreteness algorithm. However, these numbers were never actually computed.

The interpretation of the algorithm as a non-Euclidean Euclidean Algorithm gave a theorem that showed how to actually compute the $n_i$ [11].

We note that if the algorithm begins with a pair of hyperbolics, the new pair will either be a pair of hyperbolics or a hyperbolic–parabolic pair or it will stop. If it arrives at or begins will a hyperbolic-pair, the next pair will either be a hyperbolic–parabolic pair again or stop or be a parabolic–parabolic pair. A parabolic–parabolic may repeat or be a stopping pair. It repeats a given type of pair at most a finite number of times and then moves on the next type or stops. The algorithm can be extended to apply to elliptics of finite order [9] but we do not consider that case here as the definition of the terms in the $F$-sequence needs to be modified (see [23]).

**Remark 3.3.** The algorithm stops and says that $G$ is discrete at a pair $(C_t, D_t)$ when Trace $C_t \geq$ Trace $D_t \geq 2$ and Trace $C_t^{-1}D_j \leq 0$. If one orders the stopping generators in terms of their traces or the length of the corresponding curves on the quotient surface there are three possibilities

(1) Trace $C_t \geq$ Trace $D_t \geq |$Trace $C_t^{-1}D_t|$.
(2) Trace $C_t \geq |$Trace $C_t^{-1}D_t| \geq$ Trace $D_t$.
(3) $|$Trace $C_t^{-1}D_t| \geq$ Trace $C_t \geq$ Trace $D_t$.

When the algorithm stops and says that the group is discrete, there is a right angled hexagon with disjoint sides (see Section 5 for more details). For each parabolic generator, one of the six sides reduces to a point. In such a case $|$Trace $C_t^{-1}D_t| = 2$ and one or two or none of the other traces may have absolute value 2.

**Theorem 3.4.** [11] **The Non-Euclidean Euclidean Algorithm** (2013). *Let $G = \langle A, B \rangle$ where $A$ and $B$ are a coherently ordered pair of transformations in $PSL(2, \mathbb{R})$ whose axes are disjoint in the case of a pair of hyperbolics.*

*If one applies the Euclidean algorithm to the non-Euclidean translation lengths of the generators at each step and the algorithm stops at a pair of generators saying that the group is discrete, the output of the Non-Euclidean Algorithm is the F-sequence $[n_1, \ldots, n_k]$. It is obtained by* *applying the Euclidean algorithm to the non-Euclidean translation lengths of the generators at each step to the possible cases as follows:*

A. *If the initial $A$ and $B$ are hyperbolic generators with disjoint axes and the algorithm stops at a pair of hyperbolics saying that the group is discrete,*

one has

$$n_1 = \left[\frac{(|\log K_A|)/2}{(|\log K_B|)/2}\right] = \left[\frac{T_A/2}{T_B/2}\right]$$

*where* [ ] *denotes the greatest integer function.*

$$n_2 = \left[\frac{T_B/2}{T_D/2}\right] \quad \text{where } D = A^{-1}B^{n_1}$$

*and*

$$n_j = \left[\frac{T_{C_j}/2}{T_{D_j}/2}\right].$$

*Here $(A, B) = (C_0, D_0)$ and $(C_j, D_j) = (D_{j-1}^{-1}, C_{j-1}^{-1}D_{j-1}^{n_j})$ is the ordered pair of generators at step $j$, $0 \leq j \leq t$.*

B. *If one begins at any type of pair and stop at a pair that includes one or more parabolics,*

one has

(i) *If at step $j$, $(C_j, D_j)$ is hyperbolic–parabolic pair, then*

$$n_j = \left[\frac{\text{Trace}\,(C_j) - 2}{\sqrt{|\text{Trace}\,([C_j, D_j]) - 2|}}\right]$$

*or equivalently, setting $A = C_j$ and $B = D_j$,*

$$n_j = \left[\frac{2\cosh(\frac{T_A}{2}) - 2}{\sqrt{2\cosh\left(\frac{T_{[A,B]}}{2}\right) - 2}}\right].$$

(ii) *If at step $j$, $(C_j, D_j)$ is a parabolic–parabolic pair, $j = t$ and $n_t = 1$.*

*Step* (ii) continued *Further if* $t \geq 2$, *at step* $(t-1)$, $(C_{t-1}, D_{t-1})$ *is a hyperbolic–parabolic pair with*

$$n_{t-1} = \left[ \frac{\text{Trace } (C_{t-1}) - 2}{\sqrt{|\text{Trace } ([C_{t-1}, D_{t-1}]) - 2|}} \right]$$

*and*

$$n_t = 1.$$

(iii) *If the initial pair is a hyperbolic–parabolic pair, then the F-sequence is of length 2 and is* $[n_1, n_2]$ *where*

$$n_1 = \left[ \frac{2\cosh\left(\frac{T_A}{2}\right) - 2}{\sqrt{2\cosh\left(\frac{T_{[A,B]}}{2}\right) - 2}} \right]$$

*and*

$$n_2 = 1.$$

(iv) *If the initial pair is a parabolic–parabolic pair, the F-sequence is of length 1 with* $n_1 = 1$.

## 4. Preliminaries: Standard Hexagons, Notation and Facts

We recall that to any group two generator group $G = \langle A, B \rangle$ one associates a standard three generator group $3G = \langle H_L, H_{L_A}, H_{L_B} \rangle$ where $L$ is the common perpendicular to the axes of $A$ and $B$ and $L_A$ and $L_B$ are geodesics such $A = H_L H_{L_A}$ and $B = H_L H_{L_B}$. The six geodesics $Ax_A, L_A, Ax_B, L_B, A_{A^{-1}B}, L$ determine a right angled hexagon in $\mathbb{H}^2$. The hexagon will have self-intersecting sides with the exception of the hexagon determined by the stopping generators when the groups are discrete. This hexagon will be convex. In this case the projections of the axes sides will be the three shortest geodesics on the quotient [13]. We speak of the interior sides of the hexagon which are the portions of the axes of the generators between two half-turn axes and the portions of the half-turn axes between two axes. It will be clear from the context whether we are speaking about a geodesic or the portion of the geodesic interior to the hexagon.

When $G$ is discrete, we let $\pi$ be the projection of $\mathbb{H}^2$ onto $\mathbb{H}^2/G$. The quotient is a three-holed sphere or a sphere with up to three punctures.

If $G$ is generated by $W$ and $V$, then $W$ and $V$ are called *primitive associates* and $(W, V)$ is called a *primitive pair*.

The primitive elements project to closed curves on the quotient which we call the *primitive curves* and we may assume primitive curves are geodesics, as are all curves mentioned. We let the generators that stop the algorithm be $\tilde{A}_0, \tilde{B}_0, \tilde{C}_0 = \tilde{A}_0^{-1}\tilde{B}_0$ and $\alpha_0, \beta_0, \gamma_0$ their respective images on the surface. These are the three shortest curves on the surface [13] and $l_0, l_{\alpha_0}$, and $l_{\beta_0}$, the corresponding seams are the three longest seams.

We denote the length of the images of the stopping generators by $L(\alpha_0), L(\beta_0),$ and $L(\gamma_0)$. The trace convention allows us to assume that $L(\alpha_0) \geq L(\beta_0)$. We allow the possibility of parabolic stopping generators whose lengths are zero. In that case, we set $L(\hat{\alpha}_0)$ to be the length of the first non-zero winding curve, that is $L(\pi(\tilde{A}_0\tilde{B}_0))$.

## 5. Preliminaries: Essential Self-Intersections and Winding and Unwinding Sequences

Any primitive geodesic on $\mathbb{H}^2/G$ will have a certain number of self-intersections on the images of the half-turn lines. There are termed *essential self-intersections*. All other self-intersections are self-intersections that will be undone when an essential self intersection is cut. These are termed *trailing intersections*. If a primitive curve that is not a stopping generator has $F$-sequence $[n_1, \ldots n_t]$, then there is a formula for essential self intersections. This is Theorem 6.1 of [13]. Note that the formula in Theorem 7.1 part (7) has a typo but the formula in Theorem 6.1 is correct.

Before we can present the formula, we need to consider *winding* and *unwinding* sequences. Since the algorithm ends with curves that are simple, we view the algorithm as an unwinding procedure [13]. We can run the algorithm backwards as a winding procedure.

**Theorem 5.1.** [13] *Proposition* 1.6. *If F-sequence or the unwinding sequence is* $[n_1, \ldots, n_t]$, *then the winding sequence is* $[-n_t, \ldots, -n_1]$.

However, we can also write the winding sequence with positive integers, adjusting the orientation of curves carefully as in [13] so that the winding sequence is $[n_1, \ldots, n_t]$ and the corresponding continued fraction entries are $[n_0, n_1, \ldots, n_k]$ where $n_0 = 0$.

**Theorem 5.2.** [13] *If W is a primitive element in G, replacing W by its inverse if necessary, we may assume that there is a primitive associate V* *such that the ordered pair $(W, V)$ is coherently ordered. After unwinding to stopping generators and then winding, $W$ corresponds to a unique positive rational $r$ given by $p/q$ where $p$ and $q$ are relatively prime positive integers.*

In what follows, we let $\gamma_r$ denote the primitive curve on the quotient surface given by the rational $r = p/q$ that comes from winding about a pair of stopping generators. Note that curve $\gamma_r$ depends only on the conjugacy class of its pre-image.

**Theorem 5.3.** [13] *We let $I(n_1, \ldots, n_t)$ denote the number of essential self-intersections of the curve with unwinding sequence $[-n_t, \ldots, -n_1]$ so that $\gamma_r$ has winding sequence $[n_1, \ldots, n_t]$ and $r$ is the rational continued fraction expansion $[n_0, \ldots, n_t]$ and we let $r_k = [n_0, \ldots, n_k]$ denotes its k-th approximant. We set $r_k = p_k/q_k$ where $p_k$ is the numerator and $q_k$ the denominator of the approximant and $r_k$ is given in lowest terms. Then the essential self-intersection numbers are given inductively as follows:*

$$I(\alpha_0) = 0, \quad I(\beta_0) = 0, \quad I(\alpha_0^{-1}\beta_0) = 0,$$
$$I(\alpha_0\beta_0) = 1, \quad I(\alpha_0\beta_0^2) = 2$$

*and $I_{p/q}$ is determined inductively by the formulas*

$$I_{p_{k+1}/q_{k+1}} = 1 + n_{k+1} \cdot I_{p_k/q_k} + I_{p_{k-1}/q_{k-1}}$$

We write $I(r)$ and $I(p/q)$ for the number of essential self-intersections.

## 6. Length Inequalities

In this section we assume that group is discrete and free and that the stopping generators are the coherently oriented pair $(\tilde{A}_0, \tilde{B}_0)$. The corresponding hexagon is right angled and convex and we let $L(\lambda_0)$ be the largest seam length of the corresponding pair of pants. As before let $\alpha_0, \beta_0, \gamma_0$ denote the images of the sides of the hexagon on the quotient and $L(\delta)$ the length of any geodesic, $\delta$, on the quotient surface.

By [13] and the formula for $\cosh \pi(L)$ given below and with this orientation the length of the longest seam is the largest length between any associate primitive pair of curves on the surface.

Let $W \in G$ be primitive and assume that after replacing $W$ by its inverse if necessary there is a primitive $V$ such that $(W, V)$ is a coherently oriented primitive pair. There is the unwinding sequence given by $(C_j, D_j)$ where $j = 0, \ldots, t$ and $W = C_0$ and $V = D_0$ which stops at generators $\{\tilde{A}_0, \tilde{B}_0, \tilde{A}_0^{-1}\tilde{B}_0\}$ where the stopping generators are $C_t, D_t$ have been renamed taking into account any necessary normalization needed following remark 3.3. That is, we now assume that $L(\alpha_0) \geq L(\beta_0) \geq L(\gamma_0)$, and that the corresponding winding sequence begins at $\tilde{A}_0, \tilde{B}_0, \tilde{A}_0^{-1}\tilde{B}_0$ is given by $[m_1, \ldots, m_t]$ and the rational $r$ has the continued fraction entries $[0, m_1, \ldots, m_t]$.

We note that if $(W, V)$ is a primitive pair, it determines a set of stopping generators. Any other set of stopping generators will be conjugate to this set.

In the winding case where we start with three coherently oriented parabolics, we use part **B** of Theorem 3.4 backwards.

**Theorem 6.1. Winding Sequences and Translation Lengths.** *Let $W$ be any primitive element of $G = \langle A, B \rangle$ where $G$ is discrete and free and has stopping generators $(\tilde{A}_0, \tilde{B}_0)$. Assume that $[n_1, \ldots, n_t]$ is the $F$-sequence of $W$. Assume $W$ is not a stopping generator.*

*Assume that not all three of the stopping generators are parabolic. With the normalization above where $\frac{T_{\tilde{A}_0}}{2} = \frac{L(\alpha_0)}{2}$ is half the translation length of the longest hyperbolic stopping generator, we have*

$$\left(\Pi_{i=1}^t n_i\right) \cdot \frac{T_{\tilde{A}_0}}{2} < \frac{T_W}{2} \leq \left(\Pi_{i=1}^t (n_i + 1)\right) \cdot \frac{T_{\tilde{A}_0}}{2}.$$

*If all three stopping generators are parabolic, set $\hat{A}_0 = \tilde{A}_0^{-1}\tilde{B}_0$. We have*

$$\left(\Pi_{i=1}^t n_i\right) \cdot \frac{T_{\hat{A}_0}}{2} < \frac{T_W}{2} \leq \left(\Pi_{i=1}^t (n_i + 1)\right) \cdot \frac{T_{\hat{A}_0}}{2}.$$

*Equivalently, since $|m_k| = |n_{t-k}|$, the formulas can be written with $m_i$ replacing $n_i$.*

**Proof.** We first assume that the stopping generators are all hyperbolics and that $\tilde{A}_0$ is the generator with largest translation length. We note for the unwinding sequence the Non-Euclidean Euclidean Algorithm tells if we remove the greatest integer symbols, for each integer $j$ we have

$$n_j \frac{T_{C_{j+1}}}{2} \leq \frac{T_{C_j}}{2} \leq (n_j + 1)\frac{T_{C_{j+1}}}{2}$$

where the unwinding sequence is $[n_1, \ldots n_t]$ and the winding sequence taken to be positive is $[m_1, \ldots, m_t]$ where $m_k = n_{t-k}$. This gives the theorem. If the algorithm includes one or more parabolic elements before it stops, we have the same formula except we must re-interpret the initial few $m_i$'s.

That is, in the winding case where we start with three coherently oriented parabolics, we use part **B** of theorem 3.4 backwards.    □

The theorem can be rephrased as

**Theorem 6.2. Winding Sequences and Curve Lengths.** *Let $W$ be any non-parabolic primitive element of $G = \langle A, B \rangle$ where $G$ is discrete and free and has stopping generators $(\tilde{A}_0, \tilde{B}_0)$. Assume that $W$ is not conjugate to a stopping generator. Let $W$ have F-sequence $[n_1, \ldots, n_t]$ and let $L(W)$ be the length of its image on the surface. Let $L(S_0)$ be longest length of any simple curve on the quotient surface.*

$$\left( \Pi_{i=1}^t n_i \right) \cdot L(S_0) < L(W) \le \left( \Pi_{i=1}^t (n_i + 1) \right) \cdot L(S_0).$$

*Equivalently, since $|m_k| = |n_{t-k}|$, the formulas can be written with $m_i$ replacing $n_i$.*

We recall from [16] that if we begin with a coherently ordered pair of hyperbolics, then as long as we stay in the hyperbolic-hyperbolic case with all traces positive, at each small step going from replacing $X$ and $Y$ by $XY^{-1}$ and $Y$, the trace is reduced by at least $\frac{(\sqrt{2}-1)^2}{\sqrt{2}}$.

**Theorem 6.3.** [9] Counting Hyperbolic Repetitions *Assume that $G$ is discrete and free and that we unwind beginning with $(A, B)$ and ending with $(A_0, B_0)$, hyperbolics with disjoint axes. Let $[n_1, \ldots, n_t]$ be the unwinding sequence. Then*

$$\frac{(\sqrt{2}-1)^2}{\sqrt{2}} \left( \Sigma_{1=1}^t n_t \right) \ge \text{Trace } A - 2$$

*and the algorithm repeats the hyperbolic–hyperbolic case at most $q$ times where $q = \left( \Sigma_{1=1}^t n_t \right)$.*

In [9] similar bounds for $t$ were found for all other geometric types of pairs of generators.

**Theorem 6.4. Minimal curve length and essential intersections.** *Continuing with the notation above, let $\gamma_r$ denote the primitive curve on the quotient corresponding to the rational $r = p/q$. Then its length $L(\gamma_r)$ satisfies*

$$I(r) \times L(\gamma_0) \le L(\gamma_r) \le (I(r) + 1) \times L(\alpha_0)$$

*where $\gamma_0$ is the shortest stopping generator and $\alpha_0$ the longest.*

**Proof.** This follows from the fact that one plus the number of essential-self intersections correspond to winding around curves of length at most $L(\alpha_0)$ and at least $L(\gamma_0)$. Every essential self-intersection corresponds to the image of a segment that runs between either the images of $L$ and $L_A$, $L$ and $L_B$ or $L_B$ and $L_A$. Thus the images of a segment is of length at least $L(\frac{\gamma_0}{2})$. An essential self-intersection will correspond to traversing a segment between two seams twice. $\square$

We recall from Fenchel [6] that if a convex right angled hexagon has alternating sides of length $x \leq y \leq z$ then the length of the interior side $L$ opposite the shortest side $x$ satisfies:

$$\cosh L = \frac{\cosh x + \cosh y \cosh z}{\sinh y \sinh x}.$$

From this we can compute the hyperbolic length of the longest seam.

We consider the set of curves $AB^j$, $j = 1, \ldots, \infty$ in $\mathbb{H}^2$. We let $q_i$ be the point on the $L$, where $AB^i$ intersects $L$. We note from [11, 23] $q_{i+1}$ lies to the right of $q_i$ given a coherent orientation when we are at a pair of stopping generators. We set $q_0$ be the point where the axis of $A$ intersects $L$ and $\rho_i$ the hyperbolic distance from $q_{i-1}$ to $q_i$, $i = 1, \ldots \infty$.

We have

**Theorem 6.5. Seam Lengths and Intersections.** *Assume $G$ is discrete and free with stopping generators $\tilde{A}_0$ and $\tilde{B}_0$ and that $L$ is their common perpendicular and let $L(\lambda_0)$ be the length of $L$ between the axes of $\tilde{A}_0$ and $\tilde{B}_0$. If $\rho_i$ is the distance between the intersections of $\tilde{A}_0 \tilde{B}_0^{i-1}$ and $\tilde{A}_0 \tilde{B}_0^i$ with $L$, $i = 1, \ldots$, then*

$$\lim_{t \to \infty} \Sigma_{i=1}^t \rho_i = L(\lambda_0).$$

**Proof.** Assume for ease of notation that the stopping generators are $A$ and $B$. We recall that the transformation $B$ can be factored in an infinite number of ways as the product of two half-turns about geodesics perpendicular to its axis which are half its translation length apart along the axis of $B$. We write these lines

$$L_B, L_{B^2}, \ldots, L_{B^j}, \ldots$$

and their images under the half-turn about $L$ and

$$L_{\overline{B}}, L_{\overline{B}^2}, \ldots, L_{\overline{B}^j}, \ldots$$

We note that $AB^j = L_A L_{\overline{B}^j}$ and these curves will intersect $L$ between the axes of $A$ and $B$.    □

**Corollary 6.6.** *Let $l_{(C,D)}$ be a half-turn line corresponding to any primitive pair $(C, D)$ and $L(l_{(C,D)})$ the length of its image on the quotient. Then*

$$L(l_{(C,D)}) \leq \lim_{t \to \infty} \Sigma_{i=1}^{t} \rho_i = L(\lambda_0)$$

*where $\rho_i$ is as in Theorem 6.5.*

**Proof.** The image of $L$ is the longest of any of the half-turn lines.    □

**Remark 6.7.** In the case that the group is generated by hyperbolics with intersecting axes and the commutator is not elliptic, the group will be discrete and no algorithm is needed. In earlier papers we noted that following the methods of [11], one can still obtain a finite $F$-sequence which stops at a pair where lengths are shortest. One can wind and unwind. This will be addressed elsewhere as will the cases involving elliptic elements of finite order from either the intertwining case or the intersecting axes case when the group is discrete and infinite order when the group is not. For elliptic elements of finite order one has to use the extended notion of $F$-sequence developed by Malik in [23].

**Remark 6.8.** These inequalities and bounds have obvious applications to lengths of geodesics on pairs of pants in a pants decomposition for a compact surface of genus $g \geq 2$. The formulas can be stated in terms of the minimal curve length on a pair of pants with one self intersection instead of the minimal non-zero curve length on the pair of pants. This will be addressed in [12]

# References

[1]  Baibaud, C. M., *Closed geodeics on pairs of pants*, Israeli J of Math (1999) 339-347.

[2]  Basmajian, Ara, *Universal length bounds for non-simple closed geodesics on hyperbolic surfaces*, J. Topol (2013), 1–12.

[3]  Chas, M. and Phillips, T., *Self-intersection numbers of curves in the punctured torus*, Exp. Math **19** (2010) 129–148.

[4]  Cohen, M. and Lustig, M., *Paths of geodesics and geometric intersection numbers* Combinatorial groups theory and topology, Alta utah 1984 Annals of Math Studies 111, princeton U. Press (1987) 479–500.

[5]  Beardon, Alan, *The geometry of discrete groups*, Graduate Texts in Mathematics, **91** Springer-Verlag, New York, (1983).

[6]  Fenchel, W., *Elementary geometry in hyperbolic space* de Gruyter Studies in Mathematics, **11** Walter de Gruyter & Co., Berlin, (1989).

[7]  Fine, B. and Rosenberger, G., *Classification of all pair of two-generator Fuchsian groups*, Groups' 93, Galway/St. Andrews, London Math. Soc., Lecture Notes Series 211 (1995), 203–232.

[8]  Gilman, Jane, *Two-generator Discrete Subgroups of $PSL(2,\mathbb{R})$*, Memoirs of the AMS, Volume 117, No 561, 1995.

[9]  Gilman, Jane, *Algorithms, Complexity and Discreteness Criteria in $PSL(2,\mathbb{C})$*, Journal D'Analyse Mathematique, Vol. 73, (1997), 91–114.

[10]  Gilman, Jane, *Informative Words and Discreteness*, Contemp. Math. **421**, (2007) 147–155. Cont Math.

[11]  Gilman, Jane, *The Non-Euclidean Euclidean Algorithm*, Advances in Mathematics 250 (2014) 227241.

[12]  Gilman, Jane *in preparation*.

[13]  Gilman, Jane and Keen, Linda, *Word sequences and intersection numbers*. Complex manifolds and hyperbolic geometry (Guanajuato, 2001), 231–249, Contemp. Math., 311, Amer. Math. Soc., Providence, RI, 2002.

[14]  Gilman, Jane and Keen, Linda, *Discreteness Criteria and the Hyperbolic Geometry of Palindromes*, Conformal Geometry and Dynamics, 13 (2009), 76–90 arXiv:0808.3488.

[15]  Gilman, Jane and Keen, Linda, *Enumerating Palindromes and Primitives in Rank Two Free Groups*, Journal of Algebra, (2011), 1–13.

[16]  Gilman, J. and Maskit, B., *An algorithm for 2-generator Fuchsian groups*, Michigan Math. J. **38** (1991) 13–32.

[17]  Purzitsky, N., *Two generator free product*, Math Zeit. **126** (1972) 209–223.

[18]  Purzitsky, N. and Rosenberger, G., *Two generator Fuchsian groups of genus one*, Math Zeit. **128** (1972) 245–251.

[19]  Rosenberger, G., *Fuchssche Gruppen die freies Produkt zweier zyklischer Gruppen sind, und die Gleichung $x^2 + y^2 + z^2 = xyx$*. Math. Ann. 199 (1972) 213–227.

[20]  Rosenberger, G., *All generating pairs of all two-generator Fuchsian groups*, Arch. Math. 46 (1986), 198–204.

[21]  Hoban, Ryan, *Shadow Problems in Spherical and Hyperbolic Geometry* http://www-users.math.umd.edu/~rfhoban/Shadows/Shadows.html.

[22]  Jiang, Yicheng, *Polynomial complexity of the Gilman-Maskit discreteness algorithm*, Ann. Acad. Sci. Fenn. Math. 26 (2001), no. 2, 375–390.

[23]  Malik, Vidur, *Primitive words and self-intersections of curves on a surface generated by the Gilman-Maskit algorithm*, AMS Cont Math 510 (2010) 209–238.

## Chapter 9

# Drawing Inferences Under Maximum Entropy From Relational Probabilistic Knowledge Using Group Theory

Gabriele Kern-Isberner[1], Marco Wilhelm[1], and Christoph Beierle[2]

[1] *Dept. of Computer Science*
*TU Dortmund University, Germany*

[2] *Dept. of Mathematics and Computer Science*
*University of Hagen, Germany*

**ABSTRACT.** Probabilistic reasoning under the principle of maximum entropy (so-called MaxEnt principle) is a viable and convenient alternative to graph-based methodologies such as Bayesian networks that realises an idea of information economy, i.e., of being as unbiased as possible. For relational conditional knowledge, the aggregating semantics provides a semantic framework which sorts well with MaxEnt. In this paper, we exploit the group structure that underlies relational conditional knowledge bases to condense both, the aggregating semantics and the MaxEnt formalism. Based on this group structure, we present an approach to performing MaxEnt reasoning that makes use of symbolic computations. Given a knowledge base, we encode the MaxEnt optimization problem into a system of polynomial equations and then apply elimination theory to find MaxEnt inferences as points of an affine variety.

## 1. Introduction

Reasoning with probabilistic information is hard, both due to its exponential complexity[1] and to its intransparency for human reasoners. It was only after Pearl's seminal book (Pearl, 1988) that computer scientists, logicians, and psychologists got more and more interested in probabilistic reasoning. Pearl's merit was that he interlinked the numerical probabilities with a qualitative structure that expresses dependencies and also independencies.

**Keywords**: Knowledge representation, free abelian group, conditional structure, aggregating semantics, principle of maximum entropy, elimination theory.

[1] A probability distribution over $n$ binary variables has $2^n$ values.

Bayesian networks (Cowell *et al.*, 1999) arose from that idea, in which a directed acyclic graph plays the qualitative part, and tables of conditional probabilities at each node provide the numerical information. Bayesian networks are now one of the prevalent methodologies for the implementation of probabilistic reasoning.

However, Bayesian networks also show drawbacks: First, in principle a Bayesian network needs complete information in the conditional probability tables which is often hard to specify without a considerable amount of statistical work. For instance, a medical doctor might easily give the probability of a symptom given that he knows the disease is present, but what would be the right conditional probability in case the disease is not present? Second, the dominant element for structuring knowledge in a Bayesian network is the concept of conditional independency — each node is conditionally independent of its non-descendants given its parents. It is exactly this property that allows for decomposing the probability distribution into smaller local distributions and obtain the global distribution via factorization. However, extensive psychological experiments (Rottman and Hastie, 2014) have shown that the concept of conditional independence is not intelligible for human reasoners, so an essential component of Bayesian networks is still not completely accessible to human reasoners. Finally, Bayesian networks yield good results when the problem has a clearly directed, causal structure (therefore, they are often called *causal networks*) but they cannot cope well with symmetric dependencies between merely correlated variables.

An alternative to Bayesian networks that is claimed to (at least partially) solve these problems is the so-called *principle of maximum entropy (MaxEnt principle)* which is based on Shannon's entropy $\mathcal{H}(\mathcal{P}) = -\sum_{\omega \in \Omega_{\mathcal{D}}} \mathcal{P}(\omega) \log \mathcal{P}(\omega)$ (Jaynes, 1983b) reflecting the amount of indeterminateness inherent to a probability distribution $\mathcal{P}$. Given a probabilistic knowledge base consisting of conditional constraints $\mathcal{KB} = \{(B_1|A_1)[\xi_1], \ldots, (B_n|A_n)[\xi_n]\}$, the idea is to take the unique distribution that both satisfies all the constraints in $\mathcal{KB}$ and has maximal entropy as a "best" model of the knowledge base and use it for inductive reasoning from the knowledge base. With maximizing entropy, the amount of information necessarily added to the knowledge base to come up with a complete distribution is minimized, thus the MaxEnt principle selects a model of $\mathcal{KB}$ which is most conservative and cautious. Although this combination of an expressive semantical framework like probabilities with a black-box optimization procedure seems to increase the complexity of probabilistic reasoning considerably (in particular with respect to cognitive intelligibility), reasoning on maximum entropy has been shown to satisfy logical standards of uncertain reasoning very well (Kern-Isberner, 1998a,b), and to comply with commonsense reasoning in an optimal way (Paris, 1999). For an application of MaxEnt reasoning in the medical domain see e.g. (Beierle *et al.*, 2015). These results are encouraging because MaxEnt reasoning does not have any of the drawbacks mentioned above (admittedly, at somewhat higher computational costs): The knowledge base $\mathcal{KB}$ does not need to (and usually will not) be complete, no explicit conditional independence assumptions are necessary (although conditional independencies will be established automatically), and dependencies can be symmetric and cyclic (no directed acyclic graph is assumed to underly the problem).

But one question remains — beyond the abstract optimality of the MaxEnt principle, what are the qualitative structures that guide its reasoning? If no graph is needed, what are the lines along which MaxEnt reasoning works (and which hopefully can explain MaxEnt reasoning to humans)? In (Kern-Isberner, 1998a, 2001), an algebraic framework of so-called *conditional structures* was set up that was based on combinatorial group theory (Kern-Isberner and Rosenberger, 1984; Lyndon and Schupp, 1977) and was shown to provide the conditional-logical structures that can be used to characterize MaxEnt reasoning. Conditional structures can be associated with each probabilistic conditional knowledge base by exploiting the logical interactions of conditionals, but are completely independent of probabilities (hence algebraic or qualitative, respectively). They are elements of a free abelian group, and the MaxEnt distribution appertaining to a knowledge base is shown to satisfy an invariance property that can be stated in terms of a relation within the group (for more details, please see (Kern-Isberner, 1998a, 2001)). Even more — using this algebraic foundation of MaxEnt reasoning and the theory of Gröbner bases, we were able to show how MaxEnt probabilities can be computed symbolically (Kern-Isberner *et al.*, 2014a,b, 2015).

In this paper, we will now apply these ideas from (Kern-Isberner *et al.*, 2014a,b, 2015) to a first-order probabilistic framework. Actually, combining first-order logic and probability theory has been a challenge within the area of knowledge representation for at least two decades, but there is still not a standard theory. Instead, there are lots of different approaches (Bacchus *et al.*, 1996; Beierle *et al.*, 2010; De Raedt and Kersting, 2003; Getoor and Taskar, 2007; Halpern, 1990; Jaeger, 2005; Richardson and Domingos, 2006), for some overview and comparison, please refer to (Beierle *et al.*,

2011), but the topic of a general semantics in logical terms has been addressed very rarely (cf., e.g., (Kern-Isberner and Thimm, 2010)). To such a semantics, the MaxEnt principle can be applied, and the theory of conditional structures can be effective. We describe conditional structures for first-order (resp. relational[2]) probabilistic knowledge bases and show how they help structure probabilistic reasoning under the aggregating semantics (Kern-Isberner and Thimm, 2010) and simplify the view on MaxEnt reasoning in this semantical environment. Then we show that the conditionals in the knowledge base induce polynomials, and how the numerical realizations of their conditional structures can be realized within the affine variety defined by these polynomials. Also inferences are possible: Given a relational conditional, we might ask with which probability this query conditional holds in the MaxEnt model of the knowledge base. We provide a theorem that proves that the queried probability is a point in the affine variety of the Gröbner basis appertaining to an extended set of polynomials.

The rest of the paper is organized as follows: In Section 2, the basic logical notations are set and the aggregating semantics as well as the principle of maximum entropy are recalled. A closer insight into the conditional structure of the represented knowledge together with characterizations for the aggregating semantics and the principle of maximum entropy in terms of equivalence classes of possible worlds follow in Section 3. In Section 4, the relevant basics of elimination theory for applying it to MaxEnt reasoning are given. A polynomial representation of the MaxEnt distribution of relational conditional knowledge is developed in Section 5, and Section 6 finally shows how to draw MaxEnt inferences from this knowledge using symbolic computation. We illustrate the presented techniques at the end of Section 6 in an extensive example before we conclude.

## 2. Representing Probabilistic Conditional Knowledge

We consider a first-order language $\mathcal{L}_\Sigma$ over the signature $\Sigma = \langle \mathsf{Pred}, \mathcal{D} \rangle$ consisting of a finite set of predicate symbols $\mathsf{Pred}$, a finite set of constants $\mathcal{D}$, and without functions of arity greater than zero. Formulas in $\mathcal{L}_\Sigma$ shall be built in the usual way using the logical connectives negation ($\neg$), conjunction ($\wedge$), disjunction ($\vee$), and quantifiers ($\forall, \exists$). To improve

---

[2]We do not use full first-order predicate logic but build in the conditional operator which goes beyond that, so we prefer the term *relational* instead of *first-order*.

readability, we write $AB$ instead of $A \wedge B$ and $\overline{A}$ instead of $\neg A$ for formulas $A, B$. Variables are denoted with uppercase, constants with lowercase letters, and vectors of those are written in bold type. A formula $A$ is called a *closed formula* iff it does not contain free variables, i.e., all variables in $A$ are bound by quantifiers. If $A$ is just a predicate applied to a tuple of constants, $A$ is called a *ground atom*. By substituting all free variables in a formula $A$ with constants from $\mathcal{D}$, we get a *ground instantiation* of $A$.

We define a *probabilistic conditional language* by

$$(\mathcal{L}_\Sigma | \mathcal{L}_\Sigma)^{prob} = \{(B|A)[\xi] \mid A, B \in \mathcal{L}_\Sigma, \ \xi \in [0,1]\}.$$

Elements in $(\mathcal{L}_\Sigma | \mathcal{L}_\Sigma)^{prob}$ are called *(probabilistic) conditionals* and can be understood as expressions of the form "If $A$ is true, then $B$ holds with probability $\xi$". When we write $(B(\vec{X})|A(\vec{X}))[\xi]$, $\vec{X}$ is the vector of all free variables occuring in $A$ or $B$. A conditional $(B|A)[\xi]$ is called a *closed conditional* iff both $A$ and $B$ are closed. Ground instantiations of conditionals are defined analogously to those of formulas, and we denote the set of all ground instantiations of a conditional $(B|A)[\xi]$ with $\text{ground}_{\mathcal{D}}((B|A)[\xi])$. A finite set of (not necessarily closed) conditionals

$$\mathcal{KB} = \{(B_1|A_1)[\xi_1], \ldots, (B_n|A_n)[\xi_n]\} \subseteq (\mathcal{L}_\Sigma | \mathcal{L}_\Sigma)^{prob}$$

is called a *knowledge base*.

The semantics for $(\mathcal{L}_\Sigma | \mathcal{L}_\Sigma)^{prob}$ is based upon probability distributions over possible worlds. Here, possible worlds are Herbrand interpretations which are subsets of all ground atoms that can be built up from Pred and $\mathcal{D}$. We denote the set of all possible worlds with $\Omega_{\mathcal{D}}$. A possible world $\omega \in \Omega_{\mathcal{D}}$ satisfies a ground atom $A$, in symbols $\omega \models A$, iff $A \in \omega$. The satisfaction relation $\models$ is extended to closed formulas in the usual way.

Let $\mathcal{P} : \Omega_{\mathcal{D}} \to [0,1]$ be a probability distribution over the set of possible worlds $\Omega_{\mathcal{D}}$. $\mathcal{P}$ is extended to closed formulas $A$ by $\mathcal{P}(A) = \sum_{\omega \models A} \mathcal{P}(\omega)$. The entailment of closed conditionals $(B|A)[\xi]$ by $\mathcal{P}$ is straightforward by means of conditional probabilities, whereas the entailment of arbitrary conditionals by $\mathcal{P}$ allows for clearance due to the ambiguous interpretation of conditionals with free variables. The naive approach of grounding all conditionals with respect to $\mathcal{D}$ and treating them similar to closed conditionals usually results in inconsistency and does not reflect the essence of first-order logic. Instead, we refer to the aggregating semantics (Kern-Isberner and Thimm, 2010) which mimicks a statistical interpretation of the conditionals from a subjective point of view.

**Definition 1 (Aggregating Semantics).** Let    $(B(\vec{X})|A(\vec{X}))[\xi]$    $\in$ $(\mathcal{L}_\Sigma|\mathcal{L}_\Sigma)^{prob}$ be a conditional, and let $\mathcal{P}$ be a probability distribution over $\Omega_\mathcal{D}$. The entailment relation $\models_\odot$ is defined by $\mathcal{P} \models_\odot (B(\vec{X})|A(\vec{X}))[\xi]$ iff

$$\frac{\displaystyle\sum_{(B(\vec{a})|A(\vec{a}))[\xi]\in\text{ground}_\mathcal{D}((B(\vec{X})|A(\vec{X}))[\xi])} \mathcal{P}(A(\vec{a})B(\vec{a}))}{\displaystyle\sum_{(B(\vec{a})|A(\vec{a}))[\xi]\in\text{ground}_\mathcal{D}((B(\vec{X})|A(\vec{X}))[\xi])} \mathcal{P}(A(\vec{a}))} = \xi \qquad (1)$$

and    $\displaystyle\sum_{(B(\vec{a})|A(\vec{a}))[\xi]\in\text{ground}_\mathcal{D}((B(\vec{X})|A(\vec{X})[\xi])} \mathcal{P}(A(\vec{a})) > 0.$

If $(B(\vec{X})|A(\vec{X}))[\xi]$ is a closed conditional, Equation (1) reduces to the conditional probability $\mathcal{P}(B|A) = \mathcal{P}(AB)/\mathcal{P}(A) = \xi$. If $\mathcal{P}$ is the uniform distribution, i.e., $\mathcal{P}(\omega) = |\Omega_\mathcal{D}|^{-1}$ for all $\omega \in \Omega_\mathcal{D}$, the aggregating semantics just leads up to counting ground instantiations and therefore is purely statistical.

A probability distribution $\mathcal{P}$ entails a knowledge base $\mathcal{KB}$ iff $\mathcal{P}$ entails every conditional in $\mathcal{KB}$. If such a probability distribution exists, it is called a *model* of $\mathcal{KB}$ and the knowledge base $\mathcal{KB}$ is called *consistent*. For consistent knowledge bases there typically exist several models. In order to use inductively the information in $\mathcal{KB}$, it is very helpful to choose a certain model of $\mathcal{KB}$. The *principle of maximum entropy (so-called MaxEnt principle)* (Kern-Isberner, 1998b; Paris, 1994; Paris and Vencovská, 1990) provides such a unique model which is preferable from a logical point of view (Paris, 1999). The idea is to fulfill the conditionals stated in the knowledge base exactly while assuming the least amount of information, i.e., to be as cautious as possible (Jaynes, 1983a; Shore and Johnson, 1980). This is also known as the paradigm of informational economy (Gärdenfors, 1988). Formally, the so called MaxEnt distribution is the model of $\mathcal{KB}$ with maximal *entropy* $\mathcal{H}(\mathcal{P}) = -\sum_{\omega\in\Omega_\mathcal{D}} \mathcal{P}(\omega) \log \mathcal{P}(\omega)$.

**Definition 2 (MaxEnt Distribution).** Let $\mathcal{KB}$ be a consistent knowledge base. Then

$$\mathcal{P}_{\mathsf{ME}}(\mathcal{KB}) = \arg \max_{\mathcal{P}\models\mathcal{KB}} - \sum_{\omega\in\Omega_\mathcal{D}} \mathcal{P}(\omega) \log \mathcal{P}(\omega) \qquad (2)$$

is called the *MaxEnt distribution of* $\mathcal{KB}$. In particular, $\mathcal{P}_{\mathsf{ME}}(\mathcal{KB})$ is a model of $\mathcal{KB}$.

We are now able to define a nonmonotonic inference relation based on the MaxEnt distribution. A conditional $(B|A)[\xi]$ can be ME-inferred from a knowledge base $\mathcal{KB}$, written $\mathcal{KB} \mathrel{|\!\!\sim^{\mathsf{ME}}_{\odot}} (B|A)[\xi]$, iff $\mathcal{P}_{\mathsf{ME}}(\mathcal{KB}) \models_{\odot} (B|A)[\xi]$.

Determining the MaxEnt distribution according to Equation (2) requires taking every single possible world into account. The same holds true for applying the aggregating semantics since the probabilities of the possible worlds are included in Equation (1) implicitly. As the number of possible worlds grows double exponentially ($|\Omega_{\mathcal{D}}| = \prod_{A \in \mathsf{Pred}} 2^{\hat{}}(|\mathcal{D}|^{\hat{}}\mathsf{ar}(A))$ where $\mathsf{ar}(A)$ ist the arity of $A$) this makes computations expensive. In the next section, we recall how the concept of conditional structure can be used to formulate characterizations for the aggregating semantics and the principle of maximum entropy which can decrease this blow-up.

## 3. Exploiting the Conditional Structure

In (Kern-Isberner, 2001) the concept of conditional structure is introduced which reflects the interpretation of propositional conditionals within a possible world. Formally, the conditional structure of a possible world $\omega$ is given by a product in a free abelian group $\mathcal{F} \simeq \mathbb{Z}^{2n}$ with generators $a_i^+, a_i^-$ where $a_i^+$ indicates that $\omega$ verifies the $i$-th conditional in a propositional knowledge base and $a_i^-$ indicates that $\omega$ falsifies it. Thus, each conditional is associated with two algebraic symbols that allow for symbolic computations. Following the explanations in (Finthammer and Beierle, 2012; Kern-Isberner and Thimm, 2012), we show how the concept of conditional structure carries over to the first-order case, and we reformulate the aggregating semantics and the MaxEnt distribution in terms of equivalence classes of possible worlds based on their conditional structure.

**Definition 3 (Conditional Structure).** Let the knowledge base

$$\mathcal{KB} = \{(B_1(\vec{X}_1)|A_1(\vec{X}_1))[\xi_1], \dots, (B_n(\vec{X}_n)|A_n(\vec{X}_n))[\xi_n]\}$$

be given, and let $(B_i(\vec{a}_i)|A_i(\vec{a}_i))[\xi_i]$ be a proper ground instantiation of the $i$-th conditional in $\mathcal{KB}$. The *conditional structure* $\sigma_{\mathcal{KB}}(\omega)$ of a possible world $\omega \in \Omega_{\mathcal{D}}$ is a group element of $\mathcal{F}$ and is stepwise defined by

$$\sigma_{i,\vec{a}_i}^{\mathcal{KB}}(\omega) = \begin{cases} a_i^+ & \text{iff } \omega \models A_i(\vec{a}_i)B_i(\vec{a}_i) \\ a_i^- & \text{iff } \omega \models A_i(\vec{a}_i)\overline{B_i(\vec{a}_i)} \\ 1 & \text{iff } \omega \models \overline{A_i(\vec{a}_i)} \end{cases},$$

$$\sigma_i^{\mathcal{KB}}(\omega) = \prod_{\substack{\vec{a}_i:(B_i(\vec{a}_i)|A_i(\vec{a}_i))[\xi_i] \\ \in \text{ground}_{\mathcal{D}}((B_i(\vec{X}_i)|A_i(\vec{X}_i))[\xi_i])}} \sigma_{i,\vec{a}_i}^{\mathcal{KB}}(\omega),$$

$$\sigma_{\mathcal{KB}}(\omega) = \prod_{i=1}^{n} \sigma_i^{\mathcal{KB}}(\omega).$$

The conditional structure induces an equivalence relation on $\Omega_{\mathcal{D}}$. Two possible worlds $\omega$ and $\omega'$ are equivalent, written $\omega \equiv_{\mathcal{KB}} \omega'$, iff $\sigma_{\mathcal{KB}}(\omega) = \sigma_{\mathcal{KB}}(\omega')$. The equivalence class $[\omega]_{\mathcal{KB}} = \{\omega' \in \Omega_{\mathcal{D}} \mid \omega' \equiv_{\mathcal{KB}} \omega\}$ is the set of all possible worlds with the same conditional structure as $\omega$. $\Omega_{\mathcal{D}}/{\equiv_{\mathcal{KB}}} = \{[\omega]_{\mathcal{KB}} \mid \omega \in \Omega_{\mathcal{D}}\}$ denotes the set of all such equivalence classes.

The belonging of a possible world to an equivalence class is determined by tuples built of any two of the following counting functions.

**Definition 4 (Counting Functions).** The *counting functions* $v_i$, $f_i$, and $a_i$ that map possible worlds to natural numbers including zero with respect to the conditional $r_i = (B_i(\vec{X}_i)|A_i(\vec{X}_i))[\xi_i] \in (\mathcal{L}_{\Sigma}|\mathcal{L}_{\Sigma})^{prob}$ are defined by

$$v_i(\omega) = |\{(B_i(\vec{a}_i)|A_i(\vec{a}_i))[\xi_i] \in \text{ground}_{\mathcal{D}}(r_i) \mid \omega \models A_i(\vec{a}_i)B_i(\vec{a}_i)\}|,$$

$$f_i(\omega) = |\{(B_i(\vec{a}_i)|A_i(\vec{a}_i))[\xi_i] \in \text{ground}_{\mathcal{D}}(r_i) \mid \omega \models A_i(\vec{a}_i)\overline{B_i(\vec{a}_i)}\}|,$$

$$a_i(\omega) = v_i(\omega) + f_i(\omega).$$

They count the ground instantiations of $r_i$ that are verified ($v_i(\omega)$), falsified ($f_i(\omega)$), and applicable ($a_i(\omega)$) within the possible world $\omega \in \Omega_{\mathcal{D}}$.

The counting functions $v_i$ and $f_i$ lead to the notation

$$\sigma_{\mathcal{KB}}(\omega) = \prod_{i=1}^{n} (a_i^+)^{v_i(\omega)} (a_i^-)^{f_i(\omega)}$$

of the conditional structure of a possible world $\omega$ and $\omega \equiv_{\mathcal{KB}} \omega'$ holds iff $(v_i(\omega), f_i(\omega)) = (v_i(\omega'), f_i(\omega'))$ for all $i = 1, \ldots, n$. In particular, we may write $v_i([\omega_{\mathcal{KB}}]) = v_i(\omega)$ since all possible worlds within a given equivalence class verify the same number of ground instantiations of the conditionals in $\mathcal{KB}$. The same holds true for $f_i$ and $a_i$ as well.

In terms of conditional structures the entailment of a knowledge base under the aggregating semantics can now be characterized as follows.

**Proposition 1.** *Let* $\mathcal{KB} = \{(B_1|A_1)[\xi_1], \ldots, (B_n|A_n)[\xi_n]\}$ *be a knowledge base, and let* $\mathcal{P}$ *be a probability distribution over* $\Omega_{\mathcal{D}}$. *Then*

$$\mathcal{P} \models_{\odot} \mathcal{KB} \quad \textit{iff} \quad \forall i = 1, \ldots, n : \frac{\displaystyle\sum_{[\omega]_{\mathcal{KB}} \in \Omega_{\mathcal{D}}/\equiv_{\mathcal{KB}}} v_i([\omega]_{\mathcal{KB}}) \, \mathcal{P}([\omega]_{\mathcal{KB}})}{\displaystyle\sum_{[\omega]_{\mathcal{KB}} \in \Omega_{\mathcal{D}}/\equiv_{\mathcal{KB}}} a_i([\omega]_{\mathcal{KB}}) \, \mathcal{P}([\omega]_{\mathcal{KB}})} = \xi_i.$$

(3)

**Proof.** The probability distribution $\mathcal{P}$ entails $\mathcal{KB}$ iff every conditional in $\mathcal{KB}$ is entailed by $\mathcal{P}$. From the aggregating semantics (cf. Definition 1), it follows that

$$\mathcal{P} \models_{\odot} \mathcal{KB} \quad \text{iff} \quad \frac{\displaystyle\sum_{(B_i(\vec{a}_i)|A_i(\vec{a}_i))[\xi_i] \in \mathsf{ground}_{\mathcal{D}}((B_i(\vec{X}_i)|A_i(\vec{X}_i))[\xi_i])} \mathcal{P}(A_i(\vec{a}_i)B_i(\vec{a}_i))}{\displaystyle\sum_{(B_i(\vec{a}_i)|A_i(\vec{a}_i))[\xi_i] \in \mathsf{ground}_{\mathcal{D}}((B_i(\vec{X}_i)|A_i(\vec{X}_i))[\xi_i])} \mathcal{P}(A_i(\vec{a}_i))} = \xi_i$$

for $i = 1, \ldots, n$ which holds iff

$$\frac{\displaystyle\sum_{(B_i(\vec{a}_i)|A_i(\vec{a}_i))[\xi_i] \in \mathsf{ground}_{\mathcal{D}}((B_i(\vec{X}_i)|A_i(\vec{X}_i))[\xi_i])} \sum_{\omega \models A_i(\vec{a}_i)B_i(\vec{a}_i)} \mathcal{P}(\omega)}{\displaystyle\sum_{(B_i(\vec{a}_i)|A_i(\vec{a}_i))[\xi_i] \in \mathsf{ground}_{\mathcal{D}}((B_i(\vec{X}_i)|A_i(\vec{X}_i))[\xi_i])} \sum_{\omega \models A_i(\vec{a}_i)} \mathcal{P}(\omega)} = \xi_i$$

for $i = 1, \ldots, n$. Thus, we have to investigate for every ground instance of $(B_i(\vec{X}_i)|A_i(\vec{X}_i))[\xi_i]$ in which possible worlds the particular ground instance is verified or applicable, respectively. By changing perspective, we can equivalently determine for every possible world which ground instances of $(B_i(\vec{X}_i)|A_i(\vec{X}_i))[\xi_i]$ are verified or applicable within the particular possible world. Thereby, $\mathcal{P} \models_{\odot} \mathcal{KB}$ holds iff

$$\xi_i = \frac{\displaystyle\sum_{\omega \in \Omega_{\mathcal{D}}} \sum_{\substack{(B_i(\vec{a}_i)|A_i(\vec{a}_i))[\xi_i] \in \mathsf{ground}_{\mathcal{D}}((B_i(\vec{X}_i)|A_i(\vec{X}_i))[\xi_i]) \\ \omega \models A_i(\vec{a}_i)B_i(\vec{a}_i)}} \mathcal{P}(\omega)}{\displaystyle\sum_{\omega \in \Omega_{\mathcal{D}}} \sum_{\substack{(B_i(\vec{a}_i)|A_i(\vec{a}_i))[\xi_i] \in \mathsf{ground}_{\mathcal{D}}((B_i(\vec{X}_i)|A_i(\vec{X}_i))[\xi_i]) \\ \omega \models A_i(\vec{a}_i)}} \mathcal{P}(\omega)}$$

$$= \frac{\displaystyle\sum_{\omega \in \Omega_{\mathcal{D}}} v_i(\omega) \, \mathcal{P}(\omega)}{\displaystyle\sum_{\omega \in \Omega_{\mathcal{D}}} a_i(\omega) \, \mathcal{P}(\omega)} = \frac{\displaystyle\sum_{[\omega]_{\mathcal{KB}} \in \Omega_{\mathcal{D}}/\equiv_{\mathcal{KB}}} v_i([\omega]_{\mathcal{KB}}) \, \mathcal{P}([\omega]_{\mathcal{KB}})}{\displaystyle\sum_{[\omega]_{\mathcal{KB}} \in \Omega_{\mathcal{D}}/\equiv_{\mathcal{KB}}} a_i([\omega]_{\mathcal{KB}}) \, \mathcal{P}([\omega]_{\mathcal{KB}})}$$

for $i = 1, \ldots, n$ which completes the proof. $\qquad\square$

Proposition 1 shows how conditional structures can help structure probabilistic reasoning and thus leads to reducing complexity.

We conclude this section by giving a characterization for the MaxEnt distribution in terms of conditional structures. The optimization problem (2) can be solved by well-known Lagrangian multiplier techniques which results in the following proposition (Jaynes, 1983b; Kern-Isberner, 2001).

**Proposition 2.** *Let* $\mathcal{KB} = \{(B_1|A_1)[\xi_1], \ldots, (B_n|A_n)[\xi_n]\}$ *be a consistent knowledge base. Then*

$$\mathcal{P}_{\mathsf{ME}}(\mathcal{KB})(\omega) = \frac{\mathcal{P}_{\mathsf{ME}}(\mathcal{KB})([\omega]_{\mathcal{KB}})}{|[\omega]_{\mathcal{KB}}|} = \alpha_0 \prod_{i=1}^{n} \alpha_i^{v_i([\omega]_{\mathcal{KB}}) - \xi_i a_i([\omega]_{\mathcal{KB}})} \qquad (4)$$

*where*

$$\alpha_0 = \left( \sum_{[\omega]_{\mathcal{KB}} \in \Omega_D / \equiv_{\mathcal{KB}}} \prod_{i=1}^{n} \alpha_i^{v_i([\omega]_{\mathcal{KB}}) - \xi_i a_i([\omega]_{\mathcal{KB}})} \right)^{-1} \qquad (5)$$

*is a normalizing constant, and the* $\alpha_i$ *'s are so-called effects and fulfill the adjustment condition*

$$\sum_{[\omega]_{\mathcal{KB}} \in \Omega_D / \equiv_{\mathcal{KB}}} (v_i([\omega]_{\mathcal{KB}}) - \xi_i a_i([\omega]_{\mathcal{KB}})) \, |[\omega]_{\mathcal{KB}}|$$

$$\times \prod_{j=1}^{n} \alpha_j^{v_j([\omega]_{\mathcal{KB}}) - \xi_j a_j([\omega]_{\mathcal{KB}})} = 0 \qquad (6)$$

*as well as the positivity condition*

$$\alpha_i \begin{cases} > 0 & \text{iff } \xi_i \in (0,1) \\ = \infty & \text{iff } \xi_i = 1 \\ = 0 & \text{iff } \xi_i = 0 \end{cases} \qquad (7)$$

*for* $i = 1, \ldots, n$. *Here, the conventions* $\infty^0 = 1$, $\infty^{-1} = 0$, *and* $0^0 = 1$ *are taken into account.*

**Proof.** We sketch the use of Lagrangian mupliplier techniques (Boyd and Vandenberghe, 2004) in order to determine the MaxEnt distribution $\mathcal{P}_{\mathsf{ME}}(\mathcal{KB})$ by maximizing the entropy as in Equation (2). This leads to the

Lagrangian function

$$\Lambda = \sum_{\omega \in \Omega_{\mathcal{D}}} \mathcal{P}(\omega) \log \mathcal{P}(\omega) + \lambda_0 \left( \sum_{\omega \in \Omega_{\mathcal{D}}} \mathcal{P}(\omega) - 1 \right) + \sum_{i=1}^{n} \lambda_i$$

$$\times \left( \sum_{[\omega]_{\mathcal{KB}} \in \Omega_{\mathcal{D}}/\equiv_{\mathcal{KB}}} v_i([\omega]_{\mathcal{KB}}) \, \mathcal{P}([\omega]_{\mathcal{KB}}) \right.$$

$$\left. - \xi_i \sum_{[\omega]_{\mathcal{KB}} \in \Omega_{\mathcal{D}}/\equiv_{\mathcal{KB}}} a_i([\omega]_{\mathcal{KB}}) \, \mathcal{P}([\omega]_{\mathcal{KB}}) \right)$$

with the Lagrangian multipliers $\lambda_0, \lambda_1, \ldots, \lambda_n$. Here, the last sum arises from the side condition guaranteeing that $\mathcal{P}$ entails $\mathcal{KB}$, and $\sum_{\omega \in \Omega_{\mathcal{D}}} \mathcal{P}(\omega) - 1 = 0$ is the normalizing constraint for $\mathcal{P}$ to be a probability distribution. By differentiating $\Lambda$ with respect to the probability of an arbitrary possible world $\omega' \in \Omega_{\mathcal{D}}$, we get

$$\frac{\partial \Lambda}{\partial \mathcal{P}(\omega')} = \log \mathcal{P}(\omega') + 1 + \lambda_0 + \sum_{i=1}^{n} \lambda_i (v_i([\omega']_{\mathcal{KB}}) - \xi_i a_i([\omega']_{\mathcal{KB}})).$$

Thus, as $\partial \Lambda / \partial \mathcal{P}(\omega')$ has to vanish if $\mathcal{P} = \mathcal{P}_{\mathsf{ME}}(\mathcal{KB})$,

$$\mathcal{P}_{\mathsf{ME}}(\mathcal{KB})(\omega') = \exp(-(1 + \lambda_0))$$

$$\times \exp \left( -\sum_{i=1}^{n} \lambda_i (v_i([\omega']_{\mathcal{KB}}) - \xi_i a_i([\omega']_{\mathcal{KB}})) \right)$$

must hold. By substituting $\alpha_0 = \exp(-1(1 + \lambda_0))$, which will be the normalizing constant, and the effects $\alpha_i = \exp(-\lambda_i)$ for $i = 1, \ldots, n$,

$$\mathcal{P}_{\mathsf{ME}}(\mathcal{KB})(\omega') = \alpha_0 \prod_{i=1}^{n} \alpha_i^{v_i([\omega']_{\mathcal{KB}}) - \xi_i a_i([\omega']_{\mathcal{KB}})}$$

follows. The value of the normalizing constant (5) can be derived by summing up (4) over all possible worlds and setting it to 1. The positivity condition for the effects $\alpha_i$ is a consequence of $\alpha_i = \exp(-\lambda_i) > 0$ for $\lambda_i \in \mathbb{R}$ and $i = 1, \ldots, n$. Notably, the probability of possible worlds within the same equivalence class is the same, as the right hand side of (3) only depends on the equivalence class of the considered possible world and not

on the possible world itself. Hence, $\mathcal{P}_{\mathsf{ME}}(\mathcal{KB})(\omega) \, |[\omega]_{\mathcal{KB}}| = \mathcal{P}_{\mathsf{ME}}(\mathcal{KB})([\omega]_{\mathcal{KB}})$
holds for all $\omega \in \Omega_{\mathcal{D}}$. Substituting this into (3) yields

$$\xi_i = \frac{\displaystyle\sum_{[\omega]_{\mathcal{KB}} \in \Omega_{\mathcal{D}}/\equiv_{\mathcal{KB}}} v_i([\omega]_{\mathcal{KB}}) \, |[\omega]_{\mathcal{KB}}| \prod_{j=1}^{n} \alpha_j^{v_j([\omega]_{\mathcal{KB}}) - \xi_j a_j([\omega]_{\mathcal{KB}})}}{\displaystyle\sum_{[\omega]_{\mathcal{KB}} \in \Omega_{\mathcal{D}}/\equiv_{\mathcal{KB}}} a_i([\omega]_{\mathcal{KB}}) \, |[\omega]_{\mathcal{KB}}| \prod_{j=1}^{n} \alpha_j^{v_j([\omega]_{\mathcal{KB}}) - \xi_j a_j([\omega]_{\mathcal{KB}})}}$$

for $i = 1, \ldots, n$ which results in (6) after an easy conversion.    □

Hence, the nonlinear equation system (6) has to be solved in order
to observe the MaxEnt distribution $\mathcal{P}_{\mathsf{ME}}(\mathcal{KB})$. It cannot be expected that
there is an exact solution procedure in general and iterative algorithms are
used (Darroch and Ratcliff, 1972; Finthammer and Beierle, 2012). On the
contrary, it is often not necessary to derive the whole MaxEnt distribution
for answering queries but one can exploit the algebraic structure of the
knowledge base to circumscribe possible inferences. In the next sections we
will recall some basics of elimination theory and utilize them for drawing
inferences.

## 4.  Basics of Elimination Theory

In this section we follow the explanations in (Cox *et al.*, 2015) and recall the
main results of elimination theory that will be necessary to apply symbolic
computation to MaxEnt reasoning. We mainly consider the polynomial ring
$\mathbb{R}[y_1, \ldots, y_n]$ in the variables $y_1, \ldots, y_n$ over the field of real numbers $\mathbb{R}$.
For the purpose of drawing inferences based on algebraic representations of
conditional knowledge, it is useful to define a distinct variable, as we will see
in Section 6. Therefore, we sometimes make use of an extended polynomial
ring $\mathbb{R}[y_1, \ldots, y_n, y_{\mathsf{inf}}]$, too. Although we formulate the following results with
respect to $\mathbb{R}[y_1, \ldots, y_n]$, they obviously hold in $\mathbb{R}[y_1, \ldots, y_n, y_{\mathsf{inf}}]$ as well.

**Definition 5 (Affine Variety).** Let $f_1, \ldots, f_m$ be polynomials in
$\mathbb{R}[y_1, \ldots, y_n]$. The set

$$\mathcal{V}(f_1, \ldots, f_m) = \{(\gamma_1, \ldots, \gamma_n) \in \mathbb{R}^n \mid f_i(\gamma_1, \ldots, \gamma_n)$$
$$= 0 \; \forall i = 1, \ldots, m\}$$

is called the *affine variety* defined by $f_1, \ldots, f_m$.

Hence, the affine variety $\mathcal{V}(f_1, \ldots, f_m)$ is the set of all (real) solutions
of the polynomial equation system $f_i(y_1, \ldots, y_n) = 0$ for $i = 1, \ldots, m$.

**Definition 6 (Ideal).** A subset $\mathcal{I} \in \mathbb{R}[y_1, \ldots, y_n]$ is called an *ideal* if

1. $0 \in \mathcal{I}$.
2. If $f, g \in \mathcal{I}$, then $f + g \in \mathcal{I}$.
3. If $f \in \mathcal{I}$ and $h \in \mathbb{R}[y_1, \ldots, y_n]$, then $h \cdot f \in \mathcal{I}$.

Further, let $f_1, \ldots, f_m \in \mathbb{R}[y_1, \ldots, y_n]$. Then

$$\langle f_1, \ldots, f_m \rangle = \left\{ \sum_{i=1}^{m} h_i \cdot f_i \mid \forall i = 1, \ldots, m : h_i \in \mathbb{R}[y_1, \ldots, y_n] \right\} \quad (8)$$

is called the *ideal generated by* $f_1, \ldots, f_m$ (and indeed is an ideal).

The well-known Hilbert's Basis Theorem (Hilbert, 1890) guarantees that every ideal in $\mathbb{R}[y_1, \ldots, y_n]$ is of the form (8). Furthermore, it is obvious that $(f + g)(\gamma_1, \ldots, \gamma_n) = (h \cdot f)(\gamma_1, \ldots, \gamma_n) = 0$ if $f(\gamma_1, \ldots, \gamma_n) = g(\gamma_1, \ldots, \gamma_n) = 0$ and $h \in \mathbb{R}[y_1, \ldots, y_n]$. As a consequence, we may define

$$\mathcal{V}(\mathcal{I}) = \{(\gamma_1, \ldots, \gamma_n) \in \mathbb{R}^n \mid f(\gamma_1, \ldots, \gamma_n) = 0 \; \forall f \in \mathcal{I}\}$$

for ideals $\mathcal{I}$, and $\mathcal{V}(\langle f_1, \ldots, f_m \rangle) = \mathcal{V}(f_1, \ldots, f_m)$ holds. Thus, we may speak of the variety defined by an ideal. In other words, $V(g_1, \ldots, g_t) = V(f_1, \ldots, f_m)$ for any generating set $\{g_1, \ldots, g_t\}$ of $\langle f_1, \ldots, f_m \rangle$, and therefore, different generating sets correspond to different representations of the same affine variety. Before exposing Gröbner bases as generating sets with distinguished properties for further computations, we need to specify certain orderings on the set of terms

$$\mathcal{T} = \left\{ y_1^{s_1} \cdots y_n^{s_n} \mid s_1, \ldots, s_n \in \mathbb{N}_0 \right\}.$$

**Definition 7 (Term Ordering).** Let $\preceq$ be a total ordering on $\mathcal{T}$. Then $\preceq$ is a *term ordering* if the following two conditions hold:

1. $1 = y_1^0 \cdots y_n^0 \preceq t$ for all $t \in \mathcal{T}$,
2. $t_1 \preceq t_2 \Rightarrow t \cdot t_1 \preceq t \cdot t_2$ for all $t, t_1, t_2 \in \mathcal{T}$.

An important class of term orderings are the so-called *lexicographic term orderings*. A term ordering $\preceq$ on $\mathcal{T}$ is lexicographic when $y_1^{e_1} \cdots y_n^{e_n} \prec y_1^{f_1} \cdots y_n^{f_n}$ holds iff the first non-zero entry in $(f_1 - e_1, \ldots, f_n - e_n)$ is positive. Permutting the indices of the variables generates further lexicographic term orderings. Overall, there are $n!$ different lexicographic term orderings on $\mathcal{T}$. Term orderings allow for the definition of leading coefficients and leading monomials of polynomials.

**Definition 8 (Leading Coefficient/Monomial).** Let $\preceq$ be a fixed term ordering on $\mathcal{T}$, and let $f \in \mathbb{R}[y_1, \ldots, y_n]$ be a polynomial. Without loss of generality, let $f = \sum_{i=1}^{m} c_i \cdot t_i$ with coefficients $c_i \in \mathbb{R} \setminus \{0\}$ and terms $t_i \in \mathcal{T}$ for $i = 1, \ldots, m$, and let $t_1 \prec \cdots \prec t_m$. The *leading coefficient* of $f$ is the coefficient of the greatest term occuring in $f$ with respect to $\preceq$, i.e., $\mathsf{lc}_{\preceq}(f) = c_m$. The *leading monomial* of $f$ is $\mathsf{lm}_{\preceq}(f) = c_m \cdot t_m$. For a subset $\mathcal{F} \in \mathbb{R}[y_1, \ldots, y_n]$, we denote the set of the leading monomials of all polynomials in $\mathcal{F}$ by $\mathsf{lm}_{\preceq}(\mathcal{F})$.

Now we have all formal prerequisites to recall the definition of Gröbner bases (Buchberger, 1965, 2006).

**Definition 9 (Gröbner Basis).** Let $\preceq$ be a fixed term ordering. A finite subset $\mathcal{G}$ of an ideal $\mathcal{I}$ is called a *Gröbner basis* of $\mathcal{I}$ with respect to $\preceq$ if

$$\langle \mathsf{lm}_{\preceq}(\mathcal{G}) \rangle = \langle \mathsf{lm}_{\preceq}(\mathcal{I}) \rangle.$$

$\mathcal{G}$ is a *minimal Gröbner basis* if in addition for all $g \in \mathcal{G}$:

1. $\mathsf{lc}_{\preceq}(g) = 1$,
2. $\mathsf{lm}_{\preceq}(g) \notin \langle \mathsf{lm}_{\preceq}(\mathcal{G} \setminus \{g\}) \rangle$.

As a consequence of the Hilbert's Basis Theorem, every ideal $\mathcal{I} \neq \{0\}$ in $\mathbb{R}[y_1, \ldots, y_n]$ has a unique minimal Gröbner basis with respect to a given term ordering $\preceq$. We denote this Gröbner basis with $\mathcal{GB}_{\preceq}(\mathcal{I})$. The standard method to calculate Gröbner bases is *Buchberger's algorithm* (Buchberger, 1965, 2006) that is implemented in all current computer algebra systems.

From now on, we consider the extended polynomial ring $\mathbb{R}[y_1, \ldots, y_n, y_{\mathsf{inf}}]$. In the next theorem, we present an alleviated variant of the well-known Elimination Theorem (Cox *et al.*, 2015). The Elimination Theorem uses Gröbner bases to make visible the part of an ideal that consists of polynomials in certain variables only. The remaining polynomials form a Gröbner basis of a so-called *elimination ideal* in a subring of $\mathbb{R}[y_1, \ldots, y_n, y_{\mathsf{inf}}]$. As we do not need the Elimination Theorem in whole generality when we discuss how to apply it to MaxEnt reasoning in Section 6, we restrict ourself to the case where the (unique) elimination ideal $\mathcal{I}_{\mathsf{inf}}$ contains polynomials in the variable $y_{\mathsf{inf}}$ only, i.e., $\mathcal{I}_{\mathsf{inf}}$ lies in the subring $\mathbb{R}[y_{\mathsf{inf}}]$.

**Theorem 1.** *Let $\mathcal{I} \in \mathbb{R}[y_1, \ldots, y_n, y_{\mathsf{inf}}]$ be an ideal. Further, let $\preceq_{\mathsf{lex}}$ be the lexicographic term ordering with $y_{\mathsf{inf}} \preceq_{\mathsf{lex}} y_1 \preceq_{\mathsf{lex}} \cdots \preceq_{\mathsf{lex}} y_n$. Then $\mathcal{GB}_{\preceq_{\mathsf{lex}}}(\mathcal{I}) \cap \mathbb{R}[y_{\mathsf{inf}}]$ is a Gröbner basis of an ideal $\mathcal{I}_{\mathsf{inf}} \subseteq \mathbb{R}[y_{\mathsf{inf}}]$ called the elimination ideal in $y_{\mathsf{inf}}$.*

**Proof.** This theorem is a special case of the Elimination Theorem proven in (Cox *et al.*, 2015). □

Note that the calculation of Gröbner bases with respect to lexicographic term orderings may be very expensive, and Theorem 1 works fine with any elimination term ordering $\preceq_{\mathsf{elim}}$. The crucial point is, that all other variables $y_1, \ldots, y_n$ have to be preferred to $y_{\mathsf{inf}}$ by $\preceq_{\mathsf{elim}}$. We merely chose $\preceq_{\mathsf{lex}}$ for a simpler illustration.

The main benefit of Theorem 1 for our purpose is that there is a connection between the affine varieties $\mathcal{V}(\mathcal{I}_{\mathsf{inf}})$ and $\mathcal{V}(\mathcal{I})$.

**Proposition 3.** *Let $\mathcal{I} \in \mathbb{R}[y_1, \ldots, y_n, y_{\mathsf{inf}}]$ be an ideal, and let $\mathcal{I}_{\mathsf{inf}} \in \mathbb{R}[y_{\mathsf{inf}}]$ be its elimination ideal in the variable $y_{\mathsf{inf}}$. Further, let $(\gamma_1, \ldots, \gamma_n, \gamma_{\mathsf{inf}}) \in \mathcal{V}(\mathcal{I})$. Then the partial solution $\gamma_{\mathsf{inf}}$ lies in $\mathcal{V}(\mathcal{I}_{\mathsf{inf}})$.*

**Proof.** Let $\vec{\gamma} = (\gamma_1, \ldots, \gamma_n, \gamma_{\mathsf{inf}}) \in \mathcal{V}(\mathcal{I})$, and let $\preceq_{\mathsf{lex}}$ be the term ordering from Theorem 1. Obviously, $\vec{\gamma} \in \mathcal{V}(\mathcal{GB}_{\preceq_{\mathsf{lex}}}(\mathcal{I}))$ holds and likewise $\vec{\gamma} \in \mathcal{V}(\mathcal{F})$ for any subset $\mathcal{F} \subseteq \mathcal{GB}_{\preceq_{\mathsf{lex}}}(\mathcal{I})$ since $\mathcal{V}(\mathcal{GB}_{\preceq_{\mathsf{lex}}}(\mathcal{I})) \subseteq \mathcal{V}(\mathcal{F})$. In particular, $\vec{\gamma}$ is in the affine variety of the (largest) set $\mathcal{S}$ of polynomials in $\mathcal{GB}_{\preceq_{\mathsf{lex}}}(\mathcal{I})$ that do not mention any of the variables $y_1, \ldots, y_n$ but are univariate in the variable $y_{\mathsf{inf}}$. Finally $\gamma_{\mathsf{inf}} \in \mathcal{V}(\mathcal{I}_{\mathsf{inf}})$ follows by mapping the polynomials in $\mathcal{S} \subseteq \mathbb{R}[y_1, \ldots, y_n, y_{\mathsf{inf}}]$ to the appropriate polynomials in $\mathbb{R}[y_{\mathsf{inf}}]$. □

Note that the set $\mathcal{S}$ considered in the proof of Proposition 3 actually consists of a single polynomial only. This is due to the fact that $\mathbb{R}[y_{\mathsf{inf}}]$ is a principal ideal domain.

## 5. Compiling Knowledge Bases

We now extract polynomials $f_1, \ldots, f_n$ from the nonlinear equation system (6) such that the calculation of the effects $\alpha_1, \ldots, \alpha_n$ can be performed by locating the affine variety $\mathcal{V}(f_1, \ldots, f_n)$. For this, let $\mathcal{KB} = \{(B_1|A_1)[\xi_1], \ldots, (B_n|A_n)[\xi_n]\}$ be a knowledge base with given probabilities. For the rest of the paper, we assume the probabilities to be rational and non-trivial, i.e., $\xi_i \in (0,1)$ for $i = 1, \ldots, n$. Both, the rationality and the non-triviality, are not heavy restrictions. Using rational numbers is typical in practical applications and favorable for further processing on the computer. Probabilities $\xi_i \in \{0,1\}$ lead to effects $\alpha_i \in \{0, \infty\}$ and can be retained in a kind of preprocessing. We will discuss such a case in Example 1.

Due to the rationality of the probabilities $\xi_i$, there are unique, relatively prime natural numbers $p_i, q_i \in \mathbb{N}$ with $\xi_i = p_i/q_i$ for $i = 1, \ldots, n$. We exploit this fact in order to find the polynomial representation of (6) in the next proposition.

**Proposition 4.** *Let* $\mathcal{KB} = \{(B_1|A_1)[p_1/q_1], \ldots, (B_n|A_n)[p_n/q_n]\}$ *be a consistent knowledge base. There are polynomials* $f_1, \ldots, f_n \in \mathbb{R}[y_1, \ldots, y_n]$ *(in one-to-one correspondence with the conditionals in* $\mathcal{KB}$*) with* $(\gamma_1, \ldots, \gamma_n) \in \mathcal{V}(f_1, \ldots, f_n)$ *such that the effects* $\alpha_i$ *from the nonlinear equation system* (6) *are given by* $\alpha_i = \gamma_i^{q_i}$ *for* $i = 1, \ldots, n$.

**Proof.** We define

$$a_{j,i}^{\max}(\mathcal{KB}) = \max(\{a_j([\omega]_{\mathcal{KB}}) \mid [\omega]_{\mathcal{KB}} \in \Omega_\mathcal{D}/\equiv_{\mathcal{KB}}, a_i([\omega]_{\mathcal{KB}}) > 0\})$$

for $i, j = 1, \ldots, n$. Replacing the probability $\xi_i$ in (6) by $p_i/q_i$ and multiplying both sides of (6) with $q_i \prod_{j=1}^{n} \alpha_j^{\xi_j a_{j,i}^{\max}(\mathcal{KB})} \neq 0$ for $i = 1, \ldots, n$ leads to

$$\sum_{[\omega]_{\mathcal{KB}} \in \Omega_\mathcal{D}/\equiv_{\mathcal{KB}}} (q_i v_i([\omega]_{\mathcal{KB}}) - p_i a_i([\omega]_{\mathcal{KB}})) \, |[\omega]_{\mathcal{KB}}|$$

$$\times \prod_{j=1}^{n} \alpha_j^{v_j([\omega]_{\mathcal{KB}}) + \frac{p_j}{q_j}(a_{j,i}^{\max}(\mathcal{KB}) - a_j([\omega]_{\mathcal{KB}}))} = 0.$$

Substituting $\alpha_i = \gamma_i^{q_i}$ on the left hand side leads to the polynomial expression

$$f_i = \sum_{[\omega]_{\mathcal{KB}} \in \Omega_\mathcal{D}/\equiv_{\mathcal{KB}}} (q_i v_i([\omega]_{\mathcal{KB}}) - p_i a_i([\omega]_{\mathcal{KB}})) \, |[\omega]_{\mathcal{KB}}|$$

$$\times \prod_{j=1}^{n} y_j^{q_j v_j([\omega]_{\mathcal{KB}}) + p_j(a_{j,i}^{\max}(\mathcal{KB}) - a_j([\omega]_{\mathcal{KB}}))} \tag{9}$$

in the variables $y_1, \ldots, y_n$. The existence of $(\gamma_1, \ldots, \gamma_n) \in \mathcal{V}(f_1, \ldots, f_n)$ follows now from the existence of the positive and real effects $\alpha_i$ and thus from the existence of the radicals $\alpha_i^{1/q_i} \in \mathbb{R}$ for $i = 1, \ldots, n$ due to the consistency of $\mathcal{KB}$. $\qquad\square$

Note that the proof of Proposition 4 generates constructively a unique set of polynomials $\mathcal{F}(\mathcal{KB}) = \{f_1, \ldots, f_n\}$ with $f_i$ as given in (9) for $i = 1, \ldots, n$. Although the coefficients of these polynomials are integers, we consider $f_1, \ldots, f_n$ within the polynomial ring over the real numbers since we are interested in real solutions.

As already mentioned, we concentrate on knowledge bases with non-trivial probabilities, and therefore, we may simplify the polynomials $f_1, \ldots, f_n$ as long as we do not lose the point in $\mathcal{V}(f_1, \ldots, f_n)$ that corresponds to the sought effects. Obviously, we may cancel variables $y_1, \ldots, y_n$ in the polynomials since this means discarding those points $(\gamma_1, \ldots, \gamma_n) \in \mathcal{V}(f_1, \ldots, f_n)$ that have at least one zero-entry $\gamma_i = 0$ which therefore corresponds to $\alpha_i = 0$ and thus $\xi_i = 0$. Furthermore, we may cancel polynomial combinations of variables if all of the coefficients of these combinations are positive, for the reason that we are only interested in positive effects and positive effects lead to positive values $\gamma_i = \alpha_i^{1/q_i}$. In the following corollary, the polynomials $f_i^+$ and $f_i^-$ will have positive coefficients, only.

**Corollary 1.** *Let $\mathcal{KB} = \{(B_1|A_1)[p_1/q_1], \ldots, (B_n|A_n)[p_n/q_n]\}$ be a consistent knowledge base, and let $f_1, \ldots, f_n$ be the polynomials defined by (9). Further, let*

$$
f_i^+ = \sum_{[\omega]_{\mathcal{KB}} \in \Omega_\mathcal{D}/\equiv_{\mathcal{KB}}} q_i v_i([\omega]_{\mathcal{KB}}) \, |[\omega]_{\mathcal{KB}}|
$$

$$
\times \prod_{j=1}^{n} y_j^{q_j v_j([\omega]_{\mathcal{KB}}) + p_j(a_{j,i}^{\max}(\mathcal{KB}) - a_j([\omega]_{\mathcal{KB}}))}
$$

$$
f_i^- = \sum_{[\omega]_{\mathcal{KB}} \in \Omega_\mathcal{D}/\equiv_{\mathcal{KB}}} p_i a_i([\omega]_{\mathcal{KB}}) \, |[\omega]_{\mathcal{KB}}|
$$

$$
\times \prod_{j=1}^{n} y_j^{q_j v_j([\omega]_{\mathcal{KB}}) + p_j(a_{j,i}^{\max}(\mathcal{KB}) - a_j([\omega]_{\mathcal{KB}}))}
$$

*for $i = 1, \ldots, n$. Then $f_i = f_i^+ - f_i^-$, and the effects $\alpha_i$ for $i = 1, \ldots, n$ from the equation system (6) are given by $\alpha_i = \gamma_i^{q_i}$ where $(\gamma_1, \ldots, \gamma_n) \in \mathcal{V}(f_1^*, \ldots, f_n^*)$ and*

$$
f_i^* = \frac{f_i^+ - f_i^-}{\gcd(f_i^+, f_i^-)}, \quad i = 1, \ldots, n.
$$

**Proof.** This corollary follows directly from Proposition 4 and the foregoing arguments. □

Note that the task of finding the greatest common divisor of multivariate polynomials can be reduced to deriving the intersection of polynomial ideals (Cox *et al.*, 2015).

Corollary 1 makes it possible to determine the effects $\alpha_1, \ldots, \alpha_n$ by solving the polynomial equation system $f_i^* = 0$ for $i = 1, \ldots, n$ instead of the highly nonlinear equation system (6). Therefore, we call $\mathcal{F}^*(\mathcal{KB}) = \{f_1^*, \ldots, f_n^*\}$ the *set of polynomials associated with* $\mathcal{KB}$. However, we still cannot assume that there is an analytic solution. For answering queries based on the knowledge base $\mathcal{KB}$ it is therefore gainful when we do not need to calculate the $\alpha_i$'s at all. In the next section, we use elimination theory to avoid these calculations.

## 6. Drawing Inferences

A main task of reasoning is ascerting what inferences can be drawn from a given knowledge base $\mathcal{KB}$. For this purpose, we consider an additional conditional $C_{\mathsf{inf}} = (B_{\mathsf{inf}}(\vec{X})|A_{\mathsf{inf}}(\vec{X}))[\xi_{\mathsf{inf}}]$ and investigate whether $\mathcal{KB}\,|\!\sim_{\odot}^{\mathsf{ME}} C_{\mathsf{inf}}$ holds. In the sense of the aggregating semantics (cf. Definition 1), $\xi_{\mathsf{inf}}$ must satisfy

$$\xi_{\mathsf{inf}} = \frac{\displaystyle\sum_{(B_{\mathsf{inf}}(\vec{a})|A_{\mathsf{inf}}(\vec{a}))[\xi_{\mathsf{inf}}]\in\mathsf{ground}_{\mathcal{D}}((B_{\mathsf{inf}}(\vec{X})|A_{\mathsf{inf}}(\vec{X}))[\xi_{\mathsf{inf}}])} \mathcal{P}(A_{\mathsf{inf}}(\vec{a})B_{\mathsf{inf}}(\vec{a}))}{\displaystyle\sum_{(B_{\mathsf{inf}}(\vec{a})|A_{\mathsf{inf}}(\vec{a}))[\xi_{\mathsf{inf}}]\in\mathsf{ground}_{\mathcal{D}}((B_{\mathsf{inf}}(\vec{X})|A_{\mathsf{inf}}(\vec{X}))[\xi_{\mathsf{inf}}])} \mathcal{P}(A_{\mathsf{inf}}(\vec{a}))}.$$

(10)

As an apparent pitfall, we cannot easily exploit the conditional structure of the possible worlds with respect to $\mathcal{KB}$ in order to simplify (10). This is because there may exist an equivalence class in $\Omega_{\mathcal{D}}/\!\equiv_{\mathcal{KB}}$ that contains possible worlds that behave differently for a certain ground instantiation of the conditional $(B_{\mathsf{inf}}(\vec{X})|A_{\mathsf{inf}}(\vec{X}))[\xi_{\mathsf{inf}}]$. Therefore, we have to build the equivalence classes with respect to the equivalence relation $\equiv_{\mathcal{KB}\cup\{C_{\mathsf{inf}}\}}$ defined analogously to $\equiv_{\mathcal{KB}}$. However, $\Omega_{\mathcal{D}}/\!\equiv_{\mathcal{KB}\cup\{C_{\mathsf{inf}}\}}$ is just a finer partition of $\Omega_{\mathcal{D}}$ than $\Omega_{\mathcal{D}}/\!\equiv_{\mathcal{KB}}$, and it is obvious that every "new" equivalence class is a subset of an "old" one. Thus, we get $\Omega_{\mathcal{D}}/\!\equiv_{\mathcal{KB}\cup\{C_{\mathsf{inf}}\}}$ by decomposing every equivalence class in $\Omega_{\mathcal{D}}/\!\equiv_{\mathcal{KB}}$ with respect to the evaluation of $C_{\mathsf{inf}}$. We get

$$\xi_{\mathsf{inf}} = \frac{\displaystyle\sum_{[\omega]_{\mathcal{KB}\cup\{C_{\mathsf{inf}}\}}\in\Omega_{\mathcal{D}}/\equiv_{\mathcal{KB}\cup\{C_{\mathsf{inf}}\}}} v_{\mathsf{inf}}([\omega]_{\mathcal{KB}\cup\{C_{\mathsf{inf}}\}})\,\mathcal{P}([\omega]_{\mathcal{KB}\cup\{C_{\mathsf{inf}}\}})}{\displaystyle\sum_{[\omega]_{\mathcal{KB}\cup\{C_{\mathsf{inf}}\}}\in\Omega_{\mathcal{D}}/\equiv_{\mathcal{KB}\cup\{C_{\mathsf{inf}}\}}} a_{\mathsf{inf}}([\omega]_{\mathcal{KB}\cup\{C_{\mathsf{inf}}\}})\,\mathcal{P}([\omega]_{\mathcal{KB}\cup\{C_{\mathsf{inf}}\}})}.$$

(11)

From now on, to shorten formulas, we use inf instead of $\mathcal{KB}\cup\{C_{\mathsf{inf}}\}$ as a subscript whenever the latter should occur.

In order to apply symbolic computation for answering the query $\mathcal{KB} \mid \sim_\odot^{ME} C_{\inf}$, we transform (11) into a polynomial equation as well. We consider the extended polynomial ring $\mathbb{R}[y_1, \ldots, y_n, y_{\inf}]$. Using (4) and by substituting $\xi_i = p_i/q_i$ and $\alpha_i = y_i^{q_i}$ for $i = 1, \ldots, n$, Equation (10) leads to

$$f_{\inf} = \sum_{[\omega]_{\inf} \in \Omega_{\mathcal{D}} / \equiv_{\inf}} (y_{\inf} \, a_{\inf}([\omega]_{\inf}) - v_{\inf}([\omega]_{\inf})) \, |[\omega]_{\inf}|$$

$$\times \prod_{i=1}^{n} y_i^{q_i v_i([\omega]_{\inf}) + p_i (a_{i,\inf}^{\max}(\mathcal{KB} \cup \{C_{\inf}\}) - a_i([\omega]_{\inf}))} = 0. \qquad (12)$$

Here, $a_{i,\inf}^{\max}(\mathcal{KB} \cup \{C_{\inf}\})$ is defined analogously to $a_{j,i}^{\max}(\mathcal{KB})$ in the proof of Proposition 4. Note that the variable $y_{\inf}$ has a slightly different meaning than the other variables $y_1, \ldots, y_n$ since $y_{\inf}$ is a place holder for the probability of $C_{\inf}$, while $y_1, \ldots, y_n$ stand for the effects as mentioned above. As we have done in the previous section for the polynomials associated with $\mathcal{KB}$, we may cancel polynomial combinations of variables in (12) if all of the coefficients are positive. Thus, we may consider

$$f_{\inf}^* = \frac{f_{\inf}^+ - f_{\inf}^-}{\gcd(f_{\inf}^+, f_{\inf}^-)}$$

instead of $f_{\inf}$ when $f_{\inf}^+$ and $f_{\inf}^-$ fulfill the positivity condition for their coefficients and satisfy $f_{\inf} = f_{\inf}^+ - f_{\inf}^-$.

By treating $f_1^*, \ldots, f_n^*$ as polynomials within the extended polynomial ring $\mathbb{R}[y_1, \ldots, y_n, y_{\inf}]$, drawing inferences now reduces to determining the affine variety $\mathcal{V}(f_1^*, \ldots, f_n^*, f_{\inf}^*) = \mathcal{V}(f_1^*, \ldots, f_n^*) \cap \mathcal{V}(f_{\inf}^*)$. More precisely, we are only interested in the last entries of the elements $(\gamma_1, \ldots, \gamma_n, \gamma_{\inf}) \in \mathcal{V}^*$, i.e., the *partial solutions* $\gamma_{\inf}$. This circumstance is dealt with in the following theorem.

**Theorem 2.** *Let* $\mathcal{KB} = \{(B_1|A_1)[\xi_1], \ldots, (B_n|A_n)[\xi_n]\}$ *be a consistent knowledge base with rational, non-trivial probabilities,* $C_{\inf} = (B_{\inf}|A_{\inf})[\xi_{\inf}]$ *an additional conditional and* $\mathcal{KB} \mid \sim_\odot^{ME} C_{\inf}$. *Further, let* $\preceq_{\text{lex}}$ *be the lexicographic term ordering with* $y_{\inf} \prec_{\text{lex}} y_1 \prec_{\text{lex}} \cdots \prec_{\text{lex}} y_n$. *Then*

$$\xi_{\inf} \in \mathcal{V}(\mathcal{GB}_{\preceq_{\text{lex}}}(\langle f_1^*, \ldots, f_n^*, f_{\inf}^* \rangle) \cap \mathbb{R}[y_{\inf}]) \cap [0, 1].$$

**Proof.** This theorem is a direct consequence of Proposition 3 and the fact that $\xi_{\inf}$ is a probability, i.e., $\xi_{\inf} \in [0, 1]$. The consistency of $\mathcal{KB}$ guarantees that $\mathcal{V}(\mathcal{GB}_{\preceq_{\text{lex}}}(\langle f_1^*, \ldots, f_n^*, f_{\inf}^* \rangle) \cap \mathbb{R}[y_{\inf}]) \cap [0, 1]$ is not empty. $\qquad \square$

Theorem 2 provides the user with an instruction on how to restrict the search space for the sought probability $\xi_{\text{inf}}$. As mentioned in the remark after Proposition 3, $\mathcal{GB}_{\preceq_{\text{lex}}}(\langle f_1^*, \ldots, f_n^*, f_{\text{inf}}^*\rangle) \cap \mathbb{R}[\gamma_{\text{inf}}]$ consists of a single polynomial and $\xi_{\text{inf}}$ is a real root of this polynomial. The following example condenses and illustrates the essential techniques presented throughout this paper.

**Example 1.** Assume Alice knows that there is a flu epidemic and every fifth person is infected. Further, she knows that the probability of having a flu oneself increases to $p = 0.5$ if one is in contact with an infected person. We represent Alice's knowledge in the knowledge base $\mathcal{KB}_{\text{ex}} = \{r_1, r_2, r_3\}$ with

$$r_1 = (\text{Flu}(X)|\top)[0.2],$$

$$r_2 = (\text{Flu}(X)|\text{Contact}(X,Y) \wedge \text{Flu}(Y))[0.5],$$

$$r_3 = (\text{Contact}(X,X)|\top)[0].$$

The conditional $r_3$ states that having contact is an irreflexive relation which makes sense in this context since no one can infect oneself. Alice now wonders how great the risk of having a flu is when she is in contact with Bob from whom she does not know wether he has a flu or not. That is, she asks for what probability $\xi_{\text{inf}}$ the inference

$$\mathcal{KB}_{\text{ex}} \mid\sim_\odot^{\text{ME}} (\text{Flu}(alice)|\text{Contact}(alice, bob))[\xi_{\text{inf}}] \tag{13}$$

holds. We will answer this query with the help of the methods presented in this paper. To simplify calculations and to concentrate on the used techniques, we consider the smallest possible domain $\mathcal{D}_{\text{ex}} = \{alice, bob\}$. Therefore, the set of possible worlds is

$$\Omega_{\mathcal{D}_{\text{ex}}} = \mathfrak{P}(\{\text{Flu}(alice), \text{Flu}(bob), \text{Contact}(alice, alice), \text{Contact}(alice, bob),$$

$$\text{Contact}(bob, alice), \text{Contact}(bob, bob)\})$$

where $\mathfrak{P}(S)$ denotes the power set of $S$. Thus, $|\Omega_{\mathcal{D}_{\text{ex}}}| = 64$, and one can show that there are 20 different equivalence classes of possible worlds with respect to $\equiv_{\mathcal{KB}_{\text{ex}}}$. However, we will see that most of them do not have to be considered for drawing the inference (13).

As the compilation of conditional knowledge to polynomial equations only works f all of the probabilities of the stated conditionals are non-trivial, we have to investigate on how to deal with conditional $r_3$.

Obviously, this conditional forces some of the possible worlds to have zero probability, namely the possible worlds that contain the ground atoms Contact($alice, alice$) or Contact($bob, bob$). The idea is to shrink the set of possible worlds $\Omega_{\mathcal{D}_{ex}}$ and to consider only those possible worlds that do not have zero probabilities by construction of the knowledge base. This leads to the reduced set of possible worlds

$$\Omega_{\mathcal{D}_{ex}}^r = \mathfrak{P}(\{\text{Flu}(alice), \text{Flu}(bob), \text{Contact}(alice, bob),$$

$$\times \text{Contact}(bob, alice)\}.$$

None of the possible worlds in $\Omega_{\mathcal{D}_{ex}}^r$ can verify the conditional $r_3$, hence $v_3(\omega) = 0$ holds for all $\omega \in \Omega_{\mathcal{D}_{ex}}^r$. In terms of equivalence classes of possible worlds, we only consider those equivalence classes $[\omega]_{\mathcal{KB}_{ex}} \in \Omega_{\mathcal{D}}/\equiv_{\mathcal{KB}_{ex}}$ with $v_3([\omega]_{\mathcal{KB}_{ex}}) = 0$.

We now focus on the MaxEnt distribution $\mathcal{P}_{ME}(\mathcal{KB}_{ex})$. From $\xi_3 = 0$, we get $\alpha_3 = 0$. Thus, $v_3([\omega]_{\mathcal{KB}_{ex}}) - \xi_3 a_3([\omega]_{\mathcal{KB}_{ex}}) = 0 - 0 \cdot a_3([\omega]_{\mathcal{KB}_{ex}}) = 0$ and hence

$$\alpha_3^{v_3([\omega]_{\mathcal{KB}_{ex}}) - \xi_3 a_3([\omega]_{\mathcal{KB}_{ex}})} = 0^0 = 1 \tag{14}$$

for the equivalence classes under consideration. The adjustment condition with respect to $r_3$ reduces to $0 = 0$ and every factor of the form (14) in the adjustment conditions regarding to $r_1$ (respectively $r_2$) reduces to 1. As a quintessence, when we restrict our investigations to those equivalence classes of possible worlds with $v_3([\omega]_{\mathcal{KB}_{ex}}) = 0$, we may remove $r_3$ from the knowledge base $\mathcal{KB}_{ex}$ and proceed as in the case with non-trivial probabilities. The six remaining equivalence classes are shown in Table 1.

Table 1.   The equivalence classes $[\omega]_{\mathcal{KB}_{ex}}$ of the possible worlds $\omega \in \Omega_{\mathcal{D}_{ex}}$ with $v_3([\omega]_{\mathcal{KB}_{ex}}) = 0$ are shown as well as their relevant properties. Here, the ground atoms are abbreviated, e.g., $Cab$ means Contact($alice, bob$).

| $[\omega]_{\mathcal{KB}_{ex}}$ | $\lvert \cdot \rvert$ | $v_1(\cdot)$ | $f_1(\cdot)$ | $v_2(\cdot)$ | $f_2(\cdot)$ | $v_3(\cdot)$ | $f_3(\cdot)$ |
|---|---|---|---|---|---|---|---|
| $\{\{Cab\}, \{Cba\}, \{Cab, Cba\}, \emptyset\}$ | 4 | 0 | 2 | 0 | 0 | 0 | 2 |
| $\{\{Fa\}, \{Fb\}, \{Fa, Cab\}, \{Fb, Cba\}\}$ | 4 | 1 | 1 | 0 | 0 | 0 | 2 |
| $\{\{Fa, Cba\}, \{Fb, Cab\}, \{Fa, Cab, Cba\},$ $\{Fb, Cab, Cba\}\}$ | 4 | 1 | 1 | 0 | 1 | 0 | 2 |
| $\{\{Fa, Fb\}\}$ | 1 | 2 | 0 | 0 | 0 | 0 | 2 |
| $\{\{Fa, Fb, Cab\}, \{Fa, Fb, Cba\}\}$ | 2 | 2 | 0 | 1 | 0 | 0 | 2 |
| $\{\{Fa, Fb, Cab, Cba\}\}$ | 1 | 2 | 0 | 2 | 0 | 0 | 2 |

We are now able to apply symbolic computation, thus to establish the polynomials from Equation (9). We get

$$f_{1,\text{ex}} = -8y_2^2 + 12y_1^5y_2^2 + 12y_1^5y_2 + 8y_1^{10}y_2^2 + 16y_1^{10}y_2^3 + 8y_1^{10}y_2^4,$$
$$f_{2,\text{ex}} = -4y_1^5y_2 + 2y_1^{10}y_2^3 + 2y_1^{10}y_2^4.$$

We may divide the first polynomial $f_1$ by $4y_2$ and the second polynomial $f_2$ by $2y_1^5y_2$ without loss of information and obtain

$$f_{1,\text{ex}}^* = -2y_2 + 3y_1^5y_2 + 3y_1^5 + 2y_1^{10}y_2 + 4y_1^{10}y_2^2 + 2y_1^{10}y_2^3,$$
$$f_{2,\text{ex}}^* = -2 + y_1^5y_2^2 + y_1^5y_2^3.$$

In order to obtain the polynomial $f_{\text{inf,ex}}$ arising from the query (13) in addition, we have to split the equivalence classes given in Table 1 with respect to their behaviour regarding to $C_{\text{inf}} = (\text{Flu}(alice)|\text{Contact}(alice, bob))[\xi_{\text{inf}}]$. By this, we get eleven equivalence classes $[\omega]_{\text{inf}}$ with respect to $\mathcal{KB} \cap \{C_{\text{inf}}\}$. The six resulting equivalence classes with $a_{\text{inf}}([\omega]_{\text{inf}}) > 0$ are shown in Table 2. The other five equivalence classes do not make a contribution to $f_{\text{inf,ex}}$ and, therefore, are omitted. We get

$$f_{\text{inf,ex}} = 2y_{\text{inf}}\left(y_2^2 + y_1^5y_2\right) + (y_{\text{inf}} - 1)y_1^5\left(y_2^2 + y_2 + y_1^5y_2^3 + y_1^5y_2^4\right)$$

and in reduced form (by division through $y_2$)

$$f_{\text{inf,ex}}^* = 2y_{\text{inf}}\left(y_2 + y_1^5\right) + (y_{\text{inf}} - 1)y_1^5\left(y_2 + 1 + y_1^5y_2^2 + y_1^5y_2^3\right).$$

We now take Theorem 2 into account and derive a Gröbner basis $\mathcal{GB}_{\text{inf,ex}}$ of the elimination ideal $\mathcal{I}_{\text{inf,ex}}$. For this, let $\preceq_{\text{lex,ex}}$ be the lexicographic term

Table 2. The equivalence classes $[\omega]_{\text{inf,ex}}$ of the possible worlds $\omega \in \Omega_{\mathcal{D}\text{ex}}$ with $v_3([\omega]_{\text{inf,ex}}) = 0$ and $a_{\text{inf}}([\omega]_{\text{inf}}) > 0$ are shown as well as their relevant properties. Here, the ground atoms are abbreviated, e.g., $Cab$ means Contact($alice, bob$).

| $[\omega]_{\text{inf}_{\text{ex}}}$ | $|\cdot|$ | $v_1(\cdot)$ | $f_1(\cdot)$ | $v_2(\cdot)$ | $f_2(\cdot)$ | $v_3(\cdot)$ | $f_3(\cdot)$ | $v_{\text{inf}}(\cdot)$ | $f_{\text{inf}}(\cdot)$ |
|---|---|---|---|---|---|---|---|---|---|
| $\{\{Cab\}, \{Cab, Cba\}\}$ | 2 | 0 | 2 | 0 | 0 | 0 | 2 | 0 | 1 |
| $\{\{Fa, Cab\}\}$ | 1 | 1 | 1 | 0 | 0 | 0 | 2 | 1 | 0 |
| $\{\{Fb, Cab\}, \{Fb, Cab, Cba\}\}$ | 2 | 1 | 1 | 0 | 1 | 0 | 2 | 0 | 1 |
| $\{\{Fa, Cab, Cba\}\}$ | 1 | 1 | 1 | 0 | 1 | 0 | 2 | 1 | 0 |
| $\{\{Fa, Fb, Cab\}\}$ | 1 | 2 | 0 | 1 | 0 | 0 | 2 | 1 | 0 |
| $\{\{Fa, Fb, Cab, Cba\}\}$ | 1 | 2 | 0 | 2 | 0 | 0 | 2 | 1 | 0 |

ordering with $y_{\inf} \prec_{\text{lex}_{\text{ex}}} y_1 \prec_{\text{lex}_{\text{ex}}} y_2$. We get

$$\mathcal{GB}_{\inf,\text{ex}} = \mathcal{GB}_{\preceq_{\text{lex}_{\text{ex}}}}(\langle f_{1,\text{ex}}^*, f_{2,\text{ex}}^*, f_{\inf,\text{ex}}^*\rangle) \cap \mathbb{R}[y_{\inf}]$$
$$= \{-2304y_{\inf}^3 + 1824y_{\inf}^2 - 485y_{\inf} + 43\}.$$

As the only polynomial in $\mathcal{GB}_{\inf,\text{ex}}$ has just one real root, namely $\gamma_{\inf,\text{ex}} \approx 0.223$, it is

$$\mathcal{V}(\mathcal{GB}_{\preceq_{\text{lex}_{\text{ex}}}}(\langle f_{1,\text{ex}}^*, f_{2,\text{ex}}^*, f_{\inf_{\text{ex}}}^*\rangle) \cap \mathbb{R}[y_{\inf}]) \cap [0,1] = \{\gamma_{\inf,\text{ex}}\}.$$

Eventually, Alice has to assume having a flu with probability $\xi_{\inf} \approx 0.223$.

## 7. Future Work

For bigger examples than Example 1, the polynomial representations of the adjustment condition (9) and the query (12) might become large. As a consequence, the Gröbner basis calculations are expensive and the resulting polynomials the real roots of which form the sought affine variety become large, too. Hence, these polynomials have a lot of (complex) roots one would like to discount. From a theoretical point of view, an interesting question which we pursue in our ongoing work is whether the latter is possible. In other words: Are there polynomial representations of the adjustment condition and the query which are more suitable for drawing inferences symbolically than (9) and (12), i.e., lead to smaller polynomials describing the affine variety? The following seems to be a meaningful starting point for investigations following this research line: Besides the conditional structure (cf. Definition 3) that arises from the evaluation of the premises and the conclusions of the conditionals in a knowledge base $\mathcal{KB}$, there may be redundancies in $\mathcal{KB}$ caused by the probabilities of the conditionals and their connections under the MaxEnt principle. From an algebraic point of view, these MaxEnt redundancies lead to additional relations $\mathcal{R}$ between the generators of the free abelian group $\mathcal{F}$ and, therefore, to a more complicated group structure $\mathcal{F}/\mathcal{R}$. In future work, we investigate on how far our approach presented in this paper deals with this problem, and how to eliminate these redundancies from $\mathcal{KB}$. Thus, we want to clarify the relations $\mathcal{R}$ and find a knowledge base $\mathcal{KB}'$ such that $\mathcal{F}_{\mathcal{KB}}/\mathcal{R} \simeq \mathcal{F}_{\mathcal{KB}'}$.

## 8. Conclusion

In this paper, we presented an approach to drawing probabilistic inferences from relational conditional knowledge according to the MaxEnt principle by means of symbolic computation. This approach is based on our results on symbolic MaxEnt reasoning upon propositional conditional knowledge presented in (Kern-Isberner *et al.*, 2014a,b, 2015) and is lifted to the relational case, here. Previous work by Dukkipati (Dukkipati, 2009, 2012) investigated similar connections between Gröbner bases theory and MaxEnt distributions in the field of statistics but without addressing any issues of knowledge representation. In contrast to common methods for MaxEnt reasoning that are based on probabilistic networks as implemented in SPIRIT (Rödder and Meyer, 1996; Rödder *et al.*, 2006), our approach preserves accuracy as long as possible.

The algebraic structure of a relational conditional knowledge base can be represented as a free abelian group (Kern-Isberner, 2001) which is the starting point of our investigation. Based on this abstract representation we built a system of polynomial equations that encodes the information provided by the knowledge base. A solution of this equation system defines the MaxEnt distribution which fulfills the stated knowledge. Adding an additional polynomial equation with respect to a query allows for MaxEnt inferences by means of elimination theory and thus connects wide-ranging mathematical methods with inductive probabilistic reasoning.

In (Kern-Isberner *et al.*, 2014a, 2015), our approach is also used to draw generic MaxEnt inferences, i.e., inferring MaxEnt probabilities symbolically without knowing the given probabilities of the knowledge base explicitly. As the compilation of the probabilistic knowledge towards the polynomial equation system proceeds very similar in both, the propositional and the relational case, the ideas from (Kern-Isberner *et al.*, 2014a, 2015) are applicable to relational knowledge, likewise.

We intend to strengthen the connection between symbolic reasoning and the principle of maximum entropy in order to improve the understanding of the MaxEnt principle that is often constituted as a black box methodology, and make it more transparent and usable for applications.

## References

Fahiem Bacchus, Adam J. Grove, Joseph Y. Halpern, and Daphne Koller. From statistical knowledge bases to degrees of belief. *Artificial Intelligence*, 1996.

Christoph Beierle, Marc Finthammer, Gabriele Kern-Isberner, and Matthias Thimm. Automated reasoning for relational probabilistic knowledge representation. In *Proceedings of the 5th International Joint Conference on Automated Reasoning (IJCAR'2010)*, 2010.

Christoph Beierle, Marc Finthammer, Gabriele Kern-Isberner, and Matthias Thimm. Evaluation and comparison criteria for approaches to probabilistic relational knowledge representation. In *Advances in Artificial Intelligence, 34th Annual German Conference on AI (KI 2001), Proceedings*, volume 7006 of *LNCS*, pages 63–74. Springer, 2011.

Christoph Beierle, Gabriele Kern-Isberner, Marc Finthammer, and Nico Potyka. Extending and completing probabilistic knowledge and beliefs without bias. *KI*, 29(3):255–262, 2015.

Stephen Boyd and Lieven Vandenberghe. *Convex Optimization*. Cambridge University Press, 2004.

Bruno Buchberger. *Ein Algorithmus zum Auffinden der Basiselemente des Restklassenringes nach einem nulldimensionalen Polynomideal*. PhD thesis, University of Innsbruck, 1965.

Bruno Buchberger. An algorithm for finding the basis elements in the residue class ring modulo a zero dimensional polynomial ideal. *Journal of Symbolic Computation*, 2006.

Robert G. Cowell, Philip A. Dawid, Steffen L. Lauritzen, and David J. Spiegelhalter. *Probabilistic networks and expert systems*. Springer, 1999.

David A. Cox, John Little, and Donal O'Shea. *Ideals, Varieties, and Algorithms: An Introduction to Computational Algebraic Geometry and Commutative Algebra*. Springer, 2015.

John N. Darroch and Douglas Ratcliff. Generalized iterative scaling for log-linear models. *Annals of Mathematical Statistics*, 1972.

Luc De Raedt and Kristian Kersting. Probabilistic logic learning. *SIGKDD Explor. Newsl.*, 2003.

Ambedkar Dukkipati. On embedding maximum entropy models in algebraic varieties by Gröbner bases methods. In *Proceedings of the IEEE International Symposium on Information Theory (ISIT'09)*, 2009.

Ambedkar Dukkipati. On maximum entropy and minimum kl-divergence optimization by Gröbner basis methods. *Applied Mathematics and Computation*, 2012.

Marc Finthammer and Christoph Beierle. Using equivalences of worlds for aggregation semantics of relational conditionals. In *Proceedings of the 35th Conference on Advances in Artificial Intelligence (KI'2012)*, 2012.

Peter Gärdenfors. *Knowledge in Flux: Modeling the Dynamics of Epistemic States*. MIT Press, 1988.

Lise Getoor and Benjamin Taskar. *Introduction to Statistical Relational Learning*. MIT Press, 2007.

Joseph Y. Halpern. An analysis of first-order logics of probability. *Artificiul Intelligence*, 1990.

David Hilbert. Ueber die theorie der algebraischen formen. *Mathematische Annalen*, 1890.

Manfred Jaeger. A logic for inductive probabilistic reasoning. *Synthese*, 2005.

Edwin T. Jaynes. *Papers on Probability, Statistics and Statistical Physics*. D. Reidel Publishing Company, 1983a.

Edwin T. Jaynes. Where do we stand on maximum entropy? *Papers on Probability, Statistics and Statistical Physics*, 1983b.

Gabriele Kern-Isberner. Characterizing the principle of minimum cross-entropy within a conditional-logical framework. *Artificial Intelligence*, 1998a.

Gabriele Kern-Isberner. A note on conditional logics and entropy. *International Journal of Approximate Reasoning*, 1998b.

Gabriele Kern-Isberner. *Conditionals in nonmonotonic reasoning and belief revision*. Springer, 2001.

Gabriele Kern-Isberner and Gerhard Rosenberger. A note on numbers of the form $n = x^2 + Ny^2$. *Arch. Math.*, 1984.

Gabriele Kern-Isberner and Matthias Thimm. Novel semantical approaches to relational probabilistic conditionals. In *Proceedings of the 12th International Conference on Principles of Knowledge Representation and Reasoning (KR'2010)*, 2010.

Gabriele Kern-Isberner and Matthias Thimm. A ranking semantics for first-order conditionals. In *Proceedings of the 20th European Conference on Artificial Intelligence (ECAI'2012)*, 2012.

Gabriele Kern-Isberner, Marco Wilhelm, and Christoph Beierle. Probabilistic knowledge representation using Gröbner basis theory. In *Proceedings of the 13th International Symposium on Artificial Intelligence and Mathematics (ISAIM'14)*, 2014a.

Gabriele Kern-Isberner, Marco Wilhelm, and Christoph Beierle. A novel methodology for processing probabilistic knowledge bases under maximum entropy. In *Proceedings of the 27th International Conference of the Florida Artificial Intelligence Research Society (FLAIRS'14)*, 2014b.

Gabriele Kern-Isberner, Marco Wilhelm, and Christoph Beierle. Probabilistic knowledge representation using the principle of maximum entropy and Gröbner basis theory. *Annals of Mathematics and Artificial Intelligence*, 2015. doi 10.1007/s10472-015-9457-7.

Roger C. Lyndon and Paul E. Schupp. *Combinatorial group theory*. Springer, 1977.

Jeff B. Paris. *The uncertain reasoner's companion — A mathematical perspective*. Cambridge University Press, 1994.

Jeff B. Paris. Common sense and maximum entropy. *Synthese*, 1999.

Jeff B. Paris and Alena Vencovská. A note on the inevitability of maximum entropy. *International Journal of Approximate Reasoning*, 1990.

Judea Pearl. *Probabilistic Reasoning in Intelligent Systems*. Morgan Kaufmann, 1988.

Matt Richardson and Pedro Domingos. Markov logic networks. *Machine Learning Journal*, 2006.

Wilhelm Rödder and Carl-Heinz Meyer. Coherent knowledge processing at maximum entropy by SPIRIT. In *Proceedings of the 12th Conference on Uncertainty in Artificial Intelligence (UAI'96)*, 1996.

Wilhelm Rödder, Elmar Reucher, and Friedhelm Kulmann. Features of the expert-system shell spirit. *Logic Journal of the IGPL*, 2006.

Benjamin M. Rottman and Reid Hastie. Reasoning about causal relationships: Inferences on causal networks. *Psychological bulletin*, 2014.

John E. Shore and Rodney W. Johnson. Axiomatic derivation of the principle of maximum entropy and the principle of minimum cross-entropy. *IEEE Transactions on Information Theory*, 1980.

## Chapter 10

## On Some Infinite-Dimensional Linear Groups
## and the Structure of Related Modules

L. A. Kurdachenko[1] and I. Ya. Subbotin[2]

[1] *Department of Algebra, National University of Dnepropetrovsk*
*A Gagarin Prospect 72, Dnepropetrovsk 10, 49010, Ukraine*
*lkurdachenko@i.ua*

[2] *Department of Mathematics and Natural Sciences, National University*
*5245 Pacific Concourse Drive, LA, CA 90045, USA*
*isubboti@nu.edu*

*To Dennis Spellman in the occasion of his 70th birthday*

**ABSTRACT.** The authors consider some new approaches to study of infinite-dimensional linear groups.

Let $F$ be a field, $A$ be a vector space over $F$, and $\mathbf{GL}(F, A)$ be the group of all $F$-automorphisms (non-singular linear transformations) of the vector space $A$. As usual, the group $\mathbf{GL}(F, A)$ and its subgroups are called *linear groups*. If the vector space $A$ has a finite dimension, say $n$, then $\mathbf{GL}(F, A)$ is isomorphic to the group $\mathbf{GL}_n(F)$ of all non-singular matrices whose coefficients belong to the field $F$. The availability of efficient matrix instruments is one of the reasons, perhaps, the most significant reason, for the theory of linear groups to be positioned among most developed algebraic theories. If the vector space $A$ has an infinite dimension, the situation becomes quite different. A possibility of applying matrix theory is not realistic here simply because there is no more or less general theory

2000 *Mathematics Subject Classification.* AMS classification: 20G15, 20F22, 20G99.
*Key words and Phrases.* infinite linear groups, infinite-dimensional linear groups, chain conditions, $G$-invariant subspaces.

of infinite matrices. One can only find some sporadic results on infinite matrices. We also observe another significant difference. Not every group can be isomorphic to a subgroup of $\mathbf{GL}_n(F)$. At the same time, every group $G$ can be realized as a subgroup of $\mathbf{GL}(F, A)$ For some vector space $A$. In fact, let $B = \mathbb{Z}/p\mathbb{Z}$ where $p$ is a prime. Let $W = B$ **wr** $G$ be a restricted wreath product of $B$ and $G$. Then $W = A \lambda G$ where $A = \mathbf{Dr}_{g \in G} B_g$ and $B_g \cong B$ for every $g \in G$. We can think of $A$ as a vector space over a prime field $F = \mathbf{F}_p$ and $\dim_F(A) = |G|$. The equation $C_G(A) = \langle 1 \rangle$ shows that we can consider $A$ as a subgroup of $\mathbf{GL}(F, A)$. Similar arguments can be developed in the case of characteristic 0.

The above example shows that study of linear groups is possible only with imposing quite serious restrictions on the groups or on the vector space. The traditional for finitely dimensional linear groups approaches could not always be extended to infinite-dimensional case.

The restrictions on vector space A often come from outside. They arise from the specific objectives of various areas of algebra and other mathematical disciplines. Perhaps the most effective restrictions here are the finiteness conditions. Finitary linear groups were among the first objects of study in the theory of infinite-dimensional linear groups. The results related to other conditions were reflected in the surveys [3], [4], [10]. Some other restrictions are associated with the structure of vector space $A$. Thus the family of $G$-invariant subspaces plays a major role here. Figuratively speaking, "the more saturated" this family, the more information we can get regarding the structure of the corresponding linear group. For example, if every subspace is $G$-invariant, then $G$ is abelian. Of course, this is the most extreme example, but the theory of generalized soluble groups shows that this approach has good prospects. On the other hand, vector space $A$, on which a group $G$ operates, naturally could be seen as an $FG$-module. Therefore another expected problem here is the description of the structure of this $FG$-module, namely, the study of the pattern arrangement of its G-invariant subspaces. This theme was reflected in the survey [10]. In the current paper, we show some of the other results obtained relatively recently.

Let $A$ be a vector space over a field $F$ and $G$ be a subgroup of $\mathbf{GL}(F, A)$. Suppose that $D$ and $C$ are the $G$-invariant subspaces of $A$ such that $C \leq D$. Then the factor $D/C$ is called **G-chief** if for every $G$-invariant subspace $B$ with the property $C \leq B \leq D$ either $C = B$ or $B = D$. If we think of $A$ as an $FG$-module, then every $G$-chief factor $C/D$ is precisely a simple $FG$-module.

As usual, a subgroup $G$ of $\mathbf{GL}(F, A)$ is called ***irreducible*** if $A$ does not include a proper non-zero $G$-invariant subspace (in other words, the factor $A/\langle 0 \rangle$ is $G$-chief).

If $A$ has finite dimension, then $A$ has a finite series

$$\langle 0 \rangle = B_0 \le B_1 \le \cdots \le B_n = A$$

of $G$-invariant subspaces whose factors are $G$-chief. Then the factor-groups $G/C_G(B_j/B_{j-1})$ are irreducible, $1 \le j \le n$. Put

$$Z = C_G(B_1) \cap C_G(B_2/B_1) \cap \ldots \cap C_G(B_n/B_{n-1}).$$

Then we can choose the basis in $A$ in such a way that it will be isomorphic to a subgroup of the unitriangular matrices $\mathbf{UT}_n(F)$. The subgroup $\mathbf{UT}_n(F)$ is nilpotent and has many other good properties. By Remak's theorem, the factor-group $G/Z$ is isomorphic to some subgroup of the direct product

$$G/C_G(B_1) \times G/C_G(B_2/B_1) \times \cdots \times G/C_G(B_n/B_{n-1})$$

of finitely many irreducible linear groups.

If vector space $A$ has infinite dimension, then we are facing a much more complicated case. Firstly, we cannot always ensure the existence of a finite $G$-chief series in $A$. Naturally, we need to deal here with the families of $G$-invariant subspaces, which were introduced in the classical article by A.G. Kurosh and S.N. Chernikov [16] for the groups with operators. Observe that vector spaces are partial cases of the groups with operators.

Let $A$ be a vector space over a field $F$ and $G$ be a subgroup of $\mathbf{GL}_n(F)$. Further, let $\mathfrak{S}$ be linearly ordered by inclusion family of $G$-invariant subspaces of $A$. Then $S$ is called a ***complete system*** if it satisfies the following condition: *For every subfamily $\mathfrak{L}$ of $\mathfrak{S}$ the intersection and the union of all member of $\mathfrak{L}$ belong to $\mathfrak{S}$.*

It is possible to prove that *every linearly ordered by inclusion family $\mathfrak{S}$ of $G$-invariant subspaces of $A$ can be extended to a complete system.*

Let $\mathfrak{S}$ be the family of $G$-invariant subspaces of $A$. Then a pair $(B, C)$ of elements of $\mathfrak{S}$ where $B \le C$, is said to be a ***jump*** of $\mathfrak{S}$ if it satisfies the following conditions:

*For every element $D$ of $\mathbf{S}$ such that $B \le D \le C$, one of the following equalities $B = D$ or $D = C$ holds.*

If a pair $(B, C)$ is a jump of $\mathfrak{S}$, then $C/B$ is called the ***factor*** of $\mathfrak{S}$.

Suppose now that $S$ is a complete system of $G$-invariant subspaces of $A$. For each non-zero element $a \in A$, we consider the union $V_a$ of all members of $\mathfrak{S}$ containing no element a and the intersection $\Lambda_a$ of all members of $S$ containing $a$. Since $S$ is a complete system, $V_a, \Lambda_a \in S$. Since $S$ is linearly ordered, either $V_a \leq \Lambda_a$ or $\Lambda_a \leq V_a$. On the other hand, $a \in \Lambda_a$ but $a \notin V_a$ so that the second inclusion is impossible. Hence $V_a \leq \Lambda_a$ and moreover, it is not hard to see that the pair $(V_x, \Lambda_x)$ is a jump of the system $S$. Conversely, let $(B, C)$ be a jump of the system $S$. Choose an element $y \in C \setminus B$. From the definitions of the subspaces $V_y$ and $\Lambda_y$, we obtain that $B \leq V_y$ and $\Lambda_y \leq C$. Since $y \notin C$, $V_y \neq C$. Since $(B, C)$ is a jump, we obtain that $B = V_y$. Taking into account the inequality $\Lambda_y \neq B$ and the fact that $(B, C)$ is a jump, we conclude that $\Lambda_y = C$.

*Let $A$ be a vector space over a field $F$ and $G$ be a subgroup of $\mathbf{GL}_n(F)$. The family $S$ of $G$-invariant subspaces of $A$ is called a* **Kurosh-Chernikov system** *if it satisfies the following conditions*:

$(KC\ 1)$ $\langle 0 \rangle, A \in S$;
$(KC\ 2)$ *the family $S$ is linearly ordered and complete.*

If $\mathfrak{S}$, $\mathfrak{R}$ are the families of linearly ordered by inclusion $G$-invariant subgroups, then we say that $\mathfrak{R}$ is a **refinement** of $\mathfrak{S}$ if every term of $\mathfrak{S}$ is also a term of $\mathfrak{R}$ (in other words, $\mathfrak{S}$ is a subset of $\mathfrak{R}$). If $\mathfrak{S}$ is a proper subset of $\mathfrak{R}$, then $\mathfrak{R}$ is called a **proper refinement** of $S$.

Let $\mathfrak{S}$ be a Kurosh-Chernikov system. Then $\mathfrak{S}$ is called **G-chief** if it has no proper **refinement** of $\mathfrak{S}$.

It is possible to prove that *a Kurosh-Chernikov system $\mathfrak{S}$ of $G$-invariant subspaces of $A$ is $G$-chief if and only if every factor of $\mathfrak{S}$ is $G$-chief.*

*It is possible to also prove that every linearly ordered by inclusion family of $G$-invariant subspaces of $A$ can be extended to a $G$-chief Kurosh-Chernikov system.*

As usual, if $\mathfrak{S}$ is well ordered in ascending order, then we are talking about an ascending series, and if $\mathfrak{S}$ is well ordered in descending order, then talk about a descending series.

The existence of the $G$-chief families of $G$-invariant subspaces shows that the irreducible linear groups still play quite a significant role.

Let, now, $\mathfrak{S}$ be a Kurosh-Chernikov system and $\mathfrak{J}$ be a set of all jumps of $\mathfrak{S}$. For every jump $(B, C)$ of $\mathfrak{S}$, we consider a subgroup $C_G(C/B)$, and put

$$\mathbf{St}_G(S) = \cap_{(B,C) \in \mathfrak{J}} C_G(C/B).$$

The subgroup $\mathbf{St}_G(\mathfrak{S})$ is called the **stability group of system** $S$ or the **stabilizer** of $S$. A system $\mathfrak{S}$ is called **central** if $G = \mathbf{St}_G(\mathfrak{S})$. As in the finite-dimensional case, Remak's theorem implies the imbedding of factor-group $G/\mathbf{St}_G(\mathfrak{S})$ into $\mathbf{Cr}_{(B,C)\in\mathfrak{J}} G/C_G(C/B)$. In particular, if $\mathfrak{S}$ is a $G$-chief, every factor-group $G/C_G(C/B)$ is irreducible. Of course the structure of Cartesian products is much more complicated, but even in this case we are able to obtain some information. The structure of some infinite-dimensional irreducible groups was considered in the survey [15].

The case of the subgroup $\mathbf{St}_G(\mathfrak{S})$ is much more complicated. Indeed, as we noted above, in the case when $A$ has finite dimension this subgroup is nilpotent. We provide the reader with some examples showing that in the case when A has infinite dimension the subgroup $\mathbf{St}_G(\mathfrak{S})$ can be very far from being nilpotent. In the paper of C. Casolo and O. Puglisi [2], an $\mathbf{F}_pG$-module $A$ for countable non-abelian free group $G$ such that $A$ has an ascending $G$-chief series $\mathfrak{S}$ of the length $\omega$ and $G = \mathbf{St}_G(\mathfrak{S})$ has been constructed. This example shows that the productive study of subgroups $\mathbf{St}_G(\mathfrak{S})$ is only possible under certain natural restrictions. C. Casolo and O. Puglisi [2] made the first steps in the study of locally nilpotent radical of the subgroup $\mathbf{St}_G(\mathfrak{S})$, which shows how difficult it is.

Since our starting point is the case when $\dim_F(A)$ is finite, it would be natural to consider a finite series of G-invariant subspaces first. Let

$$\langle 0 \rangle = B_0 \leq B_1 \leq \cdots \leq B_n = A$$

be a finite series of $G$-invariant subspaces, and $Z$ be a stabilizer of this series. Then $B_j(g-1) \leq B_{j-1}$ for every element $g \in Z$ and for each $j$, $1 \leq j \leq n$. In particular, $A(g-1)^n = \langle 0 \rangle$, and we come to the following notion. An element $g \in G$ is called **unipotent** if $(g-1)^n = 0$ for some positive integer. A subgroup $U$ of $G$ is called **unipotent**, if every its element is unipotent. If $\dim_F(A)$ is finite, then a subgroup $H$ of $\mathbf{GL}(F, A)$ is unipotent if and only if every element of $H$ is unipotent.

However, in the case when $\dim_F(A)$ is infinite, this restriction of unipotency is not efficient. Indeed, let $p$ be a prime and $G$ be an arbitrary $p$-group, and let $A$ be an arbitrary $\mathbf{F}_pG$-module. It is not hard to see that for each element $g \in G$ the natural semidirect product $A\lambda\langle g \rangle$ is a nilpotent group. It follows that every element of $G$ is unipotent.

A. Yu. Olshanskij [18] constructed his famous example of an infinite $p$-group $G$ such that every proper subgroup of $G$ has order $p$ where $p$ is a prime and $p > 10^{75}$. This group is generated by every two of its elements and is simple. Let $A$ be an arbitrary $\mathbf{F}_pG$-module. As we mentioned above,

$A$ has the $G$-chief Kurosh-Chernikov system $S$. If $(B, C)$ is a jump of $S$, then $C_G(C/B)$ is a normal subgroup of $G$. Since $G$ is simple, then either $G = C_G(C/B)$ or $C_G(C/B) = \langle 1 \rangle$. If we suppose that $G = C_G(C/B)$ for every jump of $S$, then $G = \mathbf{St}_G(S)$. By Lemma 4 of the paper [9], in this case, $G$ must be an $SN$-group. Hence there exists a jump $(U, V)$ of $S$ such that $C_G(V/U) = \langle 1 \rangle$. In other words, we can consider $G$ an irreducible subgroup of $\mathbf{GL}(\mathbf{F}_p, V/U)$. As we have seen above, this group is also unipotent.

Consider some other concepts related to stabilizers. Denote by $\omega(FG)$ the **augmentation ideal** of the group ring $FG$, that is the two-sided ideal of $FG$ generated by the set $\{g - 1 | g \in G\}$. Submodule $A\omega(FG)$ is called the **derived submodule** of $A$. In fact, this is the minimal subspace of all $G$-invariant subspaces defining a trivial the factor-module. Farther for this submodule we will use the more familiar notation $[A, G]$. The following $G$-invariant subspace is dual to $A/[A, G]$. Put

$$\zeta_{FG}(A) = \{a \in A | a(g - 1) = 0 \text{ for all } g \in G\} = C_A(G).$$

It is not hard to see that $\zeta_{FG}(A)$ is a $G$-invariant subspace. $\zeta_{FG}(A)$ is called an **FG-center** of $A$. $FG$-center of $A$ is the maximal subspace of $G$-invariant subspaces which are annihilated by the augmentation ideal of $FG$. In some sense, $[A, G]$ and $\zeta_{FG}(A)$ are analogs of the derived subgroup and the center of a group.

In the group theory, one of the most important questions is the question of how closely related the properties of the central factor-group $G/\zeta(G)$ and the derived subgroup $[G, G]$ are. One of the classical results here is the proven by B.H. Neumann [17] theorem yielding that the finiteness of $G/\zeta(G)$ implies the finiteness of $[G, G]$. This raises the following question about the linear analogue of this theorem:

*Let $G$ be a subgroup of $\mathbf{GL}(F, A)$. Suppose that $\zeta_{FG}(A)$ has finite codimension. Is $[A, G]$ finite-dimensional?*

The answer is negative. The following simple example justifies this.

Let $A$ be a vector space over a field $F$ having countable basis $\{a_n | n \in \mathbf{N}\}$. For each positive integer $k$, define the $F$-automorphism $g_k$ of $A$ by the following rule: $a_1 g_k = a_1 + a_k$ and that $a_t g_k = a_t$ for $t > 1$. It is not hard to see that $G = \langle g_k | k \in \mathbf{N} \rangle$ is elementary abelian if $\mathbf{char}(F) = p$ is a prime, and $G$ is free abelian if $\mathbf{char}(F) = 0$. Furthermore, $\zeta_{FG}(A)$ is a subspace generated by all elements $a_n$ where $n > 1$, so that $\dim_F(A/\zeta_{FG}(A)) = 1$. At the same time $[A, G] = \zeta_{FG}(A)$, in particular, $\dim_F([A, G])$ is infinite.

However, the above example provide us with a hint to a natural restriction at which a linear analogue of B.H. Neumann's theorem is possible.

Let $p$ be a prime. We say that a group $G$ has **finite section** $p$-**rank** $\mathbf{sr}_p(G) = \mathbf{r}$ if every elementary abelian $p$-section of $G$ is finite of order at most $p^r$ and there is an elementary abelian $p$-section $A/B$ of $G$ such that $|A/B| = p^r$.

Similarly:

We say that a group $G$ has **finite section** 0-**rank** $\mathbf{sr}_0(G) = \mathbf{r}$ if every torsion-free abelian section of $G$ has finite $\mathbf{Z}$-rank at most $\mathbf{r}$ and there is an abelian torsion-free section $A/B$ of $G$ such that $\mathbf{r}_{\mathbf{Z}}(A/B) = \mathbf{r}$.

**Theorem 1.** (*M.R. Dixon, L.A. Kurdachenko, J. Otal* [6]). *Let $G$ be a subgroup of* $\mathbf{GL}(F, A)$. *Suppose that* $\mathrm{codim}_F(\zeta_{FG}(A)) = \mathbf{c}$ *is finite. Then the following assertions hold.*

(i) *If* $\mathbf{char}(F) = p > 0$ *and* $\mathbf{sr}_p(G) = \mathbf{r}$ *is finite, then* $[A, G]$ *has finite dimension.*

(ii) *If* $\mathbf{char}(F) = 0$ *and* $\mathbf{sr}_0(G) = \mathbf{r}$ *is finite, then* $[A, G]$ *has finite dimension.*

*Moreover there exists a function $\kappa$ such that* $\dim_F([A, G]) \leq \kappa(\mathbf{c}, \mathbf{r})$.

The group $\mathbf{UT}_n(F)$ show us two kinds of the central series existing in the vector space. The point of origin for the first one is the $FG$-center. Starting from the $FG$-center, we can construct the upper $FG$-central series of $A$:

$$\langle 0 \rangle = \zeta_{FG0}(A) \leq \zeta_{FG1}(A) \leq \zeta_{FG2}(A) \leq \cdots \leq \zeta_{FG\alpha}(A) \leq \zeta_{FG\alpha+1}(A)$$
$$\leq \cdots \zeta_{FG\gamma}(A),$$

defined the rule $\zeta_{FG1}(A) = \zeta_{FG}(A)$ is the center of $G$, and recursively

$$\zeta_{FG\,\alpha+1}(A)/\zeta_{FG\,\alpha}(A) = \zeta_{FG}(A/\zeta_{FG\,\alpha}(A)) \text{ for all ordinals } \alpha,$$
$$\zeta_{FG\lambda}(A) = \cup_{\mu<\lambda}\zeta_{FG\mu}(A) \text{ for the limit ordinals } \lambda,$$

and $\zeta_{FG}(A/\zeta_{FG\gamma}(A)) = \langle 0 \rangle$. The last term, $\zeta_{FG\gamma}(A) = \zeta_{FG\infty}(A)$, of this series is called the **upper FG-hypercenter** of **A**. The ordinal $\gamma$ is called the **FG-hypercentral length of a module A** and will be denoted by $\mathbf{zl}_{FG}(A)$. We observe that $[\zeta_{FG\alpha+1}(A), G] \leq \zeta_{FG\alpha}(A)$ for all $\alpha < \gamma$.

If the upper $FG$-hypercenter of $A$ coincides with $A$, then $A$ is called **FG-hypercentral**.

Starting from the derived submodule, we can construct the lower $FG$-central series of $A$:

$$A = \gamma_{FG1}(A) \geq \gamma_{FG2}(A) \geq \gamma_{FG3}(A) \geq \cdots \geq \gamma_{FG\alpha}(A)\gamma_{FG\alpha+1}(A)$$
$$\leq \cdots \gamma_{FG\gamma}(A),$$

defined by the rule $\gamma_{FG2}(A) = [A, G]$, and, recursively, $\gamma_{FG\alpha+1}(A) = [\gamma_{FG\alpha}(A), G]$ for all ordinals $\alpha$, $\gamma_{FG\lambda}(A) = \cap_{\mu<\lambda}\gamma_{FG\mu}(A)$ for the limit ordinals $\lambda$, and $\gamma_{FG\gamma}(A) = [\gamma_{FG\gamma}(A), G]$. The last term $\gamma_{FG\gamma}(A) = \gamma_{FG\infty}(A)$ of this series is called the **lower FG-hypocenter** of **A**.

Another classical result of group theory is generalization of the B.H. Neumann's theorem obtained by R. Baer in the paper [1]. R. Baer proved that the finiteness of factor-group by the $k$-th term of the upper central series of a group $G$ implies the finiteness of $\gamma_{k+1}(G)$. Under the same natural restrictions, which were used in Theorem 1, we can obtain the following linear analogue of Baer's theorem.

**Theorem 2.** (*M.R. Dixon, L.A. Kurdachenko, J. Otal* [6]). *Let $G$ be a subgroup of* **GL**$(F, A)$ *and suppose that there exists a positive integer $k$ such that* $\mathrm{codim}_F(\zeta_{FG\,k}(A)) = $ **c** *is finite. Then the following assertions hold.*

(i) *If* **char**$(F) = p > 0$ *and* **sr**$_p(G) = $ **r** *is finite, then $\gamma_{FG\,k+1}(A)$ has finite dimension.*

(ii) *If* **char**$(F) = 0$ *and* **sr**$_0(G) = $ **r** *is finite, then $\gamma_{FG\,k+1}(A)$ has finite dimension.*

*Moreover there exists a function $\kappa$ such that* $\dim_F(\gamma_{FG\,k+1}(A)) \leq \kappa_1(\mathbf{c}, \mathbf{r}, \mathbf{k})$.

This Baer's theorem has been generalized by the following way. In the paper of M. de Falco, F. de Giovanni, C. Musella, Ya.P. Sysak [8], it has been proven that if the upper hypercenter of a group $G$ has finite index, then $G$ includes a normal finite subgroup $K$ such that $G/K$ is hypercentral. L.A. Kurdachenko, J. Otal, I.Ya. Subbotin in [12] significantly enhanced this result. They proved that the order of $K$ is bounded by only the order of the factor-group by the upper hypercenter. The question of the linear analogue of the last result is natural. This analog was obtained recently.

**Theorem 3.** (*M.R. Dixon, L.A. Kurdachenko, J. Otal* [7]). *Let $G$ be a subgroup of* **GL**$(F, A)$ *and suppose that the upper FG-hypercenter of $A$ has* *finite codimension* **c**. *Then the following assertions hold.*

(i) *If* **char**$(F) = p > 0$ *and* $\mathbf{sr}_p(G) = \mathbf{r}$ *is finite, then* $A$ *includes a finite-dimensional* $G$-*invariant subspace* $K$ *such that* $A/K$ *is* $FG$-*hypercentral. Moreover there exists a function* $\kappa_2$ *such that* $\dim_F(K) \leq \kappa_2(\mathbf{c}, \mathbf{r})$.

(ii) *If* **char**$(F) = 0$ *and* $\mathbf{sr}_0(G) = \mathbf{r}$ *is finite, then* $A$ *includes a finite-dimensional* $G$-*invariant subspace* $K$ *such that* $A/K$ *is* $FG$-*hypercentral. Moreover there exists a function* $\kappa_3$ *such that* $\dim_F(K) \leq \kappa_3(\mathbf{c}, \mathbf{r})$.

Let $B$ be a subspace of a vector space $A$ and $G$ be a subgroup of $\mathbf{GL}(F, A)$. With a subspace $B$, the following natural $G$-invariant subspaces are related. First is $BFG$, the intersection of all $G$-invariant subspaces, including $B$. It is the least $G$-invariant subspace containing $B$. Second is $\mathbf{Core}_G(B)$, the sum of all $G$-invariant subspaces containing in $B$. This is the largest $G$-invariant subspace of $B$.

The dimension $\dim_F(BFG/B)$ is called the **upper measure of non** $G$-**invariance** of $B$.

The dimension $\dim_F(B/\mathbf{Core}_G(B))$ is called the **lower measure of non** $G$-**invariance** of $B$.

If $\dim_F(BFG/B) = 0$ (respectively $\dim_F(B/\mathbf{Core}_G(B)) = 0$), then $B = BFG = \mathbf{Core}_G(B)$. In other words, a subspace $B$ is $G$-invariant.

Therefore it is natural to consider the case when the upper (respectively the lower) measures of non $G$-invariance of all subspaces of $A$ are finite, and, moreover, bounded.

The case when lower measures of non $G$-invariance of every subspace is finite began to be considered in the paper of L.A. Kurdachenko, A.V. Sadovnichenko, and I.Ya. Subbotin [14].

**Theorem 4.** (*L.A. Kurdachenko, A.V. Sadovnichenko, and I.Ya. Subbotin* [14]. *Let* $G$ *be a periodic subgroup of* $\mathbf{GL}(F, A)$. *Suppose that the lower measures of non* $G$-*invariance of every subspace is finite. Then the following assertions hold.*

(i) $A$ *includes a finite-dimensional* $G$-*invariant subspace* $B$ *such that every subspace of* $A/B$ *is* $G$-*invariant.*

(ii) *A group* $G$ *has a series of normal subgroups* $E \leq H \leq G$ *where* $G/H$ *is isomorphic to a subgroup of the multiplicative group of the field* $F$, $H/E$ *is a locally finite and finite-dimensional linear group, and* $E$ *is abelian.*

*Moreover, if* **char**$(F) = p$ *is a prime, then $E$ is an elementary abelian $p$-subgroup; if* **char**$(F) = 0$, *then $E = \langle 1 \rangle$.*

The case when the lower measures of non $G$-invariance of every subspace is bounded, has been considered by L.A. Kurdachenko, J.M. Muños-Escolano, and J. Otal [11].

**Theorem 5.** (*L.A. Kurdachenko, J.M. Muños-Escolano, J. Otal* [11]). *Let $G$ be a subgroup of* **GL**$(F, A)$ *and suppose that there exists a positive integer* **b** *such that the lower measures of non $G$-invariance of every subspace is at most* **b** *finite. Then the following assertions hold.*

(i) *A includes a $G$-invariant subspace $B$ such that every subspace of $B$ is $G$-invariant and $A/B$ has finite dimension at most* $6\mathbf{b}^2 + \mathbf{b} = \psi(\mathbf{b})$.

(ii) *A group $G$ includes the normal subgroups $H$ such that*

(iia) *$H$ is abelian. Moreover, if* **char**$(F) = 0$, *then $H$ is torsion-free. If* **char**$(F) = p$ *is a prime, then $H$ is an elementary abelian $p$-subgroup;*

(iib) *$G/H$ is isomorphic to subgroup of $L \times D$, where $L$ is isomorphic to a subgroup of the multiplicative group of a field $F$ and $D$ is isomorphic to a subgroup of* **GL**$_{\psi(b)}(F)$.

If there exists a positive integer **b** such that $\dim_F(BFG/B) \leq \mathbf{b}$ for every subspace $B$ of a vector space $A$, then $\dim_F(aFG) \leq \mathbf{b} + 1$, for each element $a \in A$. The last case has been studied in the paper [5]. The main result of this paper is following.

**Theorem 6.** (*M.R. Dixon, L.A. Kurdachenko, J. Otal* [5]). *Let $G$ be a subgroup of* **GL**$(F, A)$, *and suppose that there exists a positive integer* **b** *such that every subspace $aFG$ has finite dimension at most* **b**. *Then the following assertions hold.*

(i) *A includes a $G$-invariant subspace $B$, having finite dimension, such that every subspace of $A/B$ is $K$-invariant, where $K = C_G(B)$. Moreover there exists a function $\lambda$ such that $\dim_F(B) \leq \lambda(\mathbf{b})$.*

(ii) *$K$ includes a $G$-invariant abelian subgroup $T$ such that $[A, T] \leq B$ and the factor $K/T$ is isomorphic to some subgroup of the multiplicative group of a field $F$;*

(iii) *if* **char**$(F) = 0$, *then $T$ is torsion-free, if* **char**$(F) = p$ *is a prime, then $T$ is an elementary abelian $p$-subgroup.*

*In particular, $G$ is an extension of a metabelian group by a finite-dimensional linear group.*

And the last result of the current survey is about a vector space, whose subspaces have finite upper measures of non $G$-invariance.

**Theorem 7.** (*L.A. Kurdachenko, A.V. Sadovnichenko* [13]). *Let $G$ be a subgroup of* $\mathbf{GL}(F, A)$, *and suppose that the upper measures of non $G$-invariance of every subspace is finite. Then $A$ includes a finite-dimensional $G$-invariant subspace $B$ such that every subspace of $A/B$ is $G$-invariant.*

## References

[1]  R. Baer. *Endlichkeitskriterien für Kommutatorgruppen.* Math. Annalen-124(1952), 161–177.

[2]  C. Casolo and O. Puglisi. *Hirsch-Plotkin radical of stability groups.* Journal Algebra **370** (2012), 133–151.

[3]  M.R. Dixon and L.A. Kurdachenko *Linear groups with infinite central dimension*, Groups St Andrews 2005. Volume 1. London Mathematical Society. Lecture Note Series 339. Cambridge Univ. Press. 2007, 306–312.

[4]  M.R. Dixon, L.A. Kurdachenko, J.M. Munoz-Escolano, and J. Otal *Trends in infinite-dimensional linear groups.* Groups St Andrews 2009 in Bath. Volume 1. London Mathematical Society. Lecture Note Series 387. Cambridge Univ. Press. 2011, 271–282.

[5]  M.R. Dixon, L.A. Kurdachenko, and J. Otal *Linear groups with finite-dimensional orbits.* Ischia Group Theory 2010, Proceedings of the conference in Group Theory, World Scientific, 2012, 131–145.

[6]  M.R. Dixon, L.A. Kurdachenko, and J. Otal *Linear analogues of theorem of Schur, Baer and Hall.* International Journal of Group Theory 2(2013), No. 1, 79–89.

[7]  M.R. Dixon, L.A. Kurdachenko, and J. Otal *On the structure of some infinite-dimensional linear groups.* B Meyath*.

[8]  M. De Falco, F. de Giovanni, C. Musella, and Ya.P. Sysak *On the upper central series of infinite groups.* Proc. Amer. Math. Soc. **139** (2011), 385–389.

[9]  P. Hall and B. Hartley *The stability group of a series of subgroups.* Proc. London Math. Soc. **16** (1966), 1–39.

[10]  L.A. Kurdachenko *On some infinite-dimensional linear groups.* Note di Matematica, **30** (2010), no. 1, 21–36.

[11]  L.A. Kurdachenko, J.M. Muños-Escolano, and J. Otal *Groups acting on vector spaces with a large family of invariant subspaces.* Linear and Multilinear Algebra, **60** (2012), Issue 4, 487–498.

[12]  L.A. Kurdachenko, J. Otal, and I. Ya. Subbotin *On a generalization of Baer Theorem.* Proc. Amer. Math. Soc., **141** (3013), 2597–2602.

[13]  L.A. Kurdachenko and A.V. Sadovnichenko *On some linear groups, having a big family of $G$-invariant subspaces.* Algebra and Discrete Mathematics **16** (2013), number 2, 217–225.

[14]    L.A. Kurdachenko, A.V. Sadovnichenko, and I.Ya. Subbotin. *Infinite-dimensional linear groups with a large family of G-invariant subspaces.* Comment. Math. Univ. Carolin. **51**(2010), 551–558.

[15]    L.A. Kurdachenko and I.Ya. Subbotin. *On some infinite-dimensional linear groups.* Groups St. Andrews 2001, Volume II, London Mathematical Society Lecture Note Series. 305, Cambridge University Press 2003, 377–384.

[16]    A.G. Kurosh and S.N. Chernikov *Soluble and nilpotent groups.* Uspehi Mat. Nauk-**2** (1947) no. 3, 18–59. Amer. Math. Soc. Transl. No. 80 (1953).

[17]    B.H. Neumann. *Groups with finite classes of conjugate elements.* Proc. London Math. Soc. — 1(1951), 178–187.

[18]    A.Yu. Olshanskij *Groups of bounded period with subgroups of prime order.* Algebra and logic **21** (1982), no. 5, 369–418.

Chapter 11

# On New Analogs of Some Classical Group Theoretical Results in Lie Rings

L. A. Kurdachenko[1], A. A. Pypka[2] and I. Ya. Subbotin[3]

[1] *Department of Geometry and Algebra, Dnipropetrovsk National University*
*72 Gagarin Av., Dnipropetrovsk, Ukraine 49010*
*lkurdachenko@gmail.com*

[2] *Department of Geometry and Algebra, Dnipropetrovsk National University*
*72 Gagarin Av., Dnipropetrovsk, Ukraine 49010*
*Pypka@ua.fm*

[3] *Department of Mathematics and Natural Sciences, National University*
*5245 Pacific Concourse Drive, LA, CA 90045, USA*
*isubboti@nu.edu*

*To Gerhard Rosenberger in the occasion of his 70th birthday.*

**ABSTRACT.** The paper discuss the relationship between the factors of the upper and lower series in the Lie rings. As a result some analogs of well known important group theoretical results were enhanced to Lie rings.

## 1. Introduction

There are two important subgroups in a group, namely the center and the derived subgroup, which indicate how far or close the group is from being abelian. Clearly, if the central factor-group $G/\zeta(G)$ is trivial, then the derived subgroup $[G,G]$ is trivial, and conversely. The question about the extent of the influence of both $G/\zeta(G)$ and $[G,G]$ on the structure of the group $G$ naturally arises. In particular, it is interesting to find a case when the $G/\zeta(G)$ and $[G,G]$ possess the same properties. In this regard, we note the following classical result of the theory of infinite groups.

1991 *Mathematics Subject Classification.* 17B05, 17B30, 17B65.
*Key words and Phrases.* Lie ring, Derived ideal, Schur's Theorem, Baer's Theorem

**Theorem ZD.** *Let $G$ be a group, $C$ a subgroup of the center $\zeta(G)$ such that $G/C$ is finite. Then the derived subgroup $[G, G]$ is finite.*

This result in this formulation appears for the first time in the paper of B.H. Neumann [7]. But at the conclusion of this paper, B.H. Neumann stated that R. Baer wrote to him that this theorem is a consequence of a more general result, which was proved by R. Baer in his paper [2]. In fact, in Theorem 3 of the mentioned paper, it has been proven that if a normal subgroup $H$ of a group $G$ has finite index, then the factor $([G, G] \cap H)/[H, G]$ is also finite. Nevertheless, R. Baer in his later work [3] shows Theorem ZD in its regular form supplying it with a new proof. In his famous lectures on nilpotent groups, P. Hall obtained the generalization of Theorem ZD ([5], Theorem 8.7). In these lectures, P. Hall calls this theorem *Schur's Theorem* (note that P. Hall made no specific references). Following P. Hall, many algebraists began calling Theorem ZD the Schur's theorem, usually giving the reference on the paper of I. Schur [10]. The paradox is that I. Schur was not engaged in infinite groups. In this old classic work, I. Schur (only for finite groups!) introduces the concept of the group, which is now called *Schur's multiplicator* or *Schur's multiplier*. He obtained some properties of this group. In modern terminology, Schur's multiplier $\mathbf{M}(G)$ of a group $G$ is exactly the second cohomology group $\mathbf{H}^2(G, \mathbf{U}(\mathbb{C}))$. It is important that Theorem ZD has a natural extension, which just based of I. Schur's essential use. Naturally the question arises: *whether the orders of $G/\zeta(G)$ and $[G, G]$ are related, and if it is, how tight is this relation?* B.H. Neumann raised this question in [7] and obtained the first bounds for $|[G, G]|$. The best bound yet was obtained by J. Wiegold. In his paper [14], he proved that if $t = |G/\zeta(G)|$, then $|[G, G]| \leq t^m$ where $m = \frac{1}{2}(log_p t - 1)$ and $p$ is the least prime dividing $t$. Also J. Wiegold has proven that this bound is attained if and only if $t = p^n$ where $p$ is a prime [14]. When $t$ has more than one prime divisor, one can do better, though here the picture is less clear.

In [3] R. Baer obtained the following important generalization of Theorem ZD. He proved that the finiteness of $G/\zeta_n(G)$ implies the finiteness of $\gamma_{n+1}(G)$. This topic is very interesting and important. It points to the deep connection between the factors of the upper and lower central series. This topic has a more general nature, it is not specific only to group theory. It has continuations to some other algebraic structures. Thus in the article [4], Theorem ZD has been extended to $n$-groups. One can find some close results in Lie algebras (see, for example, [12]) and in their generalizations [9].

Between groups and Lie algebras, there is a certain analogy. Thus many results on groups managed to get analogues in the Lie algebras, and vice versa. Moreover, in some Lie algebras cases, it is possible to get an even deeper promotion (of course, in the case, when a Lie algebra is considered over a field) (see, for example, [11]).

For Lie algebras, an analog of Theorem ZD is well-known. Some of its extension has been obtained in the paper [12]. Some analog of Baer's theorem also holds for Lie algebras. Moreover, in the paper [6], its significant generalization was obtained. Namely, it was shown that if the upper hypercenter of Lie algebra $L$ has finite codimension, then $L$ includes a finite dimensional ideal $K$ such that $L/K$ is hypercentral.

An analog of Theorem ZD has been extended to $n$-Lie algebras in the paper [9].

It is natural to expect that some analog of Theorem ZD will hold for Lie rings. Indeed, making relatively small efforts we can be obtain the following result. Note that non-traditional notation (such as $\Pi([L,L])$ and $\varepsilon(k)$) will be explained in the main body of the article.

**Theorem A.** *Let $L$ be a Lie ring. Suppose that the factor-ring $L/\zeta(L)$ is finite. Then the derived ideal $[L,L]$ is also finite. Moreover, $\Pi([L,L]) \subseteq \Pi(L/\zeta(L))$ and $|[L,L]| \leq k^{\varepsilon(k)} \leq k^{log_2(k)}$ where $k = |L/\zeta(L)|$.*

We note, that if the derived ideal is finite, then the center $\zeta(L)$ of a Lie ring $L$ is not necessary of finite index. The same situation holds for groups. A characterization of the groups, whose derived subgroup is finite, has been obtained by B.H. Neumann [8]. It arises from Theorem ZD in the following way. If $|G/\zeta(G)| = b$, then $|G : C_G(a)| \leq b$ for each element $a \in G$. B.H. Neumann [8] considered the groups $G$ in which $|G : C_G(a)| \leq b$ for all $a \in G$. He proved that such a group has a finite derived subgroup. For Lie algebras, an analog of this theorem of B.H. Neumann was considered by M.R. Vaughan-Lee [13]. He proved that if $\mathbf{dim}_F(L/Ann_L(a)) \leq b$ for each element $a$ of a Lie algebra $L$, then $\mathbf{dim}_F([L,L]) \leq \frac{1}{2}b(b+1)$. In the current article, we obtained the following analog of this B.H. Neumann's theorem for Lie rings.

**Theorem B.** *Let $L$ be a Lie ring and suppose that there exists a positive integer $b$ such that $|L/Ann_L(a)| \leq b$ for every element $a \in L$. Then the derived ideal $[L,L]$ is finite. Moreover, there exists a function $\nu$ such that $|[L,L]| \leq \nu(b)$. Conversely, if $[L,L]$ is finite, then $|L/Ann_L(a)| \leq |[L,L]|$ for every element $a \in L$.*

As we mentioned above, R. Baer in [3] has obtained a generalization of Theorem ZD, showing that the finiteness of $G/\zeta_n(G)$ implies the finiteness of the $(n + 1)$-term of the lower central series of $G$. For a Lie ring $L$ we can introduce the notion of the *lower central series* in the following way. Put $L_1 = \gamma_1(L) = L$, $\gamma_2(L) = [L, L]$ and recursively $\gamma_{\alpha+1}(L) = [\gamma_\alpha(L), L]$ for all ordinals $\alpha$ and $\gamma_\lambda(L) = \bigcap_{\mu<\lambda} \gamma_\mu(L)$ for the limit ordinals $\lambda$. In the current work we obtained the following analog of this R. Baer's theorem for Lie rings.

**Theorem C.** *Let $L$ be a Lie ring. Suppose that the factor-ring $L/\zeta_k(L)$ is finite and has order $d$. Then $\gamma_{k+1}(L)$ is finite. Moreover, $\Pi(\gamma_{k+1}(L)) \subseteq \Pi(L/\zeta_k(L))$ and $|\gamma_{k+1}(L)| \leq d^{\varepsilon(d)+k-1} \leq d^{\log_2(d)+k-1}$.*

Note that finiteness of $\gamma_{n+1}(L)$ implies that $L$ has a least ideal $K$ such that the factor-ring $L/K$ is nilpotent. In other words, the nilpotent residual of $L$ is finite. Therefore the question about the order of the nilpotent residual naturally rises. One bound of this order was obtained in Theorem C. In the following theorem we obtained a significantly better bound.

**Theorem D.** *Let $L$ be a Lie ring and $R$ be a nilpotent residual of $L$. Suppose that there exists a positive integer $n$ such that $L/\zeta_n(L)$ is finite and has order $d$. Then $L/R$ is nilpotent and $R$ is finite. Moreover, $\Pi(R) \subseteq \Pi(L/\zeta_n(L))$ and $|R| \leq d^{\varepsilon(d)+1} \leq d^{\log_2(d)+1}$.*

## 2. Proof of Theorem A

Let $L$ be a Lie ring, and

$$\zeta(L) = \{x \in L | [x, y] = 0 \text{ for each element } y \in L\}$$

be the center of $L$. Clearly, $\zeta(L)$ is an ideal of ring $L$. Now define the *upper central series*

$$\langle 0 \rangle = \zeta_0(L) \leq \zeta_1(L) \leq \cdots \zeta_\alpha(L) \leq \zeta_{\alpha+1}(L) \leq \cdots \zeta_\gamma(L) = \zeta_\infty(L)$$

of Lie ring $L$ by the following rule: $\zeta_1(L) = \zeta(L)$ is the center of $L$, and recursively $\zeta_{\alpha+1}(L)/\zeta_\alpha(L) = \zeta(L/\zeta_\alpha(L))$ for all ordinals $\alpha$, and $\zeta_\lambda(L) = \bigcup_{\mu<\lambda} \zeta_\mu(L)$ for limit ordinals $\lambda$. The last term $\zeta_\infty(L)$ of this series is called the *upper hypercenter* of $L$. Denote by $zl(L)$ the length of the upper central series of $L$. If $L = \zeta_\infty(L)$, then $L$ is said to be a *hypercentral* Lie ring.

If $X$ and $Y$ are subsets of $L$, then $[X, Y]$ denote the additive subgroup, generated by all elements $[x, y]$, where $x \in X$, $y \in Y$. We note that if $X$ is an ideal of $L$ and $Y$ is a subring of $L$, then $[X, Y]$ is a subring of $L$. If $X$, $Y$ are the ideals of $L$, then $[X, Y]$ is an ideal of $L$.

Let $M$ is a non-empty subset of $L$ and $H$ is a subring of $L$. Put

$$Ann_H(M) = \{x \in H | [x, y] = 0 \text{ for all } y \in M\}.$$

The subset $Ann_H(M)$ is called the *annihilator* (or *centralizer*) of $M$ in a subring $H$. It is not hard to see that $Ann_H(M)$ is a subring of $L$. Moreover, if $M$ is an ideal in $L$, then $Ann_H(M)$ is an ideal in $H$.

If $M$ is a subring of $L$ and $A$, $B$ are ideals of $M$ such that $B \leq A$, then define $Ann_M(A/B)$ by the rule

$$Ann_M(A/B) = \{x \in M | [x, a] \in B \text{ for all } a \in A\}.$$

In other words, $Ann_M(A/B)/B = Ann_{M/B}(A/B)$. Clearly, $Ann_M(A/B)$ is an ideal of $M$.

**Lemma 2.1.** *Let $L$ be a Lie ring. Suppose that the additive group of the factor-ring $L/\zeta(L)$ is generated by subset $\{a_\lambda + \zeta(L) | \lambda \in \Lambda\}$. Then the additive group of the derived ideal $[L, L]$ is generated by $[a_\lambda, L]$, $\lambda \in \Lambda$.*

**Proof.** In fact, arbitrary element $d$ of $L$ has the form $d = \sum_{\mu \in M} k_\mu a_\mu + z$ where $M$ is a finite subset of $\Lambda$, $k_\mu \in \mathbb{Z}$, $\mu \in M$, $z \in \zeta(L)$. For another arbitrary element $x \in L$ we have

$$[d, x] = \left[ \sum_{\mu \in M} k_\mu a_\mu + z, x \right] = \sum_{\mu \in M} k_\mu [a_\mu, x] \in \sum_{\mu \in M} [a_\mu, L].$$

Since $[L, L]$ as an additive group is generated by the elements $[d, x]$, $d, x \in L$, we proved the result. $\square$

**Corollary 1.** *Let $L$ be a Lie ring. Suppose that the additive group of the factor-ring $L/\zeta(L)$ is generated by the subset $\{a_\lambda + \zeta(L) | \lambda \in \Lambda\}$. Then the additive group of the derived ideal $[L, L]$ is generated by the subsets $\{[a_\lambda, a_\mu] | \lambda, \mu \subset \Lambda\}$.*

**Corollary 2.** *Let $L$ be a Lie ring. Suppose that the additive group of the factor-ring $L/\zeta(L)$ is generated by the subset $\{a_j + \zeta(L) | 1 \leq j \leq n\}$. Then the additive group of the derived ideal $[L, L]$ is generated by the subsets $\{[a_j, a_m] | 1 \leq j < m \leq n\}$.*

**Proof.** Indeed, it follows from $[a_j, a_m] = -[a_m, a_j]$ that $[a_j, a_j] = 0$. $\square$

Let $G$ be a group and $g$ be an element of $G$. If $g$ has a finite order $n$, then put $\Pi(g) = \{p_1, \ldots, p_k\}$ where $n = p_1^{t_1} \cdot \ldots \cdot p_k^{t_k}$ is a primary decomposition of a natural number $n$. If $g$ has infinite order, then put $\Pi(g) = \varnothing$.

Put $\Pi(G) = \bigcup_{g \in G} \Pi(g)$.

**Proposition 1.** *Let $L$ be a Lie ring. Suppose that the additive group of the factor-ring $L/\zeta(L)$ is periodic. Then the additive group of $[L, L]$ is also periodic and $\Pi([L, L]) \subseteq \Pi(L/\zeta(L))$.*

**Proof.** Let $u$ be an arbitrary element of $[L, L]$. Lemma 2.1 shows that $u \in [a_1, L] + \cdots + [a_n, L]$ for some elements $a_1, \ldots, a_n \in L$. For each element $d \in L$ the mapping $\xi_d : L \to L$ defined by the rule $\xi_d(x) = [d, x]$ for each $x \in A$ is an endomorphism of the additive group of $L$. Furthermore, $Ker(\xi_d) = Ann_L(d)$, $Im(\xi_d) = [d, L]$, so that we have the following additive isomorphism $L/Ann_L(d) = L/Ker(\xi_d) \cong Im(\xi_d) = [d, L]$. Clearly, $\zeta(L) \leq Ann_L(d)$, which follows that the additive group $L/Ann_L(d)$ is periodic, and $\Pi(L/Ann_L(d)) \subseteq \Pi(L/\zeta(L))$. Hence the additive group $[d, L]$ is periodic, and $\Pi([d, L]) \subseteq \Pi(L/\zeta(L))$. This is valid for each $[a_j, L]$, $1 \leq j \leq n$. Therefore $\Pi(u) \subseteq \Pi(L/\zeta(L))$. It follows that $\Pi([L, L]) \subseteq \Pi(L/\zeta(L))$. □

For every finitely generated group $G$ we denote by $d(G)$ the least number of generators of $G$. Now let $G$ be an additive finite abelian group, then

$$G = P_1 \oplus \cdots \oplus P_k$$

where $P_j$ is the Sylow $p_j$-subgroup of $G$, $1 \leq j \leq k$. Let $|P_j| = p_j^{d_j}$, $1 \leq j \leq k$. Since $P_j$ is a direct sum of cyclic subgroups, $|P_j| \leq d_j$, $1 \leq j \leq k$. Then $d(G) \leq d_1 + \cdots + d_k$. If $n$ is a natural number and $n = p_1^{d_1} \cdot \ldots \cdot p_k^{d_k}$ is the primary decomposition of $n$, then put $\varepsilon(n) = d_1 + \cdots + d_k$. We have:

$$\varepsilon(n) = d_1 + \cdots + d_k = \log_{p_1}(p_1^{d_1}) + \cdots + \log_{p_k}(p_k^{d_k})$$
$$\leq \log_2(p_1^{d_1}) + \cdots + \log_2(p_k^{d_k}) =$$
$$= \log_2(p_1^{d_1} \cdot \ldots \cdot p_k^{d_k}) = \log_2(n).$$

Hence $d(G) \leq \varepsilon(|G|) \leq \log_2(|G|)$.

**Proof of Theorem A.** Let $\{a_j + \zeta(L) | 1 \leq j \leq d\}$ be a set of generators of $L/\zeta(L)$ such that $d = d(L/\zeta(L))$. The inclusion $\Pi([L, L]) \subseteq \Pi(L/\zeta(L))$ follows from Proposition 1. Lemma 2.1 implies that $[L, L] = [a_1, L] + \cdots + [a_d, L]$. As in the proof of Proposition 1, we obtain the additive isomorphism

$L/Ann_L(a_j) \cong [a_j, L]$, which shows that $||[a_j, L]|| \leq k$, $1 \leq j \leq d$. Then $||[L, L]|| \leq k^d$. By the remarked above

$$d = d(L/\zeta(L)) \leq \varepsilon(k) \leq \log_2(k). \qquad \square$$

## 3. Proof of Theorem B

**Lemma 3.1.** *Let $L$ be a non-abelian Lie ring, and suppose that there exists a positive integer $b$ such that $||[x, L]|| \leq b$ for each element $x \in L$. Let $v$ be an element of $L$ such that $||[v, L]||$ is the greatest, $B = \langle v^L \rangle$, $H = Ann_L(B)$. Then $||[x + B, H/B]|| < ||[v, L]||$ for each element $x \in H$.*

**Proof.** Without loss of generality we can suppose that $||[x, L]|| = b$. Since $L$ is not abelian, $b \neq 1$. Suppose the contrary, assume that $H$ contains an element $c$ such that $||[c+B, H/B]|| = b$. Since $[c+B, H/B] = ([c, H]+B)/B$, it follows that $||[c, H]|| = b$. In turn, it follows that $[c, H] \cap B = \langle 0 \rangle$. On the other hand, by the choice of $b$ we have $||[c, L]|| = b$, which proves that $[c, H] = [c, L]$ and, in particular, $[c, L] \cap B = \langle 0 \rangle$.

Let $z \in Ann_L(v+c)$. Then $0 = [v+c, z] = [v, z]+[c, z]$. Since $[v, z] \in B$, $[c, z] \in B$. The equality $[c, L] \cap B = \langle 0 \rangle$ implies that $[c, z] = 0$, so that $z \in Ann_L(c)$. Furthermore, $[v, z] = [v + c - c, z] = [v + c, z] - [c, z] = 0$, which shows that $z \in Ann_L(v)$. It proves the equality $Ann_L(v + c) = Ann_L(v) \cap Ann_L(c)$. On the other hand,

$$|L/Ann_L(v + c)| = ||[v + c, L]|| \leq b = ||[v, L]|| = |L/Ann_L(v)|.$$

Taking into account the inclusion $Ann_L(v + c) \leq Ann_L(v)$, we obtain that $Ann_L(v + c) = Ann_L(v)$, hence $Ann_L(v) \leq Ann_L(c)$. Clearly, $Ann_L(B) \leq Ann_L(v)$, so that $H = Ann_L(B) \leq Ann_L(c)$. It implies that $[c, H] = \langle 0 \rangle$ and $||[c + B, H/B]|| = 1 \neq b$. Thus we obtain contradiction which proves the lemma. $\qquad \square$

**Lemma 3.2.** *Let $L$ be a Lie ring and $a_1, \ldots, a_k$ be elements of $L$ such that $|L/Ann_L(a_j)| = b_j$ is finite, $1 \leq j \leq k$. Let $A$ be a subring, generated by the elements $a_1, \ldots, a_k$. Then $[A, A]$ is finite and has order at most $b^{\varepsilon(b)}$ where $b = b_1 \cdot \ldots \cdot b_k$.*

**Proof.** Put $C = \bigcap_{1 \leq j \leq k} Ann_L(a_j)$, then $L/C$ is finite and has order at most $b_1 \cdot \ldots \cdot b_k$. Then $A/(A \cap C)$ has finite order at most $b_1 \cdot \ldots \cdot b_k$. Clearly, $\zeta(A)$ includes $A \cap C$, and we can apply Theorem A. By this Theorem $[A, A]$ is finite and has order at most $b^{\varepsilon(b)}$ where $b = b_1 \cdot \ldots \cdot b_k$. $\qquad \square$

Let $L$ be a ring, $X$ and $Y$ subsets of $L$. As usual, we say that $X$ is a $Y$-invariant subset, if $[x, y] \in X$ for every elements $x \in X$, $y \in Y$. Denote by $X^Y$ the smallest $Y$-invariant additive subgroup of $L$ including $X$. It is not hard to prove that $X^Y$ is the sum of $X$ and $[X, {}_n Y]$, $n \in \mathbb{N}$.

**Lemma 3.3.** *Let $L$ be a Lie ring and suppose that there exists a positive integer $b$ such that $|L/Ann_L(a)| \leq b$ for every element $a \in A$. Let $A = \langle a^L \rangle$. Then $A$, as a subring, is generated by at most $b+1$ elements, $[A, A]$ is finite and has order at most $b^m$ where $m = (b+1)^2 \varepsilon(b)$.*

**Proof.** Put $C = Ann_L(a)$, and let $T$ be the transversal to $C$ in $L$. Let $T = \{x_1, \ldots, x_k\}$ where $k \leq b$. It is not hard to prove that the subring, generated by the elements $a, [a, x_1], \ldots, [a, x_k]$, is $L$-invariant and then it is an ideal of $L$. It follows that this subring coincides with $A$. By Lemma 3.2, $[A, A]$ is finite and has order at most $b^m$ where $m = (b+1)^2 \varepsilon(b)$.    $\square$

Now we are in a position to prove Theorem B.

**Proof of Theorem B.** We will apply induction on $b$. If $b = 1$, then $L = Ann_L(a)$ for each element $a \in L$. This means that the ring $L$ is abelian, so that $[L, L] = \langle 0 \rangle$ and $\nu(1) = 1$. Suppose that $b > 1$ and we have already proven our conjecture for all Lie ring $M$ such that $|M/Ann_M(x)| \leq b-1$ for each element $x \in M$. Choose an element $v$ with the property $|L/Ann_L(v)| = b$. Put $B = \langle v^L \rangle$ and $H = Ann_L(B)$. Then $H = Ann_L(\langle v \rangle + [v, L])$, so that $|L/H| \leq b^{b+1}$. Let $T$ be the transversal to $H$ in $L$. Put $V = B + \langle T^L \rangle$. By this choice, $L = V + H$. Using Lemma 3.3, we obtain that $D = [V, V]$ is finite and has order at most $b^{ms}$ where $m = (b+1)^2 \varepsilon(b)$ and $s = 1 + b^{b+1}$. Furthermore, $V/[V, V]$ as an additive group is generated by at most $s$ elements.

The ideal $V/D$ of the factor-ring $L/D$ is abelian and, as an additive subgroup, it is generated by the elements $v_1, \ldots, v_t$ where $t \leq s$. In the factor-ring $L/D$, we also have $|(L/D)/Ann_{L/D}(v_j + D)| \leq b$ for each $j$, $1 \leq j \leq t$, so that $|(L/D)/Ann_{L/D}(V/D)| \leq b^t \leq b^s$. Let $T/D$ be the transversal to $Ann_{L/D}(V/D)$ in $L/D$. Then $|T/D| \leq b^s$. For every element $x + D \in T/D$ we have $|[V/D, x + D]| \leq b$. Let $U/D = \sum_{x+D \in T/D} [V/D, x + D]$. Then, clearly, $U/D \leq [V/D, L/D]$, which follows that $V/U \leq \zeta(L/U)$. We note that $|U/D| \leq b^{|T|} \leq b^{m_1}$ where $m_1 = b^s$. Thus $|U| = |U/D||D| \leq b^{ms} b^{m_1} = b^k$ where $k = ms + b^s$.

The equality $L/U = V/U + (H + U)/U$ and the inclusion $V/U \leq \zeta(L/U)$ imply that $[L/U, L/U] = [(H + U)/U, (H + U)/U]$. Since $B \leq U$, Lemma 3.1 together with the induction hypothesis shows that $[(H + U)/U, (H + U)/U]$

is finite and has order at most $\nu(b-1)$. Hence

$$\|[L,L]\| = \|[L/U, L/U]\|\|U\| \le b^k \nu(b-1) = \nu(b).$$    $\square$

## 4. Proof of Theorem C

**Lemma 4.1.** *Let $L$ be a Lie ring. If $j$ and $k$ are positive integer such that $k \ge j$, then $[\gamma_j(L), \zeta_k(L)] \le \zeta_{k-j}(L)$.*

**Proof.** We will use induction on $j$. For $j = 1$ we have $[\gamma_1(L), \zeta_k(L)] = [L, \zeta_k(L)] \le \zeta_{k-1}(L)$.

Suppose now that $j > 1$ and we have already proved the inclusions $[\gamma_m(L), \zeta_k(L)] \le \zeta_{k-m}(L)$ for all $m < j$. We have $\gamma_j(L) = [\gamma_{j-1}(L), L]$. Choose the arbitrary elements $x \in \gamma_{j-1}(L)$, $y \in L$, and $z \in \zeta_k(L)$. We have

$$[[x,y],z] = -[[z,x],y] - [[y,z],x].$$

Since $[z,x] = -[x,z] \in [\gamma_{j-1}(L), \zeta_k(L)]$, the induction hypothesis implies that $[z,x] \in \zeta_{k-j+1}(L)$, so that $[[z,x],y] \in [\zeta_{k-j+1}(L), L] \le \zeta_{k-j}(L)$. Furthermore $[y,z] \in [L, \zeta_k(L)] \le \zeta_{k-1}(L)$. Since $k-1 > j-1$, using the induction hypothesis, we obtain that

$$[[y,z],x] = -[x,[y,z]] \in [\gamma_{j-1}(L), \zeta_{k-1}(L)] \le \zeta_{k-1-j+1}(L) = \zeta_{k-j}(L).$$

It follows that $[[x,y],z] \in \zeta_{k-j}(L)$.    $\square$

**Lemma 4.2.** *Let $L$ be a Lie ring and $H$ be an ideal of $L$. Suppose that $L/Ann_L(H)$ has finite order $d$. Let $E$ be a transversal to $Ann_L(H)$ in $L$. If $H/Ann_H(E)$ has finite order $t$, then $[L,H]$ has finite order at most $dt$ and $\Pi([L,H]) \subseteq \Pi(H/Ann_H(E))$.*

**Proof.** Let $E = \{e_1, \ldots, e_d\}$. If $x$ is an arbitrary element of $L$, then $x = e_j + c$ for some $j$, $1 \le j \le d$, where $c \in Ann_L(H)$. If $h$ be an arbitrary element of $H$, then

$$[x,h] = [e_j + c, h] = [e_j, h].$$

It follows that $[L,H] \le [e_1, H] + \cdots + [e_d, H]$. Let $a$ be an arbitrary element of $L$. Consider the mapping $\xi_a : H \to H$ defined by the rule $\xi_a(h) = [a,h]$, $h \in H$. This mapping is an additive endomorphism of $H$, $Im(\xi_a) = [a, H]$, $Ker(\xi_a) = Ann_H(a)$. Hence

$$[a, H] = Im(\xi_a) \cong H/Ker(\xi_a) = H/Ann_H(a).$$

The inclusion $Ann_H(E) \leq Ann_H(e_j)$ implies that $H/Ann_H(e_j)$ is finite and has order at most $t$. Hence $|[e_j, H]| \leq t$. Moreover, $\Pi([e_j, H]) \subseteq \Pi(H/Ann_H(E))$. This is true for every $j$, $1 \leq j \leq d$. It follows that $[L, H]$ has finite order at most $dt$ and $\Pi([L, H]) \subseteq \Pi(H/Ann_H(E))$. $\quad\square$

**Corollary 3.** *Let $L$ be a Lie ring and $H$ be an ideal of $L$. Suppose that $L/Ann_L(H)$ has finite order $d$. If $H/(H \cap \zeta(L))$ has finite order $t$, then $[L, H]$ has finite order at most $dt$ and $\Pi([L, H]) \subseteq \Pi(H/(H \cap \zeta(L)))$.*

**Proof.** Indeed, let $E$ be a transversal to $Ann_L(H)$ in $L$. Since $Ann_H(E) \geq H \cap \zeta(L)$, $H/Ann_H(E)$ has finite order dividing $t$, and $\Pi(H/Ann_H(E)) \subseteq \Pi(H/(H \cap \zeta(L)))$. Hence we can apply Lemma 4.2. $\quad\square$

Now we can prove Theorem C.

**Proof of Theorem C.** Let

$$\langle 0 \rangle = Z_0 \leq Z_1 \leq \cdots \leq Z_{k-1} \leq Z_k = Z$$

be an upper central series of $L$. We will apply induction on $k$. If $k = 1$, then $L/Z_1$ has finite order $d$, and Theorem A implies that $\gamma_2(L) = [L, L]$ has finite order at most $d^{\varepsilon(d)}$ and $\Pi(\gamma_2(L)) \subseteq \Pi(L/\zeta_1(L))$.

Assume now that $k > 1$ and we have already proven that $\gamma_k(L/Z_1)$ is finite and has order $t$ such that $t \leq d^{\varepsilon(d)+k-2}$ and $\Pi(\gamma_k(L/Z_1)) \subseteq \Pi(L/\zeta_k(L))$. We have $\gamma_k(L/Z_1) = (\gamma_k(L) + Z_1)/Z_1$. Put $K/Z_1 = \gamma_k(L/Z_1)$, $T = \gamma_k(L)$. Then $T \leq K$. Furthermore, the isomorphism $T/(T \cap Z_1) \cong (T + Z_1)/Z_1 \leq K/Z_1$ shows that $\Pi(T/(T \cap Z_1)) \subseteq \Pi(L/\zeta_k(L))$ and $|T/(T \cap Z_1)| \leq t$. By Lemma 4.1, $Z_k \leq Ann_L(T)$, so that the order of $L/Ann_L(T)$ divides $d$ and $\Pi(L/Ann_L(T)) \subseteq \Pi(L/\zeta_k(L))$. By Corollary 3, $[L, T]$ has finite order dividing $dt$, and $\Pi([L, T]) \subseteq \Pi(T/(T \cap Z_1)) \subseteq \Pi(L/\zeta_k(L))$. Put $D = [L, T]$. By the construction of $D$ we have $[L/D, T/D] \leq T/D$. This inclusion shows that $T/D \leq \zeta(L/D)$. Since $L/T$ is nilpotent of nilpotency class $k - 1$, the last inclusion shows that $L/D$ is nilpotent of nilpotency class at most $k$. It implies the inclusion $\gamma_{k+1}(L) \leq D$, which in its turn implies that $\gamma_{k+1}(L)$ is finite of order at most $dt \leq dd^{\varepsilon(d)+k-2} = d^{\varepsilon(d)+k-1}$ and $\Pi(\gamma_{k+1}(L)) \subseteq \Pi(L/\zeta_k(L))$. $\quad\square$

## 5. Proof of Theorem D

**Proposition 2.** *Let $L$ be a Lie ring and $H$ be an ideal of $L$. Then the factor-ring $L/Ann_L(H)$ is isomorphic to a subring of the Lie ring of all derivations of $H$.*

The proof of this Proposition almost word-to-word repeats the proof of its analog for Lie algebras (see, for example, [1], Chapter 1, Lemma 4.2]).

Let $L$ be a Lie ring and let

$$\langle 0 \rangle = K_0 \le K_1 \le \cdots \le K_n = L$$

be a series of ideals of $L$. Let $\Delta = Der(L)$ be the Lie ring of all derivations of $L$. Consider the following subset of $\Delta$:

$$N_\Delta(K_j) = \{\phi \in \Delta | \phi(x) \in K_j \text{ for every element } x \in K_j\}.$$

It is not hard to see that $N_\Delta(K_j)$ is a subring of $\Delta$. Then the intersection $N = \bigcap_{1 \le j \le n} N_\Delta(K_j)$ is also a subring of $\Delta$ and every ideal $K_j$ is $N$-invariant, $0 \le j \le n$. Consider the centralizers now

$$C_N(K_j/K_{j-1})$$
$$= \{\phi \in N | \phi(x + K_{j-1}) = x + K_{j-1} \text{ for every element } x \in K_j\}$$

of the factors of this series, $1 \le j \le n$. Put $\Sigma = \bigcap_{1 \le j \le n} C_N(K_j/K_{j-1})$. The subring $\Sigma$ of $N$ is called the *stabilizer* of the series $\{K_j | 1 \le j \le n\}$. Every ideal $K_j$ is $\Sigma$-invariant and each element $\sigma \in \Sigma$ acts trivially in every factor $K_j/K_{j-1}$, $1 \le j \le n$.

The following result is well-known, but unfortunately we were not able to find the corresponding reference. In [6], we gave its analog for Lie algebras. For Lie rings the proof is practically identical.

**Proposition 3.** *Let $L$ be a Lie ring and*

$$\langle 0 \rangle = K_0 \le K_1 \le \cdots \le K_n = L$$

*be a series of ideals of $L$. Then the stabilizer $\Sigma$ of this series is a nilpotent subring of the Lie ring $Der(L)$. Moreover, $\Sigma$ is nilpotent of class at most $n - 1$.*

Let $L$ be a Lie ring, $B$ and $C$ be ideals of $L$ such that $B \le C$. The factor $C/B$ is called $L$-central, if $Ann_L(C/B) = L$. If $A$ is an ideal of $L$, then we define *the upper $L$-central series*

$$\langle 0 \rangle = \zeta_{0,L}(A) \le \zeta_{1,L}(A) \le \cdots \zeta_{\alpha,L}(A) \le \zeta_{\alpha+1,L}(A) \le \cdots \zeta_{\gamma,L}(A)$$
$$= \zeta_{\infty,L}(A)$$

of an ideal $A$ by the following rule $\zeta_{1,L}(A) = A \cap \zeta(L)$, and recursively

$$\zeta_{\alpha+1,L}(A)/\zeta_{\alpha,L}(L) = \zeta(L/\zeta_{\alpha,L}(A)) \cap A/\zeta_{\alpha,L}(A)$$

for all ordinals $\alpha$ and $\zeta_{\lambda,L}(A) = \bigcup_{\mu<\lambda} \zeta_{\mu,L}(A)$ for limit ordinals $\lambda$. The last term $\zeta_{\infty,L}(A)$ of this series is called *the upper $L$-hypercenter* of $A$. From this definition we can see that every term $\zeta_{\alpha,L}(A)$ of the upper $L$-central series of $A$ is an ideal of $L$.

The factor $C/B$ is called *$L$-eccentric*, if $Ann_L(C/B) \neq L$. An ideal $C$ of $L$ is said to be $L - hypereccentric$, if it has an ascending series

$$\langle 0 \rangle = C_0 \leq C_1 \leq \cdots C_\alpha \leq C_{\alpha+1} \leq \cdots C_\gamma = C$$

of ideals of $L$ such that each factor $C_{\alpha+1}/C_\alpha$ is an $L$-eccentric and $L$-chief for every $\alpha < \gamma$.

We say that the ideal $A$ of $L$ has the *$Z$-decomposition*, if

$$A = \zeta_{\infty,L}(A) \oplus \eta_{\infty,L}(A)$$

where $\eta_{\infty,L}(A)$ is the maximal $L$-hypereccentric ideal of $A$. Note that in this case, $\eta_{\infty,L}(A)$ includes every $L$-hypereccentric ideal of $L$. In particular, it is unique.

**Lemma 5.1.** *Let $L$ be a Lie ring and $A$ be an ideal of $L$. Suppose that $A$ satisfies the following conditions:*

(i)   *$A$ is abelian;*
(ii)  *$L/Ann_L(A)$ is hypercentral;*
(iii) *$A$ includes an ideal $C$ of $L$ such that $C \leq \zeta(L)$ and $A/C$ is an $L$-chief factor;*
(iv)  *$Ann_L(A/C) \neq L$.*

*Then $A$ includes an ideal $D$ of $L$ such that $A = C \oplus D$. In particular, $D$ is $L$-eccentric and $L$-chief.*

**Proof.** Let $Y = Ann_L(A)$ and $Y \neq z+Y \in \zeta(L/Y)$. Consider the mapping $\xi_z : A \to A$ defined by the rule $\xi_z(a) = [a, z]$ for each $a \in A$. Clearly this mapping is an additive endomorphism of $A$, $Ker(\xi_z) = Ann_A(z)$, $Im(\xi_z) = [A, z]$, and we have an additive isomorphism $A/Ker(\xi_z) \cong Im(\xi_z) = D$.

Let $x \in L$ and $c \in Ann_A(z)$. Then by the choice of $z$, $[x, z] \in Ann_L(A)$. We have $[[c, x], z] = -[[z, c], x] - [[x, z], c]$. Since $c \in Ann_A(z)$, $[z, c] = 0$. Since $[x, z] \in Ann_L(A)$ and $c \in A$, $[[x, z], c] = 0$. So that $[[c, x], z] = 0$, and $[c, x] \in Ann_A(z)$. It follows that $Ann_A(z)$ is an ideal of $L$.

Let, again, $x \in L$, $a \in A$. We have $[[a, z], x] = -[[x, a], z] - [[z, x], a]$. Since $[z, x] \in Ann_L(A)$, $[[z, x], a] = 0$. Since $A$ is an ideal, $[x, a] \in A$, so that $[[x, a], z] \in [A, z]$, which implies that $[A, z]$ is also an ideal.

By (iii), $C \leq Ker(\xi_z)$. Since $z \notin Ann_L(A)$, $Ker(\xi_z) \neq A$. Then the fact that $A/C$ is an $L$-chief factor implies that $C = Ker(\xi_z)$. Let $K$ be an ideal of $L$ such that $K \leq D$ and $K_1$ be a preimage of $K$ at the mapping $\xi_z$. If $a \in K_1$, $x \in L$, then

$$\xi_z([a, x]) = [[a, x], z] = -[[z, a], x] - [[x, z], a].$$

The choice of $z$ implies that $[x, z] \in Ann_L(A)$, therefore $[[x, z], a] = 0$. Since $a \in K_1$, $[a, z] \in K$. The fact that $K$ is an ideal implies that

$$\xi_z([a, x]) = -[[z, a], x] = [[a, z], x] \in K.$$

In turn it follows that $[a, x] \in K_1$, and $K_1$ is an ideal of $L$. If now $K \neq \langle 0 \rangle$, then $K_1 \neq Ker(\xi_z) = C$. If $K \neq D$, then $K_1 \neq A$. Hence if we assume that $D$ is not $L$-chief, then $A/C$ is also not $L$-chief, and we obtain a contradiction with the condition (iii). This contradiction shows that $D$ is $L$-chief.

Suppose now that $D$ is $L$-central. For arbitrary elements $a \in A$ and $x \in L$ we obtain $[[a, z], x] = [\xi_z(a), x] = 0$. On the other hand

$$[[a, z], x] = -[[x, a], z] - [[z, x], a] = -[[x, a], z],$$

because $[[z, x], a] = 0$. Thus $[[x, a], z] = 0$, which implies that $[x, a] \in Ker(\xi_z) = C$. However this means that factor $A/C$ is $L$-central, which contradicts the condition (iv). This contradiction shows that $D$ is $L$-eccentric.

Since $D$ is $L$-chief, we have the following two possibilities: either $D \cap C = D$ or $D \cap C = \langle 0 \rangle$. In the first case, $D \leq C$, which is impossible since $D$ is $L$-eccentric. Hence $D \cap C = \langle 0 \rangle$. In this case, $(D + C)/C \cong D$ is a non-zero ideal of $L/C$, and the condition (iii) implies that $(D + C)/C = A/C$. Thus $D + C = A$. □

**Corollary 4.** *Let $L$ be a Lie ring and $A$ be an ideal of $L$. Suppose that $A$ satisfies the following conditions:*

(i) *$A$ is abelian;*
(ii) *$L/Ann_L(A)$ is hypercentral;*
(iii) *$A$ has a series*

$$\langle 0 \rangle = C_0 \leq C_1 \leq \cdots \leq C_n = C \leq A$$

*of ideals of $L$ such that the factors $C_j/C_{j-1}$ are $L$-central, $1 \leq j \leq n$, and $A/C$ is $L$-eccentric and $L$-chief factor.*

*Then $A$ includes an ideal $D$ of $L$ such that $A = C \oplus D$. In particular, $D$ is $L$-eccentric and $L$-chief.*

**Lemma 5.2.** *Let $L$ be a Lie ring and $A$ be an ideal of $L$. Suppose that $A$ satisfies the following conditions:*

(i) *$A$ is abelian;*
(ii) *$L/Ann_L(A)$ is hypercentral;*
(iii) *$A$ includes an ideal $C$ of $L$ such that $Ann_L(C) \neq L$ and $C$ is $L$-chief;*
(iv) *$A/C \leq \zeta(L/C)$.*

*Then $A$ includes an ideal $D$ of $L$ such that $A = C \oplus D$. In particular, $D$ is $L$-central.*

**Proof.** Let $Y = Ann_L(A)$ and $Y \neq z + Y \in \zeta(L/Y)$. Consider the mapping $\xi_z : A \to A$, defined by the rule $\xi_z(a) = [a, z]$ for each $a \in A$. Clearly, this mapping is an additive endomorphism of $A$, $Ker(\xi_z) = Ann_A(z)$, $Im(\xi_z) = [A, z]$ and we have an additive isomorphism $A/Ker(\xi_z) \cong Im(\xi_z) = D$. As in Lemma 5.1 we can prove that $Ann_A(z)$ and $[A, z]$ are ideals of $L$.

By our condition, $[A, z] \leq C$. Since $C$ is $L$-chief, then either $[A, z] = C$ or $[A, z] = \langle 0 \rangle$. However, in the second case, $z \in Ann_L(A)$, and we obtain contradiction with the choice of element $z$. Thus $Im(\xi_z) = C$.

Suppose that $[C, z] = \langle 0 \rangle$. Then we obtain the inclusion $C \leq Ker(\xi_z)$. Since $A/C$ is $L$-central, $A/Ker(\xi_z)$ is also $L$-central. Put $K = Ker(\xi_z)$. Let $c$ be an arbitrary element of $C$. Then $c = [a, z]$ for some element $a \in A$. Since $A/K$ is $L$-central, $K = [a + K, x + K]$ for every element $x \in L$. The equation $[a + K, x + K] = [a, x] + K$ implies that $[a, x] \in K = Ker(\xi_z)$. In its turn, it implies the equation $[[a, x], z] = 0$. Then

$$[c, x] = [[a, z], x] = -[[x, a], z] - [[z, x], a] = [[a, x], z] + [[x, z], a].$$

Since $[x, z] \in Ann_L(A)$, $[[x, z], a] = 0$. Therefore $[c, x] = [[a, x], z] = 0$. However, it means that $C$ is $L$-central, and we obtain contradiction with the choice of $C$. This contradiction shows that $[C, z] \neq \langle 0 \rangle$. Let $b$ be an arbitrary element of $[C, z]$ and $x$ be an arbitrary element of $L$. Then $b = [u, z]$ for some element $u \in C$. We have

$$[b, x] = [[u, z], x] = -[[x, u], z] - [[z, x], u] = [[u, x], z] + [[x, z], u].$$

Since $[x, z] \in Ann_L(A)$, $[[x, z], u] = 0$. Then $[b, x] = [[u, x], z]$. The fact that $C$ is an ideal of $L$ implies that $[u, x] \in C$, and therefore $[[u, x], z] \in [C, z]$. This proves that $[C, z]$ is an ideal of $L$. Since $[C, z] \neq \langle 0 \rangle$, the condition

(iii) implies that $[C, z] = C = [A, z]$. Then for every element $a \in A \setminus C$, there exists an element $v \in C$ such that $[a, z] = [v, z]$. It follows that $[a - v, z] = 0$. The choice of the element $a$ yields that $a - v \neq 0$. The equation $a = v + (a - v)$ shows that $A = C + Ann_A(z)$. We noted above that $Ann_A(z)$ is an ideal of $L$. Finally, $C \cap Ann_A(z)$ is an ideal of $L$, which means that either $C \cap Ann_A(z) = \langle 0 \rangle$ or $C \cap Ann_A(z) = C$. As we observed above, the equation $C \cap Ann_A(z) = C$ is impossible. Hence

$$A = C \oplus Ann_A(z) = C \oplus D. \qquad \square$$

**Corollary 5.** *Let $L$ be a Lie ring and $A$ be an ideal of $L$. Suppose that $A$ satisfies the following conditions:*

(i) *$A$ is abelian;*
(ii) *$L/Ann_L(A)$ is hypercentral;*
(iii) *$A$ has a series*

$$\langle 0 \rangle = C_0 \leq C_1 \leq \cdots \leq C_n = C \leq A$$

*of ideals of $L$ such that the factors $C_j/C_{j-1}$ are $L$-eccentric and $L$-chief, $1 \leq j \leq n$, and $A/C$ is an $L$-central factor.*

*Then $A$ includes an ideal $D$ of $L$ such that $A = C \oplus D$.*

Corollaries 4 and 5 imply

**Proposition 4.** *Let $L$ be a Lie ring and $A$ be an abelian ideal of $L$. Suppose that $A$ has a finite series of ideals of $L$, whose factors are either $L$-central or $L$-eccentric and $L$-chief. If the factor-ring $L/Ann_L(A)$ is nilpotent, then $A$ has the $Z$-decomposition.*

**Corollary 6.** *Let $L$ be a Lie ring and $A$ be an abelian ideal of $L$. Suppose that $A$ is finite. If the factor-ring $L/Ann_L(A)$ is nilpotent, then $A$ has the $Z$-decomposition.*

In fact, since $A$ is finite, it has a finite $L$-chief series of ideals of $L$, and we can apply Proposition 4.

Now we are ready to prove Theorem D.

**Proof of Theorem D.** Put $Z = \zeta_n(L)$ and $C = Ann_L(Z)$. Then Proposition 2 shows that the factor-ring $L/C$ is isomorphic to some subring of $Der(Z)$. Furthermore, $L/C$ is isomorphic to some subring of the stabilizer of the series

$$\langle 0 \rangle = \zeta_0(L) \leq \zeta_1(L) \leq \zeta_2(L) \leq \cdots \leq \zeta_n(L) = Z.$$

Using Proposition 3 we obtain that $L/C$ is nilpotent. The intersection $B = Z \cap C$ is an ideal of $L$ (in particular, of $C$), and $B \leq \zeta(C)$. Using this inclusion and the isomorphism $C/(Z \cap C) \cong (C + Z)/Z$ we obtain that $\Pi(C/\zeta(C)) \subseteq \Pi(L/Z)$ and $C/\zeta(C)$ is finite and has order dividing $d$. The application of Theorem A shows that the derived ideal $[C, C] = D$ of $C$ has finite order at most $d^{\varepsilon(d)}$, and $\Pi([C, C]) \subseteq \Pi(C/\zeta(C)) \subseteq \Pi(L/Z)$. The factor-ring $C/D$ is abelian. We remark that $D$ is an ideal of $L$. Indeed, let $x, y$ be arbitrary elements of $C$ and $z$ be an arbitrary element of $L$. Then $[[x, y], z] = -[[z, x], y] - [[y, z], x]$. The fact that $C$ is an ideal of $L$ implies that $[z, x], [y, z] \in C$, which implies that $[[x, y], z] \in D$. Using ordinary induction, we obtain that $[u, z] \in D$ for every element $u \in D$.

Since $C/D$ is abelian, $C/D \leq Ann_{L/D}(C/D)$. It follows that

$$(L/D)/Ann_{L/D}(C/D)$$

is a nilpotent Lie ring. Then by Proposition 4, $C/D$ has the $Z$-decomposition: $C/D = \zeta_{\infty, L/D}(C/D) \oplus \eta_{\infty, L/D}(C/D)$. From the choice of $Z$ we obtain the inclusion $(Z + D)/D \cap C/D \leq \zeta_{\infty, L/D}(C/D)$, which shows that $\Pi((C/D)/\zeta_{\infty, L/D}(C/D)) \subseteq \Pi(L/Z)$, $(C/D)/\zeta_{\infty, L/D}(C/D)$ is finite and has order dividing $d$. As a result, it follows that the ideal $E/D = \eta_{\infty, L/D}(C/D)$ has finite order at most $d$, and $\Pi(E/D) \subseteq \Pi(L/Z)$. The isomorphism

$$C/E \cong (C/D)/(E/D) \cong \zeta_{\infty, L/D}(C/D)$$

shows that $C/E \leq \zeta_{\infty, L/E}(C/E)$. Since $L/C$ is nilpotent, $L/E$ is also nilpotent. It follows that $R \leq E$. Finally, $|E| = |D||E/D| \leq d^{\varepsilon(d)}d = d^{\varepsilon(d)+1} \leq d^{log_2(d)+1}$ and $\Pi(R) \subseteq \Pi(E) \subseteq \Pi(L/Z)$. $\qquad\square$

## References

[1] R. Amayo and I. Stewart. Infinite Dimensional Lie Algebras, Leyden, The Netherlands, Noordhoff International Publishing, 1974.

[2] R. Baer. Representations of groups as quotient groups. II. Minimal central chains of a groups, Trans. Amer. Math. Soc., 1945, **58**, 348–389.

[3] R. Baer, Endlichkeitskriterien für Kommutatorgruppen, Math. Ann., 1952, **124**, 161–177.

[4] A.M. Galmak. On theorem of Schur for $n$-are groups, Ukrainian Math. J, 2006, **58**, 730–741.

[5] P. Hall. Nilpotent groups, Notes of lectures given at the Canadian Mathematical Congress, summer seminar, University of Alberta, Edmonton, 12–30 August, 1957.

[6]   L.A. Kurdachenko, A.A. Pypka, and I.Ya. Subbotin. On some relations between the factors of the upper and lower central series in Lie algebras, Serdica Math. J., 2015, **41**, 293–306.

[7]   B.H. Neumann, Groups with finite classes of conjugate elements, Proc. Lond. Math. Soc., 1951, **1**, 178–187.

[8]   B.H. Neumann, Groups covered by permutable subsets, J. London Math. Soc., 1954, **29**, 236–248.

[9]   F. Saeedi and B. Veisi. On Schur's theorem and its converses for $n$-Lie algebras, Linear Multilinear Algebra, 2014, **62**, 1139–1145.

[10]  I. Schur. Über die Darstellungen der endlichen Gruppen durch gebrochene lineare substitutionen, J. Reine Angew. Math., 1904, **127**, 20–50.

[11]  I. Stewart. Infinite-dimensional Lie algebras in the spirit of infinite group theory, Compos. Math., 1970, **22**, 313–331.

[12]  L. Stitzinger and R. Turner. Concerning derivations of Lie algebra, Linear Multilinear Algebra, 1999, **45**, 329–331.

[13]  M.R. Vaughan-Lee. Metabelian BFC $p$-groups, J. London Math. Soc., 1972, **5**, 673–680.

[14]  J. Wiegold. Multiplicators and groups with finite central factor-groups, Math. Z., 1965, **89**, 345–347.

## Chapter 12

## Log-Space Complexity of the Conjugacy Problem in Wreath Products

Alexei Myasnikov,[1] Svetla Vassileva,[2] and Armin Weiss[3]

[1] *Department of Mathematics, Stevens Institute*
*Hoboken, New Jersey 07039, United States*
*amiasnikov@gmail.com*

[2] *Department of Mathematics, McGill University*
*Montreal, Quebec H3A 0B9, Canada*
*svassileva@math.mcgill.ca*

[3] *Department of Mathematics, Stevens Institute*
*Hoboken, New Jersey 07030, United States*
*weiss@stevens.edu*

*Dedicated to Gerhard Rosenberger and Dennis Spellman*
*on their Seventieth Birthdays*

**ABSTRACT.** We show a transfer result from individual groups to wreath products. Namely, we prove that the conjugacy problem in the wreath product $A \wr B$ of two groups $A$ and $B$ is log-space decidable, provided the factor groups $A$ and $B$ both have log-space decidable conjugacy problem and $B$ has log-space computable power problem. If, additionally, $A$ and $B$ have bounded torsion and $A$ has log-space computable power problem, we show that the iterated wreath product $A \wr^n B$ also has log-space decidable conjugacy problem. We apply these general results to show that free solvable groups and wreath products of two abelian groups (in particular, the lamplighter group, $\mathbb{Z}_2 \wr \mathbb{Z}$) have log-space decidable conjugacy problem.

## 1. Introduction

In this paper we prove that the conjugacy problem in free solvable groups, as well as in iterated wreath products of finitely generated abelian groups, has log-space complexity. The main technical result of the paper is that if $A$ and $B$ are finitely generated groups with log-space decidable conjugacy problems and the power problem in $B$ is log-space decidable then the conjugacy problem in the wreath product $A \wr B$ is also log-space decidable.

The log-space complexity is an important part of the modern theory of complexity of computation. We do not discuss basics of the log-space complexity here instead referring to the books [1, 9, 11], where various aspects of log-space complexity are thoroughly laid out. However, it is worthwhile to mention here that the log-space complexity shows up naturally and plays an important part in several areas of the theory of computation: in space complexity, parallel computations, circuit complexity, first-order definable languages, etc. It is also important to note that log-space (denoted by $\mathbb{L}$ in the sequel) implies polynomial time ($\mathbb{P}$) and that a standard conjecture in complexity theory states that $\mathbb{L} \subsetneq \mathbb{P}$. Nowadays, it seems that $\mathbb{L}$, as a complexity measure, is more natural and important, than $\mathbb{P}$. Surprisingly, there are not many results on log-space complexity of algorithmic problems in algebra, in particular in group theory. One of the most influential results in this area claims that the word problem in linear groups is in $\mathbb{L}$ (Lipton and Zalcstein [7]). Notice, that this is the only known proof that the word problem in such groups is in $\mathbb{P}$. Computing normal forms is a powerful technique in combinatorial and geometric group theory, it solves the word problem in groups and gives an approach to other group-based algorithmic problems. Elder, Elston, and Ostheimer showed in [4] that the direct product, free product, and wreath product of two groups with log-space computable normal forms all have log-space computable normal forms. They also showed that solvable Baumslag-Solitar groups have log-space computable normal forms. Furthermore, Diekert, Kausch and Lohrey proved in [2] that Coxeter groups, and in particular right-angled Artin groups, have log-space computable normal forms. These results provide a solid foundation for current research of log-space complexity of general algorithmic problems in groups. Following this route Vassileva in her PhD thesis [10] studied the space complexity of some algorithmic problems in groups, including the conjugacy and power problems in wreath products. In [3] Elder and Kalka proved that the conjugacy problem in some classes of Garside groups is also in $\mathbb{L}$. Recently, it was shown by Macdonald, Miasnikov, Nikolaev, and Vassileva in [5] that most of the standard algorithmic problems (not including the isomorphism problem) in finitely generated nilpotent groups are decidable in log-space.

It would be interesting to verify if other well-known algorithmic problems in groups which are decidable in polynomial time, are in fact decidable in log-space.

## 2. Preliminaries

### 2.1. *Log-space computability*

Let us begin by introducing the model of computation. Intuitively, it is a deterministic Turing machine, called *c*-transducer, on three tapes: input, work and output tape and the work tape has a number of cells logarithmic in the size of the input tape. More precisely, a *c-transducer* is a tuple $(S, s_0, \Sigma, \beta, \delta, c)$, where

- $S$ is a finite set, called the set of *states*,
- $s_0 \in S$ is the start state,
- $\Sigma$ is a finite alphabet,
- $\beta \in \Sigma$ is a special 'blank' symbol,
- $\delta : S \times \Sigma \times \Sigma \rightarrow (S \cup \{\text{halt}\}) \times \Sigma \times \Sigma \times \{1, 0, -1\} \times \{1, 0, -1\} \times \{0, 1\}$
  is called the *transition function* and
- $c > 0$ is a constant.

A *tape* is an infinite sequence $X = \{x_n\}_n$ with $x_n \in \Sigma$. Intuitively, we think of it as a one-ended infinite array of cells, each cell holding an element of this sequence. To every tape $X$ we associate a pointer $h_X$, called the *head*, which holds the index of the current cell (head position).

A *configuration*, $C_i$, of the c-transducer consists of a tuple $\left(s_i, (X_i, h_X^{(i)}), (Y_i, h_Y^{(i)}), (Z_i, h_Z^{(i)})\right)$ which records the current state and the three tapes $X_i, Y_i, Z_i$ (input, work, and output) with their corresponding head positions. The tapes $Y_0$ and $Z_0$ are set to be the constant sequence $\{\beta\}_n$. A *run* of the c-transducer is defined to be a (finite) sequence $\{C_i\}_{i=0}^N$ of configurations such that each configuration in the sequence is obtained from the previous one by an application of the transition function in the following manner. Given $C_i$, let $\delta(s_i, x_{h_X^{(i)}}, y_{h_Y^{(i)}}) = (s_j, a, b, \epsilon_X, \epsilon_Y, \epsilon_Z)$. Then $C_{i+1}$ is given by the tuple $\left(s_{i+1}, (X_{i+1}, h_X^{(i+1)}), (Y_{i+1}, h_Y^{(i+1)}), (Z_{i+1}, h_Z^{(i+1)})\right)$, where

- $s_{i+1} = s_j$
- $X_{i+1} = X_i$ (this means that the input tape is read-only)
- $h_X^{(i+1)} = h_X^{(i)} + \epsilon_X$
- $Y_{i+1} = y_0, \ldots, y_{h_Y^{(i)}-1}, a, y_{h_Y^{(i)}+1}, \ldots$
- $h_Y^{(i+1)} = h_Y^{(i)} + \epsilon_Y$

- $Z_{i+1} = z_0, \ldots, z_{h_Z^{(i)}-1}, b, z_{h_Z^{(i)}+1}, \ldots$
- $h_Z^{(i+1)} = h_Z^{(i)} + \epsilon_Z$

In casual language, the machine reads (according to the head position and the current state), a letter from the input tape, and a letter on the work tape, then prints a letter at the current position of the work tape, changes the state, moves the heads and prints a letter on the output tape. Furthermore, in order for a run to be valid, we require that $h_X, h_Y, h_Z \geq 0$. This is to prevent the head from "falling off" of the tape.

We call the sequence of non-blank symbols on $X_0$ the *input* and the sequence of non-blank symbols on $Z_N$ the *output*. The input and the output may be viewed as words in $\Sigma$ (without the blank symbol).

Observe that the $c$-transducer is deterministic: once a specific $\delta$ is fixed (i.e., once the machine is specified) every sequence of configurations is completely determined by the initial configuration. In other words, the output of a given machine is uniquely determined by its input.

We say that a $c$-transducer is a *$c$-log-space transducer* if the following holds for all possible runs $\{C_i\}_{i=0}^n$,

$$\max_{0 \leq i \leq n} \left\{ \max\{j \mid (Y_i)_j \neq \beta\} \right\} \leq c \log \left( \max\{i \mid X_i \neq \beta\} \right).$$

That is, the maximum number of cells used on the work tape during any run is logarithmically bounded by the number of cells used on the input tape. A c-transducer is called simply a *log-space transducer* (henceforth abbreviated LST) if it is a c-log-space transducer for some constant $c > 0$.

Given two finite alphabets $X_1$ and $X_2$, we say that a function $f : X_1^* \to X_2^*$ is *log-space computable* if there is a LST which for every input $w \in X_1^*$ produces $f(w)$ on the output tape. A decision problem, which we regard as a subset of $X_1^*$, is *log-space decidable* if its characteristic function is log-space computable. Note that in order to discuss multi-variable functions, we simply add a 'special symbol', $\alpha$, to the alphabet and separate the input words by this symbol. We then regard a decision problem as a subset of $(X_1 \cup \{\alpha\})^*$ and the above definitions apply.

Any function that is log-space computable is also computable in polynomial time. Indeed, since the machine is deterministic and any run is de facto generated by successive applications of the transition function on the initial configuration, there cannot be two configurations $C_i$ and $C_j$ such that $s_i = s_j$, $h_X^{(i)} = h_X^{(j)}$, $Y_i = Y_j$ and $h_Y^{(i)} = h_Y^{(j)}$, as otherwise the machine will enter an infinite loop. Since a run is by definition finite, this is impossible. Thus, the length of any given run (and hence the time complexity of the machine) is bounded by $|S| \cdot n \cdot c \log n \cdot |\Sigma|^{c \log n} \in O(n^{c+2})$, where $n$ is the length of the input.

Note that the type of computations that can be done in log-space are quite limited. For example, we can count up to the size of the input and we can maintain pointers to the input. In order to count up to a number $M$, we need $\log M$ bits; so if we denote by $n$ the length of the input string, we can count up to $n^k$, where $k$ is any constant. Similarly, a pointer holds a value between $0$ and $n$ (where $n$ is the length of the input string), so any pointer can be stored using at most $\log n$ cells. We purposefully do not specify the base of the logarithm: if the encoding is binary, use $\log_2$; the complexity is unaffected as long as we do not use a unary encoding.

Moreover, one can compose log-space transducers (i.e., the class of log-space computable functions is closed under composition). To show this, given machines for computing two functions $f$ and $g$, we build a new machine, which, on input $x = x_1 \ldots x_n$ produces $f(g(x))$ on its output tape. First, we slightly modify the transducer for $g$ by adding a counter, $p$ (initialized to 1), and a cell, $y$, to its work tape. The counter will only count up to $n$, so it uses $\log n$ cells (thus, we are adding $\log n + 1$ cells to the original work tape). To compute the $i^{\text{th}}$ letter, $g(x)[i]$, of the output of $g$, run the machine for $g$. Whenever the machine would write on the output tape, (over)write in $y$ instead, increment the counter and continue in this manner until the counter reaches $i$; then stop running $g$. The cell $y$ now holds $g(x)[i]$. Now, run $f$ on $x$ and whenever it requests the $i^{\text{th}}$ letter of the input, compute and use $g(x)[i]$ instead. Thus $f$ will write $f(g(x))$ on its output tape. Note that the machine that we built still uses logarithmic space, since it was built out of the machines for $f$ and $g$ and we only added the counter $p$ and a cell.

We gave a description of log-space decidable problems in terms of LST. In practice, we work with algorithms which correspond to transducers. In fact, one can think of a transducer as a formal unambiguous way to encode an algorithm. Our algorithms are presented in pseudo-code with the following special key words to distinguish between which tapes we are working with.

**return** Prints on the output tape and terminates the execution of the algorithm.

**print** Prints on the output tape and moves the head to the first empty position.

As discussed above, given any log-space algorithm $\mathcal{A}$, we can modify it to obtain a log-space algorithm $\mathcal{A}_i$, which returns the $i^{\text{th}}$ letter, $\mathcal{A}[i]$, of the output of $\mathcal{A}$. We will use this notation in subsequent algorithms.

## 2.2.  *Wreath products and representing their elements*

For two groups $A$ and $B$, $A^{(B)}$ denotes the set of functions of finite support from $B$ to $A$, and forms a group under point-wise multiplication. Mapping $a \in A$ to a function

$$a(b) = \begin{cases} a & \text{if } b = 1, \\ 1 & \text{otherwise,} \end{cases}$$

gives an embedding of $A$ into $A^{(B)}$, termed the canonical embedding. In what follows we often identify $A$ with its canonical image in $A^{(B)}$. The *wreath product* $A \wr B$ of $A$ and $B$ is defined as the semi-direct product $B \ltimes A^{(B)}$, where the action of $b \in B$ on a function $f \in A^{(B)}$ is defined by $f^b(x) = f(xb^{-1})$. We identify $B$ and $A^{(B)}$ (hence $A$) with their canonical images in $A \wr B$.

Let $X$ and $Y$ be some fixed generating sets of $A$ and $B$, correspondingly. It is easy to see that $A \wr B$ is generated by $X \cup Y$, to which we refer as the generating set of $A \wr B$. The relation $[a^b, (a')^{b'}] = 1$ holds for any $a, a' \in A$ and $b, b' \in B$. Using the relations above a given word $w$ in the generators of $A \wr B$ can be rewritten in the "standard form"

$$w = b A_1^{B_1} \dots A_k^{B_k}, \tag{1}$$

where $b \in B$, $A_1, \dots, A_k$ are non-trivial elements of $A$, and $B_1, \dots, B_k$ are pair-wise distinct elements of $B$. Here the product $A_1^{B_1} \dots A_k^{B_k}$ can be viewed as a function $f : B \to A$, namely $f = \{(B_1, A_1), \dots, (B_k, A_k)\}$. Indeed, since $A_i^{B_i}$ and $A_j^{B_j}$ commute, the order of the factors $A_1^{B_1}, \dots, A_k^{B_k}$ in equation (1) is immaterial. In this case we write $w = bf$.

**Lemma 1.** *Let $A$ and $B$ be finitely generated groups such that $B$ has log-space decidable word problem. Then there is a log-space algorithm which, given a product $x = x_1 \dots x_m$ of generators of $A \wr B$, outputs a pair $(b, f)$ such that $x = bf$ as an element of $A \wr B$.*

**Proof.** Given a word $w$ in generators $X \cup Y$, it can be considered as a product of the form $a_1 b_1 \dots a_n b_n$, where each $a_i$ is a word in generators $X$ and each $b_j$ is a word in generators $Y$. Clearly, there is a log-space algorithm which computes for each $1 \le i \le n$ the starting and ending index of $a_i$ ($b_i$ respectively) in $w$. Consequently, the subsequent algorithms will refer to the factors $a_i$ and $b_i$.

Given an algorithm, $\mathcal{A}_{\mathrm{WB}}$ for the word problem in $B$, we can modify it to take as input a product of the type $a_1 b_1 \dots a_n b_n$ by instructing it to treat any $a_j$ as the identity of $B$. We will refer to this algorithm as $\overline{\mathcal{A}}_{\mathrm{WB}}$.

Given a word $w$ in generators of $A \wr B$, we write

$$w = a_1 b_1 \dots a_n b_n = b_1 b_1^{-1} a_1 b_1 \dots a_n b_n = b_1 a_1^{b_1} a_2 b_2 \dots a_n b_n$$
$$= b_1 b_2 (a_1^{b_1} a_2)^{b_2} \dots a_n b_n = b_1 b_2 (a_1^{b_1 b_2} a_2^{b_2}) \dots a_n b_n$$
$$= b_1 \dots b_n \cdot a_1^{b_1 \dots b_n} \dots a_n^{b_n}$$

Then, we ensure that the exponents are distinct: whenever two exponents $b_i \dots b_n$, $b_j \dots b_n$ are equal as elements of $B$, use the commutativity relations of the wreath product to combine the corresponding terms into $(a_i a_j)^{b_i \dots b_n}$. Thus, we can rewrite $w$ as the product $b \cdot A_1^{B_1} \dots A_k^{B_k}$, where $A_1, \dots, A_k \in A$ are non-trivial, $b \in B$ and $B_1, \dots, B_k \in B$ are distinct. Since $B_1, \dots, B_k$ are distinct and $A_1, \dots, A_k$ are non-trivial, the product $A_1^{B_1} \dots A_k^{B_k}$ corresponds to a function $f : B \to A$ which is given by $f = \{(B_1, A_1), \dots, (B_k, A_k)\}$. In other words, a word $w$ in generators of $A \wr B$ can be rewritten as a pair $bf$.

Algorithm 1 below computes this pair. More precisely, it prints $b$ followed by $f$, represented as a list of input-output pairs. The algorithm runs in logarithmic space since it only relies on counting and running an algorithm for the word problem in $B$ (which we assumed is in log-space). $\square$

### 2.3. *Useful algorithms*

We begin by describing an algorithm for sorting in log-space. Let $X$ be an array of integers in binary form and denote by $N$ the length of $X$. To find and print the smallest element of the array we run through the array and keep a pointer to our best candidate so far. In the end, we simply print that candidate as the smallest element of the array. The next task is to find the next smallest element. We proceed in a similar manner — we run through the array, and at every step we check whether the element we are looking at is between the latest printed minimal one and our current candidate. If it is, we update the candidate. Otherwise, we continue. In the end, we print this candidate. Running through the array $N$ times in this fashion produces a sorted version of the array on the output tape. This algorithm is clearly in log-space, since at any point we only store three

---

**Input:** A product $a_1b_1 \ldots a_nb_n$, where the $a_i, b_j$ are words in the
   generators of $A$ and $B$, respectively.
**Output:** A pair $bf = \big(b, \{(B_1, A_1), \ldots, (B_k, A_k)\}\big)$ such that
   $a_1b_1 \ldots a_nb_n = bf$ as an element of $A \wr B$.
**print** $b_1 \ldots b_n$;
**for** $i = 1, \ldots, n$ **do**
    **for** $j = 1, \ldots, i-1$ **do**
        **if** $\overline{\mathcal{A}}_{\mathrm{WB}}(b_j \ldots b_{i-1}) =$ '*Yes*' **then**
            AlreadyPrinted := 'Yes';
    **if** AlreadyPrinted $=$'*No*' **then**
        **print** $\big(b_i \ldots b_n, a_i$;
        **for** $j = i+1, \ldots, n$ **do**
            **if** $\overline{\mathcal{A}}_{\mathrm{WB}}(b_i \ldots b_j) =$'*Yes*' **then**
                **print** $a_j$;
        **print** );

**Algorithm 1:** Rewriting $a_1b_1 \ldots a_nb_n$ as an element of $A \wr B$.

pointers (and each pointer has at most $\log N$ bits). Observe that we sorted
the array in increasing order of its elements, but a similar algorithm can
produce a sorting in decreasing order.

The *power problem* asks to determine whether or not for two group
elements $x$ and $y$, $x^n = y$ for some $n$ and if so, find the *smallest* non-zero
such $n$. We assume that the return value $n$ is encoded in binary. The power
problem is akin to the membership search problem for a cyclic subgroup.
The restriction that $n$ be the smallest power which works is a technicality
which is only relevant in Proposition 9. This requirement is not used by
any of the other algorithms presented in this paper. The reason for this
extra condition is an idiosyncrasy of log-space computation: if one has an
algorithm which finds an $n$ such that $x^n = y$ in logarithmic space in the
lengths of $x$ and $y$, it is not *a priori* clear how to find the *smallest* $N$ such
that $x^N = y$ in logarithmic space. The obvious algorithm which consists of
checking every $N < n$ is not in log-space, since one cannot store $n$ in space
logarithmic in the lengths of $x$ and $y$.

Let $G$ be a group with log-space decidable power problem and let $H = \{h_1, \ldots, h_n\}$ be an ordered finite subset of $G$. Fix an element $g \in G$. We
are interested in algorithmically partitioning $H$ into $\sim_g$ equivalence classes
where $\sim_g$ is the equivalence relation defined by $h_i \sim_g h_j$ if $h_i\langle g\rangle = h_j\langle g\rangle$.

Denote by $T = \{t_1, \ldots, t_k\}$ the (ordered) set of representatives of these equivalence classes. They are, in fact, $\langle g \rangle$-coset representatives in $G$.

To compute them, we simply run through $H$ checking for each $i$ whether $h_i$ falls in the same coset as some $h_j$ with $j < i$. If not, we print $h_i$. If yes, we move on. We will denote this algorithm by $\mathcal{A}_{\text{coset}}(g, H)$. Observe that by adding a counter which we augment every time we print on the output tape, we can obtain the number of distinct coset representatives (also in log-space).

In order to decide the conjugacy of two elements $bf, cg$ in a wreath product $A \wr B$ we will study the effect of $f$ and $g$ on cosets of $\langle b \rangle$ in $B$. Let $S = \{t_i\}$ be a set of $\langle b \rangle$-coset representatives in $B$. Then, for $\gamma, b \in B$ and $f \in A^{(B)}$, define

$$
\pi_{t_i}^{(\gamma)}(f) = \begin{cases} \displaystyle\prod_{j=0}^{N-1} f(t_i b^j \gamma^{-1}) & \text{if order}(b) = N < \infty, \\[2em] \displaystyle\prod_{j=-\infty}^{\infty} f(t_i b^j \gamma^{-1}) & \text{if order}(b) = \infty. \end{cases}
$$

We denote $\pi_{t_i}^{(1)}(f)$ by $\pi_{t_i}(f)$. These functions are essential to the solution of the conjugacy problem in wreath products.

The definition of the maps $\pi_{t_i}$ depends on the order of the element $b$. First, observe that in the case when the order of $b$ is infinite the product is finite, since the function $f$ is of finite support. In fact, it is the product of all possible non-trivial factors of the form $f(t_i b^j \gamma^{-1})$ multiplied in increasing order of $j$. The same is true in the case when the order of $b$ is finite. So in order to compute $\pi_{t_i}^{(\gamma)}$, we need to find all the elements of the form $t_i b^j \gamma^{-1}$ for which $f$ is non-trivial, order them in increasing order of $j$ and multiply them out. Algorithm 2 computes a list of pairs $(j, v_p)$ for which $t_i b^j \gamma^{-1} = v_p \in \text{supp}(f)$.

---

**Input:** $f = \{(v_1, u_1), \ldots, (v_k, u_k)\}, t_i, b, \gamma$
**Output:** $L = \{(j_1, v_p^{(1)}), \ldots, (j_t, v_p^{(t)})\}$, where $(j_q, v_p^{(q)})$ satisfy
$\qquad t_i b^{j_q} \gamma^{-1} = v_p^{(q)}$
**for** $p = 1, \ldots, k$ **do**
$\qquad$ **if** $\mathcal{A}_{\text{PB}}(b, t_i^{-1} v_p \gamma) = j$ **then**
$\qquad\qquad$ **print** $(j, v_p)$;

---

**Algorithm 2:** $\mathcal{A}_{\text{aux}}$

Since we assume that $\mathcal{A}_{\mathrm{PB}}$ is log-space computable, and $|(b, t_i v_p \gamma^{-1})| \leq |\mathrm{Input}(\mathcal{A}_{\mathrm{aux}})|$, we only use a logarithmic number of cells on the work tape at every step.

Finally, it is easy to see that the algorithm $\mathcal{A}_\pi = \mathcal{A}_{\mathrm{sort}} \circ \mathcal{A}_{\mathrm{aux}}$ computes the map $\pi_{t_i}$. Since $\mathcal{A}_{\mathrm{sort}}$ and $\mathcal{A}_{\mathrm{aux}}$ are log-space computable, so is $\mathcal{A}_\pi$. We will only use the algorithm $\mathcal{A}_\pi$ in the case where $\gamma = 1$, so in order to simplify the notation, we will assume it takes as input only three arguments: $f$, $t_i$ and $b$. Moreover, from the definition of $\pi_{t_i}$, it follows that for any function $f$,

$$|\pi_{t_i}^{(\gamma)}(f)| \leq |\mathrm{supp}(f)|.$$

### 2.4.  *On the conjugacy problem for $A \wr B$*

In [6], Matthews provides a criterion for testing whether two elements of a wreath product are conjugate.

**Theorem 2 ([6]).** *Let $A$ and $B$ be groups. Two elements $x = bf$ and $y = cg$ in $A \wr B$ are conjugate if and only if there exists $d \in B$ such that*

- *$db = cd$ in $B$ and*
- *if $\mathrm{order}(b)$ is finite, $\pi_{t_i}(f)$ is conjugate to $\pi_{t_i}^{(d)}(g)$ for all $t_i \in T$,*
- *if $\mathrm{order}(b)$ is infinite, $\pi_{t_i}(f)$ is equal to $\pi_{t_i}^{(d)}(g)$ for all $t_i \in T$.*

We will denote by $\tilde{\pi}_{s_i}$ the functions defined the same way as $\pi_{t_i}$, but with respect to a set of $\langle c \rangle$-coset representatives $\{s_i\}_i$.

**Lemma 3.** *In the notation above, given any $c \in B$, let $\{s_i\}_{i \in I}$ and $\{\bar{s}_i\}_{i \in I}$ be two sets of $\langle c \rangle$-coset representatives. Then for every $i \in I$ and any function $g \in A^{(B)}$, if $j$ is such that $s_i \langle c \rangle = \bar{s}_j \langle c \rangle$, then*

$$\tilde{\pi}_{s_i}(g) \sim \tilde{\pi}_{\bar{s}_j}(g).$$

**Proof.** Since $s_i \langle c \rangle = \bar{s}_j \langle c \rangle$, there is some integer $p_i$ for which $s_i = \bar{s}_j c^{p_i}$ and hence

$$\tilde{\pi}_{s_i}(g) = \prod_k g(s_i c^k) = \prod_k g(\bar{s}_j c^{p_i} c^k) = \prod_k g(\bar{s}_j c^{p_i + k}).$$

The last product in the above equation is a cyclic permutation of the factors in $\prod_k g(\bar{s}_j c^k) = \tilde{\pi}_{\bar{s}_j}(g)$ and hence is conjugate to $\tilde{\pi}_{\bar{s}_j}(g)$.    $\square$

We are now ready to show that the conjugacy problem in wreath products is decidable in log-space. Denote by $\mathcal{A}_{CA}$ and $\mathcal{A}_{CB}$ two algorithms solving the conjugacy problem in $A$ and $B$ respectively. We use $\mathcal{A}_{CA}(x, 1)$ to determine whether a word $x$ is trivial and we denote this by $\mathcal{A}_{WA}(x)$.

We begin by giving a description of the algorithm for deciding the conjugacy problem in $A \wr B$. We argue the correctness and complexity of the algorithm in Theorem 4 below.

Let $x = bf$ and $y = cg$ be two elements of $A \wr B$. Using $\mathcal{A}_{CB}$, we check whether $b$ and $c$ are conjugate in $B$. If they are not, then $x$ and $y$ are not conjugate in $A \wr B$ and we terminate. In the event $b$ and $c$ are conjugate in $B$, we consider the following three cases.

1. Suppose $g = 1$. Then $x \sim y$ if and only if $\pi_{t_i}(f) = 1$ for all $i$.
2. Suppose $g \neq 1$, but $\pi_{t_i}(f) = 1$ for all $i$. Then $x \sim y$ if and only if $\tilde{\pi}_{s_j}(g) = 1$ for all $j$.
3. Suppose $g \neq 1$ and there exists an $i$ such that $\pi_{t_i}(f) \neq 1$. Denote $\mathrm{supp}(g) = \{\beta_1, \ldots, \beta_m\}$. Then $x \sim y$ if and only if one of $d = \beta_1^{-1} t_i, \ldots, \beta_m^{-1} t_i$ satisfies $db = cd$ and $\pi_{t_i}(f) = \pi_{t_i}^{(d)}(g)$ if $\mathrm{order}(b) = \infty$, or $\pi_{t_i}(f) \sim \pi_{t_i}^{(d)}(g)$ if $\mathrm{order}(b)$ is finite.

**Theorem 4.** *Let $A$ and $B$ be finitely generated groups for which the conjugacy problem is decidable and denote by $\mathcal{A}_{CA}$ and $\mathcal{A}_{CB}$ the corresponding decision algorithms. Given two elements $x$ and $y$ of $A \wr B$, the algorithm outlined above*

(i) *correctly determines whether they are conjugate in $A \wr B$ or not, and*
(ii) *runs in space logarithmic in the length of $x$ and $y$, provided the conjugacy problems in $A$ and $B$ are log-space decidable and the power problem in $B$ is log-space decidable.*

**Proof.**

(i) Let $x = bf$ and $y = cg$ be two elements of $A \wr B$. We begin by verifying that $b$ and $c$ are conjugate. If they are not, then $x$ and $y$ are certainly not conjugate. If they are, the algorithm proceeds according to the following three cases.

- Suppose $g = 1$. Then $\pi_{t_i}^{(d)}(g) = 1$ for all $i$ and therefore, by Theorem 2, in order to check that $x \sim y$, it is enough to check that $\pi_{t_i}(f) = 1$ for all $i$.

- Suppose $g \neq 1$, but $\pi_{t_i}(f) = 1$ for all $i$. We already established that $b \sim c$, therefore by Theorem 2, $x \sim y$ if and only if $\pi_{t_i}^{(d)}(g) = 1$ for all $i$. Now,

$$\pi_{t_i}^{(d)}(g) = \prod_j g(t_i b^j d^{-1}) = \prod_j g(t_i d^{-1} c^j) = \tilde{\pi}_{t_i d^{-1}}(g),$$

and, since by Lemma 3, $\tilde{\pi}_{t_i d^{-1}}(g) \sim \tilde{\pi}_{s_j}(g)$ for any other set of $\langle c \rangle$-coset representatives $\{s_j\}$, it follows that

$$\tilde{\pi}_{t_i d^{-1}}(g) = 1 \text{ for all } i \Leftrightarrow \tilde{\pi}_{s_j}(g) = 1 \text{ for all } j.$$

Thus, to establish that $x \sim y$, it is enough to check that $\tilde{\pi}_{s_j}(g) = 1$ for all $j$.

- Finally, suppose that $g \neq 1$ and $\pi_{t_i}(f) \neq 1$ for some $i$. Denote $\mathrm{supp}(g) = \{\beta_1, \ldots, \beta_m\}$. If $\pi_{t_i}(f)$ is to be conjugate (or equal) to $\pi_{t_i}^{(d)}(g)$, there must be an element $d \in B$ such that $\pi_{t_i}^{(d)}(g) \neq 1$. Find it. Thus we have

$$\exists d \in B, \ \pi_{t_i}^{(d)}(g) \neq 1 \implies \exists l \in \mathbb{Z}, \ g(t_i b^l d^{-1}) \neq 1$$

$$\implies \exists l \in \mathbb{Z}, \ t_i b^l d^{-1} = \beta_p,$$

$$\text{for some} 1 \leq p \leq m,$$

$$\implies \exists l, d = \beta_p^{-1} t_i b^l \text{ for some } 1 \leq p \leq m.$$

$$\implies d = \beta_1^{-1} t_i b^l, \ldots, \beta_m^{-1} t_i b^l.$$

Now observe that we can assume, without loss of generality, that $l = 0$. This is due to the fact that if we can write $d_1 = d_2 b^l$, then

- $d_1 b = c d_1$ if and only $d_2 b = c d_2$ and
- if $\mathrm{order}(b) < \infty$, $\pi_{t_i}(f) \sim \pi_{t_i}^{(d_1)}(g)$ if and only if $\pi_{t_i}(f) \sim \pi_{t_i}^{(d_2)}(g)$,
- if $\mathrm{order}(b) = \infty$, $\pi_{t_i}(f) = \pi_{t_i}^{(d_1)}(g)$ if and only if $\pi_{t_i}(f) = \pi_{t_i}^{(d_2)}(g)$.

Indeed,

- $d_2 b = d_1 b^{-l} b = d_1 b^{1-l} = c^{1-l} d_1 = c \cdot c^{-l} d_1 = c d_2,$
- if $\mathrm{order}(b) = N < \infty$,

$$\pi_{t_i}^{(d_2)}(g) = \prod_{j=0}^{N-1} g(t_i b^j d_2^{-1}) = \prod_{j=0}^{N-1} g(t_i d_2^{-1} c^j)$$

$$= \tilde{\pi}_{t_i d_2^{-1}}(g) \sim \tilde{\pi}_{t_i d_1^{-1}}(g) \sim \pi_{t_i}(f), \quad \text{and}$$

∘ if order$(b) = \infty$,

$$\pi_{t_i}^{(d_2)}(g) = \prod_{j=-\infty}^{\infty} g(t_i b^j d_2^{-1}) = \prod_{j=-\infty}^{\infty} g(t_i b^{j+l} d_1^{-1})$$

$$= \prod_{j=-\infty}^{\infty} g(t_i b^j d_1^{-1}) = \pi_{t_i}^{(d_1)}(g) = \pi_{t_i}(f)$$

Thus, it is enough to check for the values $d = \beta_1^{-1} t_i, \ldots, \beta_m^{-1} t_i$ that $db = cd$ and $\pi_{t_i}(f) \sim \pi_{t_i}^{(d)}(g)$ or $\pi_{t_i}(f) = \pi_{t_i}^{(d)}(g)$ (depending on whether the order of $b$ is finite or infinite).

(ii) We show that under the given assumptions the algorithm is in log-space. Denote by $L$ the length of the input, that is $L = |x| + |y| = |b| + |c| + |f| + |g|$, where the functions $f$ and $g$ are represented as input-output pairs. Recall that all coset representatives $(t_i, s_j, \tilde{s}_j)$ are chosen from supp$(f)$ or supp$(g)$, hence both the length and number of coset representatives is bounded by $L$. Moreover, the algorithm $\mathcal{A}_\pi$ computes the map $\pi_i$ in space logarithmic in $L$.

To check whether $b$ and $c$ are conjugate in $B$ is a conjugacy problem in $B$, which we assume is log-space decidable. To determine whether case 1 occurs, we need to check whether $g = 1$, which is done by simply checking that $|\text{supp}(g)| \neq 0$. If $g = 1$, we compute $\mathcal{A}_\pi(f, t_i, b, 1)$ and solve $|\text{supp}(f)| < L$ word problems in $A$, which can be done in log-space.

If $g \neq 1$, we need to solve $|\text{supp}(f)| < L$ word problems to determine whether case 2 occurs, and if yes, we compute $\mathcal{A}_\pi(g, s_j, c, 1)$ solve another $|\text{supp}(g)| < L$ word problems in $A$ to check that each $\tilde{\pi}_{s_j}(g) = 1$.

If case 3 occurs, we search through supp$(f)$ for an element $t_i$ such that $\pi_{t_i}(f) \neq 1$ and save a pointer to this element in the input. Next, we use the algorithm for the power problem (asking whether there is an integer $N$ such that $b^N = 1$) to determine whether the order of $b$ is infinite or not. In either case, we try $m = |\text{supp}(g)| < L$, possibilities for $d$, and for each possibility we use $\mathcal{A}_{\text{CP}}$ to check the equality (or conjugacy) of $\pi_{t_i}(f)$ and $\pi_{t_i}^{(d)}(g)$.

Observe that we do not save anything in memory, except for two counters (one to keep track of the series of word/conjugacy problems we solve and another to go through the support of $g$ in case 3), and a pointer to the coset representative, $t_i$, that we found at the beginning of case 3.

Finally, all the algorithms that we use as subroutines run in log-space. Thus, the algorithm for deciding the conjugacy problem runs in log-space. □

**Theorem 5.** *Let $A$ and $B$ be finitely generated groups with log-space decidable conjugacy problems and let the power problem in $B$ be log-space decidable. Then the conjugacy problem in $A \wr B$ is also log-space decidable.*

**Proof.** Given two words in generators of $A$ and $B$, we compose Algorithm 1 from Lemma 1 with the algorithm outlined in this section. The former takes two words and converts them to the form which the input of the latter is expecting, and the latter decides whether the elements it is given are conjugate. The conditions on $A$ and $B$ are the conditions required for the second algorithm to run in log-space. □

### 2.5.  *On the conjugacy problem in wreath products of abelian groups*

**Lemma 6.** *The power problem for any finitely generated abelian group is decidable in log-space.*

**Proof.** Let $\mathbb{Z}^r \times \mathbb{Z}_{m_1} \times \ldots \times \mathbb{Z}_{m_k} = \langle x_1, \ldots, x_r \rangle \times \langle z_1 \rangle \times \ldots \times \langle z_k \rangle$ be a cyclic decomposition of a finitely generated abelian group, here $\langle x_1, \ldots, x_r \rangle$ is a free abelian group with basis $x_1, \ldots, x_r$, and $\langle z_i \rangle$ is a cyclic group of order $m_i$ generated by $z_i$. Given elements $x$ and $y$ as words in $\{x_1 \ldots x_r, z_1, \ldots, z_k\}^{\pm 1}$ we can rewrite them as

$$x = x_1^{a_1} \ldots x_r^{a_r} z_1^{a_{r+1}} \ldots z_k^{a_{r+k}} \quad \text{and} \quad y = x_1^{b_1} \ldots x_r^{b_r} z_1^{b_{r+1}} \ldots z_k^{b_{r+k}}$$

in log-space by writing the exponents in binary. Thus, it remains to reduce the exponent $a_{r+i}$ (resp. $b_{r+i}$) of $z_i$ modulo $m_i$ for $1 \leq i \leq k$ in order to obtain unique normal forms for $x$ and $y$ (see also [Murray] how to find normal forms for a free abelian group).

If $r \neq 0$, we begin by checking that the infinite parts of the words match. For each $i = 1, \ldots, r$, there are three possibilities that we need to consider.

(1) $a_i = 0$. In this case, either $b_i$ is also equal to zero, or $b_i \neq 0$ for some $i$ and so $y$ cannot be a power of $x$. If $b_i = 0$, proceed.
(2) $a_i \neq 0$. In this case, we can set $n_i = \frac{b_i}{a_i}$. If this is not an integer, then $y$ cannot be a power of $x$. If it is an integer, we save it. Note that the size

of $n_i$ is perforce logarithmic in the size of $x$, so we can indeed record it on our work tape.

In the case where $a_i = 0$ for all $1 \leq i \leq r$, $x$ and $y$ have no infinite part, so we proceed in the same way as when $r = 0$.

Otherwise, for every $i, j$ such that $a_i, a_j \neq 0$, we check whether $n_i = n_j$. If this is false, then $y$ is not a power of $x$. If, however, it is true for all such $i$ and $j$, we denote this common power by $n$ and proceed to ascertain whether $na_i = b_i \mod m_i$ for $r < i \leq r + k$. If this holds, then $x^n = y$. Otherwise, $y$ is not a power of $x$.

If $r = 0$, then we simply have $x = z_1^{a'_1} \dots z_k^{a'_k}$ and $y = z_1^{b'_1} \dots z_k^{b'_k}$. Since the group $\mathbb{Z}_{m_1} \times \cdots \times \mathbb{Z}_{m_k}$ is finite, we use brute force to find the smallest $N$ (if it exists) such that $x^N = y$. Namely, list all integers $N$ such that $1 \leq N \leq m_1 \cdots m_k$ and check for all $1 \leq i \leq k$ whether $Na'_i \equiv b'_i (\mathrm{mod}\, m_i)$. If there is no such $N$, then $y$ is not a power of $x$. Otherwise, the first $N$ that we find is the (minimal) solution to the power problem. There are $km_1 \cdots m_k$ congruences to check, but since the group is fixed, all of $k, m, \dots, m_k$ are constant and therefore finding $N$ can be done with a constant number of (and hence logarithmically many) operations. □

**Corollary 7.** *The conjugacy problem in the wreath product of two finitely generated abelian groups is decidable in log-space.*

**Proof.** Let $A$ an $B$ be two abelian groups. By Lemma 6 the power problem in $B$ is decidable in log-space. In an abelian group, the conjugacy problem reduces to the word problem. In [Murray] the authors show that the word problem in abelian groups is decidable in log-space, so the conjugacy problem in $A$ and $B$ is decidable. Thus, all the conditions of Theorem 5 are satisfied and the conjugacy problem in $A \wr B$ is decidable in log-space. □

In particular, the conjugacy problem in the much studied lamplighter group $\mathbb{Z}_2 \wr \mathbb{Z}$ is decidable in log-space.

### 2.6. *On the conjugacy problem in iterated wreath products*

In order to solve the conjugacy problem iterated wreath products, we need to reduce the power problem of $A \wr B$ to the power problems of $A$ and $B$. Therefore, we need some notation: for $N \in \mathbb{N}$, $b \in B$, and $f \in A^{(B)}$, we define $f^{(b,N)}$ by

$$f^{(b,N)}(c) = (f^{b^{N-1}} \cdots f^b f)(c) \quad \text{for } c \in B.$$

With this notation we have $(b, f)^N = (b^N, f^{(b,N)})$.

**Lemma 8.** *Let $e_i \in \mathbb{Z}$ with $e_i < e_{i+1}$ and $a_i \in A$ for $i = 1, \ldots, n$, and $t, b \in B$ with $b$ of infinite order. Furthermore, let the function $f$ be defined by $f(tb^{e_i}) = a_i$ for $i = 1, \ldots, n$ and $f(c) = 1$ for all other $c \in B$. Then we have*

$$f^{(b,N)}(tb^k) = a_i \cdots a_{j-1}$$

*for $\max\{e_{j-1}, e_{i-1} + N\} \le k \le \min\{e_i + N - 1, e_j - 1\}$ and $1 \le i \le j \le n+1$. Here, we set $e_0 = -\infty$ and $e_{n+1} = \infty$. Note that $a_i \cdots a_{j-1}$ is possibly the empty product.*

**Proof.**

$$f^{(b,N)}(tb^k) = f(tb^{k-(N-1)}) \cdots f(tb^{k-1}) f(tb^k)$$

$$= \prod_{\ell=i}^{j-1} f(tb^{e_\ell}) \quad \text{because all other } f(tb^i) \text{ are trivial}$$

for $i = \min\{\ell \mid e_\ell \ge k - (N-1)\}$ and $j = \max\{\ell \mid e_\ell \le k\} + 1$. Thus, $e_{i-1} < k - (N-1) \le e_i$, i.e. $e_{i-1} + N \le k \le e_i + N - 1$ and likewise $e_{j-1} \le k \le e_j - 1$. $\qquad\square$

**Proposition 9.** *If the power problems in two finitely generated groups $A$ and $B$ are log-space computable, and the torsion elements of $A$ and $B$ have uniformly bounded order, then the power problem in $A \wr B$ is computable in log-space.*

Observe that the power problem in $A$ being log-space computable is a necessary condition, since $A$ embeds in $A \wr B$ (as discussed in section 2.2) and so the computability of the power problem in $A \wr B$ implies the computability of the power problem in $A$.

**Proof.** Denote by $M$ the bound on the order of the torsion elements of $A$ and $B$. Let $bf, cg \in A \wr B$ and let $\mathcal{A}_{\mathrm{PA}}$ and $\mathcal{A}_{\mathrm{PB}}$ be log-space algorithms which decide the power problem in $A$ and $B$ respectively. Since the word problems in $A$ and $B$ can be reduced to their respective power problems, we can assume we are also given algorithms $\mathcal{A}_{\mathrm{WA}}$ and $\mathcal{A}_{\mathrm{WB}}$ which decide the word problem in $A$ and $B$ respectively in logarithmic space. We will describe a log-space algorithm as a constant-length sequence of log-space tranductions.

We begin by applying $\mathcal{A}_{\mathrm{PB}}$ to find the smallest integer $N$ such that $b^N = c$. We can view this as a log-space transduction and henceforth assume that $N$ is part of the input. If there is no such integer, then $bf$ cannot be a power of $cg$ in $A \wr B$. We distinguish three cases: $b \neq 1$ of infinite order, $b \neq 1$ of finite order, and $b = 1$. One can determine which case to follow in logarithmic space, since it is simply a constant number of calls to the word problem in $B$.

First, we tackle the case where $b \neq 1$. As already outlined, we have

$$(bf)^N = b^N \cdot f^{b^{N-1}} \ldots f^b f = f^{(b,N)}.$$

We first consider the case in which $b$ has infinite order. We begin by giving an algorithm, $\mathcal{A}_{\text{prod}}$, for computing a word equivalent to the expression $f^{b^{N-1}} \ldots f^b f(x)$ for $x \in B$. To this end, we introduce the following auxiliary algorithms.

---

**Input:** $f = \{(u_1, v_1), \ldots, (u_k, v_k)\}$, $b$, $x$
**Output:** $L_x = \{(j_{i_1}, v_{i_1}), \ldots, (j_{i_m}, v_{i_m})\}$
for $i = 1, \ldots, k$ **do**
    if $\mathcal{A}_{PB}(b, u_i^{-1} x) = j$ **then**
        **print** $(j, v_i)$;

---

**Algorithm 3:** $\mathcal{A}_{\text{MatchToSupp}}$

For given $f$, $b$ and $x$, the algorithm $\mathcal{A}_{\text{MatchToSupp}}$ produces a list consisting of non-trivial elements of the form $f^{b^j}(x)$, indexed by $j$. The algorithm $\mathcal{A}_{\text{sort}} \circ \mathcal{A}_{\text{MatchToSupp}}(f, b, x)$ returns this list sorted by decreasing values of $j$ (here $\mathcal{A}_{\text{sort}}$ sorts the pairs according to their first component). Next, we need to determine which of these factors we need to use. The following algorithm, denoted by $\mathcal{A}_{\text{ListToProd}}$, produces a word representing the product $f^{b^{N-1}} \ldots f^b f(x)$.

---

**Input:** $L_x = \{(j_1, v_{i_1}), \ldots, (j_m, v_{i_m})\}$ with $j_1 > j_2 > \ldots > j_m$, $b$, $c$
**Output:** $f^{b^{K-1}} \ldots f^b f(x)$, where $K = \max\{j_i | j_i \leq \mathcal{A}_{PB}(b, c)\}$
for $p = 1, \ldots, m$ **do**
    if $0 \leq j_p < \mathcal{A}_{PB}(b, c)$ **then**
        **print** $v_{i_p}$;

---

**Algorithm 4:** $\mathcal{A}_{\text{ListToProd}}$

Observe that $b^{N-i} \neq 1$ for $i = 1, \ldots, N-1$. This implies that for any $x \in B$, $xb^{N-i} \neq xb^{N-j}$ for $1 \leq i < j < N$. In particular, this means that for every $x \in B$, at most $|\mathrm{supp}(f)|$ factors of the form $f(xb^{-(N-i)})$ are non-trivial. From the proof of Lemma 8 it follows that, on input $(f, b, c, x)$, $\mathcal{A}_{\mathrm{prod}} = \mathcal{A}_{\mathrm{ListToProd}} \circ \mathcal{A}_{\mathrm{sort}} \circ \mathcal{A}_{\mathrm{MatchToSupp}}$ produces a word equivalent to $f^{(b,N)}(x)$ in $A$. Moreover, $\mathcal{A}_{\mathrm{prod}}$ runs in log-space.

We now proceed to solve the power problem in $A \wr B$ when $b \neq 1$ has infinite order. In this case, $\mathcal{A}_{\mathrm{PB}}(b, c)$ is the unique integer $N$ such that $b^N = c$. In order to check that $(bf)^N = cg$, we must check whether $f^{(b,N)} = g$. To this end, we use Lemma 8 (and its notation) to study the function $f^{(b,N)}$ on the cosets of its support. We first show how to check this equality when $\mathrm{supp}(f), \mathrm{supp}(g) \subseteq t \langle b \rangle$ for a single $t \in B$. Then we show how the general case follows from this.

Let $n = |\mathrm{supp}(f)|$ and, without loss of generality, assume $n \geq |\mathrm{supp}(g)|$. For all $1 \leq i \leq j \leq n+1$, denote by $a_i \cdots a_{j-1}$ the product defined in Lemma 8. By this lemma it is equal to $f^{(b,N)}\left(tb^{\min\{e_i+N-1, e_j-1\}}\right)$. As a first step, we use the algorithm $\mathcal{A}_{\mathrm{prod}}$ to compute all the $a_i \cdots a_{j-1}$ for $1 \leq i \leq j \leq n+1$. Moreover, also using the power problem for $B$, a list of all $k$ with $g(tb^k) \neq 1$ is computed. As before this computation can be viewed as a log-space transducer, what allows us to assume in the following that these products are part of the input.

We need to be careful when working with $f^{(b,N)}$, as its support may exponentially large. Still using Lemma 8 we can check equality of $f^{(b,N)}$ and $g$ in log-space. First, we check whether $f^{(b,N)}(tb^k) = g(tb^k)$ for all $k$ such that $tb^k \in \mathrm{supp}(g)$. These are at most $n$ applications of the word problem for $A$ to inputs of the form $g(tb^k)^{-1}a_i \cdots a_{j-1}$. Since $\mathcal{A}_{\mathrm{WA}}$ uses logarithmic space and all the executions of $\mathcal{A}_{\mathrm{WA}}$ can be done reusing always the same space in a separate part of the work tape, the entire procedure uses logarithmic space. After that, we still have to make sure that $\mathrm{supp}(f) \subseteq \mathrm{supp}(g)$. In order to do so, we do the following for all $1 \leq i \leq j \leq n+1$:

- If we have $\min\{e_i + N - 1, e_j - 1\} - \max\{e_{j-1}, e_{i-1} + N\} \leq n$, then for each $k$ such that $\min\{e_i + N - 1, e_j - 1\} \leq k \leq \max\{e_{j-1}, e_{i-1} + N\}$ we use $\mathcal{A}_{\mathrm{WA}}$ to check whether $g(tb^k) = a_i \cdots a_{j-1}$ in $A$. If this is not the case for all values of $k$, then $f^{(b,N)} \neq g$.
- Otherwise, we use $\mathcal{A}_{\mathrm{WA}}$ to test whether $a_i \cdots a_{j-1} = 1$ in $A$. If this is not the case, then $|\mathrm{supp}(f^{(b,N)})| > n \geq |\mathrm{supp}(g)|$, and therefore $f^{(b,N)} \neq g$.

If none of the above cases refutes that $f^{(b,N)} = g$, then we know that indeed $f^{(b,N)} = g$. There are at most $\binom{n}{2} \sim n^2$ different values for $i$ and $j$. Each case of these at most $n$ instances of the word problems of $A$ on inputs $g(tb^k)^{-1}a_i \cdots a_{j-1}$ resp. $a_i \cdots a_{j-1}$ have to be solved. As above, this can be done in logarithmic space. This concludes the special case where $\mathrm{supp}(f), \mathrm{supp}(g) \subseteq t\langle b \rangle$ for a single $t \in B$.

The general case follows because by using the power problem, one can find a finite system of coset representatives $t_j$ such that $\mathrm{supp}(f), \mathrm{supp}(g) \subseteq \bigcup_j t_j \langle b \rangle$. More precisely, for any two elements $e, d \in \mathrm{supp}(f) \cup \mathrm{supp}(g)$, one can use the power problem to determine in logarithmic space whether $e$ and $d$ lie in the same coset (by checking whether $\mathcal{A}_{\mathrm{PB}}(e, d)$ returns an integer or 'No'). Whenever an element is not in the same coset as any of the previous elements, it is chosen as coset representative. If some element $b_i$ is not a coset representative, then the solution of the power problem returns the exponent $e_i$ such that $t_j b^{e_i} = b_i$ (where $t_j$ is the respective coset representative, i.e. the first element which lies in the same coset as $b_i$ – note that this computation is similar to how the functions $\pi_i$ were computed – see Section 2.3). Thus, the coset representatives and the exponents $e_i$ can be computed by a log-space transduction. With another log-space transduction the exponents for each coset can be sorted. Thus, we have completed the case that $b$ has infinite order.

Now, let $b$ have finite order. The order $d$ can be determined with a constant number of calls to $\mathcal{A}_{\mathrm{WB}}$ by checking for each $1 \leq d \leq M$ whether $b^d = 1$. Likewise some $N'$ with $b^{N'} = c$ can be determined because if there is such $N'$, then there is also some $N'$ with $1 \leq N' \leq M$. Since $b^d = 1$ in $B$, we have $(b, f)^d \in A^{(B)}$. Moreover, $b^N = c$ if and only if $N \equiv N' \mod d$. Thus, it remains to check whether there is some $L$ such that $(b, f)^{N'}((b, f)^d)^L = (c, g)$. In other words it remains to solve the power problem for $(b, f)^d$ and $(b, f)^{-N'}(c, g)$ in $A^{(B)}$. Since $N'$ and $d$ are bounded by a constant, $(b, f)^d$ and $(b, f)^{-N'}(c, g)$ can be computed by a log-space transduction. Thus, we have reduced this case to the case where both $b$ and $c$ are trivial.

Let $b = c = 1$. Denote $\mathrm{supp}(f) \cup \mathrm{supp}(g)$ by $\{u_1, \ldots, u_k\}$. The goal is to find a common power, $N$, such that for all $i$, $f(u_i)^N = g(u_i)$. If, for some $i$, $f(u_i)$ has infinite order in $A$, then there is either a unique $N$ which satisfies this equation or there is no solution. Therefore, we begin by determining whether there is an element of infinite order among the $f(u_i)$ and if yes, we ensure that the $N$ obtained is a solution to the equations $f(u_j)^N = g(u_j)$ for all $1 \leq j \leq k$.

In the case where there are no infinite order elements, it is enough to check for every $1 \leq N \leq M$ and for every $1 \leq i \leq k$, that $f(u_i)^N = g(u_i)$ in $A$. That is, we need to solve $kM$ word problems, each of which is log-space decidable. Since $M$ is a constant, and $k$ is bounded by the size of the input, finding $N$ can be done in logarithmic space.  $\square$

Our main interest is in the case when $A$ and $B$ are both torsion-free groups. In the statement of Proposition 9 we require $A$ and $B$ to have uniformly bounded torsion. If they are torsion-free, this requirement is satisfied trivially and we obtain the following corollary of Proposition 9.

**Corollary 10.** *If $A$ and $B$ are two finitely generated torsion-free groups with power problems that are log-space computable, then $A \wr B$ also has log-space computable power problem.*

**Definition 11.** We define the *left iterated wreath product*, $A^{\,n}\wr B$, of two groups $A$ and $B$ inductively as follows.

- $A^{\,1}\wr B = A \wr B$
- $A^{\,n}\wr B = A \wr (A^{\,n-1}\wr B)$

Similarly, we define the *right-iterated wreath product* $A \wr^n B$ of two groups $A$ and $B$ by

- $A \wr^1 B = A \wr B$
- $A \wr^n B = (A \wr^{n-1} B) \wr B$

**Lemma 12.** *Suppose torsion elements in $A$ and $B$ have uniformly bounded order. Then torsion elements in $A \wr B$ have uniformly bounded order.*

**Proof.** Let $d \in \mathbb{N} \setminus \{0\}$ (resp. $e$) be such that $a^d = 1$ (resp. $b^e = 1$) for all torsion elements $a \in A$ (resp. $b \in B$) – that is $d$ is the lcm of all orders of torsion elements of $A$. First consider a torsion element $f \in A^{(B)}$. Then we have $f^d = 1$ since $f(b)$ is a torsion element for all $b \in B$. Next consider some arbitrary torsion element $bf \in A \wr B$. Then $b$ is torsion in $B$. Therefore, $b^e = 1$, and consequently $(bf)^e \in A^{(B)}$. Hence, $(bf)^{ed} = 1$.  $\square$

**Corollary 13.** *Given two finitely generated groups $A$ and $B$, where each has log-space decidable conjugacy and power problems and uniformly bounded torsion elements, the conjugacy problem in $A^{\,n}\wr B$ is decidable in log-space.*

**Proof.** The proof done inductively on the length, $n$, of the iteration. If $n = 1$, then by Theorem 5, the conjugacy problem in $A^{\,1}\wr B = A \wr B$ is decidable in logarithmic space. Now assume that the conjugacy problem is log-space decidable in $A^{\,n-1}\wr B$. The power problem in $A^{\,n-1}\wr B$ is log-space decidable by induction using Proposition 9 and Lemma 12; therefore, it follows from Theorem 5 that the conjugacy problem in $A \wr (A^{n-1} \wr B) = A^n \wr B$ is decidable in log-space. □

**Corollary 14.** *Given two finitely generated groups $A$ and $B$, where each has log-space decidable conjugacy and power problems and $A$ has uniformly bounded torsion elements, he conjugacy problem in $A^n \wr B$ is decidable in log-space.*

**Proof.** The proof done inductively on the length, $n$, of the iteration. If $n = 1$, then by Theorem 5, the conjugacy problem in $A^1 \wr B = A \wr B$ is decidable in logarithmic space. Now assume that the conjugacy problem is log-space decidable in $A^{n-1} \wr B$. The power problem in $A^{n-1} \wr B$ is log-space decidable by Proposition 9 and therefore it follows from Theorem 5 that the conjugacy problem in $A \wr (A^{n-1} \wr B) = A^n \wr B$ is decidable in log-space. □

**Corollary 15.** *Let $A$ and $B$ be groups with log-space decidable conjugacy problems and let $B$ have log-space decidable power problem. Then the conjugacy problem in $A \wr^n B$ is also decidable in logarithmic space.*

**Proof.** We proceed by induction on the length of iteration. When $n = 1$, the conjugacy problem in $A \wr^1 B = A \wr B$ is decidable by Theorem 5, since both $A$ and $B$ have log-space decidable conjugacy problems and $B$ has log-space decidable power problem. Now suppose that the conjugacy problem in $A \wr^{n-1} B$ is decidable in log-space. Then, since the power problem in $B$ is decidable in log-space, we can apply Theorem 5 again to obtain that the conjugacy problem in $(A \wr^{n-1} B) \wr B$ is decidable in logarithmic space. That is, $A \wr^n B$ has log-space decidable conjugacy problem. □

Iterated wreath products of free abelian groups are crucial in the study of free solvable groups. The following corollary is particularly interesting from this point of view.

**Corollary 16.** *The conjugacy problem in a fixed free solvable group, $S_{d,r}$, of degree $d$ and rank $r$ is decidable in logarithmic space.*

**Proof.** The proof is essentially the same as of Corollary 14: we proceed by induction over the solvability degree $d$ and show that both power and conjugacy problem of free solvable groups are log-space decidable. If $d = 1$, we are in the abelian case: the power problem in $\mathbb{Z}^r$ is decidable in log-space (Lemma 6) and the conjugacy problem is just the word problem, which, of course, is also in log-space.

For the inductive step, we use the Magnus embedding $S_{d,r} \to \mathbb{Z}^r \wr S_{d-1,r}$. Remeslennikov and Sokolov [8] showed that two elements are conjugate in $S_{d,r}$ if and only if their images are conjugate in $\mathbb{Z}^r \wr S_{d-1,r}$. By induction, we can assume that $S_{d-1,r}$ has log-space decidable conjugacy and power problem. Thus, by Theorem 5 the conjugacy problem in $\mathbb{Z}^r \wr S_{d-1,r}$ is decidable in log-space and, since $S_{d-1,r}$ is torsion-free, the power problem of $\mathbb{Z}^r \wr S_{d-1,r}$ is decidable in log-space by Proposition 9. By [8], it follows that the conjugacy problem of $S_{d,r}$ is decidable in log-space. Log-space decidability of the power problem of $S_{d,r}$ simply follows because $S_{d,r}$ is a subgroup of $\mathbb{Z}^r \wr S_{d-1,r}$.    □

## References

[1]  S. Arora and B. Barak. *Computational complexity. A modern approach.* Cambridge University Press, 2009.

[2]  V. Diekert, J. Kausch, and M. Lohrey. Logspace computations in Coxeter groups and graph groups. In *Computational and combinatorial group theory and cryptography*, volume 582 of Contemp. Math., page 7794. Amer. Math. Soc., Providence, RI, 2012.

[3]  M. Elder and A. Kalka. Logspace computations for Garside groups of spindle type. preprint. Available at http://arXiv:1310.0933 [math.GR], 2013.

[4]  M. Elder, G. Elston, and G. Ostheimer. On groups that have normal forms computable in log-space. *Journal of Algebra*, 381:260281, 2013.

[5]  R. Macdonald, A. Myasnikov, A Nikolaev, and S. Vassileva. Logspace and compressed word computations in nilpotent groups. Preprint. Available at http://arXiv:1503.03888 [math.GR], 2015.

[6]  J. Mathews. The conjugacy problem in wreath products and free metabelian groups. *Transactions of the American Math Society*, 121:329339, 1966.

[7]  R. Lipton and Y. Zalstein. Word problems solvable in logspace. *J. Assoc. Comput. Mach.*, 24:522526, 1977.

[8]  V.N. Remeslennikov and V.G. Sokolov. Certain properties of the magnus embedding. *Algebra i logika*, 9(5):566578, 1970.

[9]  M. Sipser. *Introduction to the Theory of Computation.* PWS Publishing, 1997.

[10]  S. Vassileva. *Space and Time Complexity of Algorithmic Problems in Groups.* PhD thesis, McGill University, 2013.

[11]  H. Vollmer. *Introduction to Circuit Complexity.* Springer, Berlin, 1999.

## Chapter 13

## Group Presentations, Cayley Graphs and Markov Processes

Peter Olszewski

*Department of Mathematics, Penn State Erie,*
*The Behrend College, Erie, PA, 16563 United States*

*Dedicated to Gerhard Rosenberger and Dennis Spellman on their*
*Seventieth Birthdays*

**ABSTRACT.** The purpose of this paper is to examine the relationship between groups, group presentations, their Cayley graphs, and associated Markov Processes. In particular, we prove that for a finite group derived from an ergodic Markov Process, a process we describe in the paper, the long range equilibrium vector is uniform on the group elements, as to be expected. We also prove a theorem giving a complete characterization of finitely generated free groups in terms of their associated Markov Processes.

## 1. Introduction

Given a group $G$, a presentation for $G$ consists of a set of generators $X$ for $G$ together with a set of defining relations $R$ on $X$. We say then that $G = \langle X, R \rangle$. Roughly, this means that every element of $G$ can be written as a word or expression in the generators, $X$, and in principle the full group table can be constructed from the relations $R$. Any countable group has a (many) group presentations. Writing a group in terms of its presentation is the most compact and simplest way of describing a countable group. The books by Magnus, Karrass and Solitar [MKS], Lyndon and Schupp [LS], Baumslag [B], and Johnson [J] are standard references for the theory of group presentations.

*AMS Subject Classification:* Primary 20A05; Secondary 08C10, 20E26
*Keywords:* Free group, Group Presentation, Cayley Graph, Markov Process

Given a group presentation we can associate to it a diagraph called its **Cayley Color Graph**. Sabidussi's Theorem (see Section 3) gives a characterization as to when a graph is the Cayley graph for some group. In addition, a random walk on a Cayley graph defines a Markov Process with the group elements as the states. Conversely, given a Markov Process with certain properties, we can build a Cayley graph. In this case, the structure of a finitely generated free group is determined by the Markov Process.

The outline of this paper is as follows. In Section 2, we describe group presentations, their Cayley graphs, give examples, and describe the isomorphism problem. In Section 3, we describe Sabidussis Theorem, which gives a characterization of when a graph $\Gamma$ is actually the Cayley graph of a group. In Section 4, we talk about choosing random elements in a group and random walks on the Cayley graph. In Section 5, we describe Markov Processes and show how to construct a Cayley graph for a Markov Process and then prove that for a finite group, the long-range equilibrium state vector must be uniform on the group elements as expected. Finally, in Section 6, we prove a theorem about the structure of finitely generated free groups in terms of the associated Markov Process.

## 2. Group Presentations and their Cayley Graphs

The basic concept in group presentations and Cayley graphs is that of a free group. Let $A$ be a set. Then, a group $F$ is free on $A$ if every mapping $f : A \to G$, where $G$ is a group, can be extended to a unique homorphism of $F$ onto $G$. We denote a free group on a set $A$ by $F[A]$. A group is free if it is free on some set $A$. It can be proved that given a set $A$, there exists a free group on $A$ and further if two sets $A_1$ and $A_2$ have the same size then the corresponding free groups $F[A_1]$ and $F[A_2]$ are isomorphic. If $A = \{x_1, \ldots, x_n\}$ is a finite set, we say that $F[A]$ is a free group of rank $n$ and denote this by $F_n$.

If $F$ is free on $\{x_1, \ldots, x_n\}$ then the elements of $F_n$ can be considered as reduced words on the alphabet $\{x_1, \ldots, x_n\}$ and their formal inverses $\{x_1^{-1}, \ldots, x_n^{-1}\}$. The identity element is considered as the empty word. Reduced means that we can cancel any occurrences of $x_i x_i^{-1}$ or $x_i^{-1} x_i$. References for relevant material on free groups are the books by Magnus, Karrass, and Solitar [MKS], Lyndon and Schupp [LS], Baumslag [B], and Johnson [J].

Any group $G$ is a homorphic image of a free group. Consider $G$ as a set. Then consider the free group on $G$. The identity map $i : G \to G$ can then be extended to a unique homorphsim $f : F[G] \to G$. Hence for a group $G$

we have $F/N \cong G$ where $F$ is a free group and $N$ is a normal subgroup. From this we get group presentations.

**Definition 2.1.** Let $A$ be a set and let $R$ be the set of words in the free group $F[A]$ on $A$. Let $N[R]$ be the least normal subgroup of $F[A]$ containing $R$. An isomorphism $\phi$ of $F[A]/N[R]$ onto a group $G$ is a presentation of $G$.

We express this presentation as $G = \langle A; R \rangle$. The set $A$ is a set of generators for $G$ and each $r \in R$ is a defining relator. An equation $r = 1$ with $r \in R$ is a relation. A finite presentation is one in which both $A$ and $R$ are finite sets.

Roughly a group presentation consists of a set of generators and a set of rules or relations among the generators from which the whole group table in principle can be determined.

A free group is the most general type of group. For a free basis there are no relations. If we start with a finite alphabet, $\{x_1, \ldots, x_n\}$ and take their formal inverses and call them $\{x_1^{-1}, \ldots, x_n^{-1}\}$ then we will consider all words on $\{x_1, \ldots, x_n\}$ and $\{x_1^{-1}, \ldots, x_n^{-1}\}$. We can have two words representing the same element of the free group. We say two words are equivalent if one can be changed into the other by insertions and deletions of the trivial relators $x_i x_i^{-1}$ and $x_i^{-1} x_i$. We say a word is reduced if it has no occurrences of any of these subwords $x_i x_i^{-1}$ and $x_i^{-1} x_i$. Consider for example the word $x_1 x_2 x_2^{-1} x_2 x_2^{-1} x_1 x_1$. Then by reduction we obtain $x_1 x_2 x_2^{-1} x_2 x_2^{-1} x_1 x_1 = x_1^3$. It is clear that each word has a unique reduced form. If we let $F$ be the set of equivalence classes of words, two words are equivalent if they can be reduced to the same word. Each equivalence class has a unique reduced word. On the set of words let us define an empty word that has no symbols in it and call it one. On the set of reduced words, define multiplication by concatenation and then reduction. In other words, we put the words next to each other and then perform any possible cancellations. This set of words then forms a group $F$ which is free on the original alphabet.

We now give some very straightforward examples of group presentations and how to present groups.

**Example 2.1.** Consider the cyclic group $\mathbb{Z}_4$ of order 4. This has a single generator $g$ with one relation $g^4 = 1$. Hence $\mathbb{Z}_4$ has a presentation

$$\mathbb{Z}_4 = \langle g : g^4 = 1 \rangle.$$

Notice that the elements of this group are $\{1, g, g^2, g^3\}$. Here $g^3$ can also be taken as a generator.

**Example 2.2.** For our next example, we consider the group of symmetries of an equilateral triangle. This group is called $D_3$. By symmetry we mean a Euclidean congruence motion that leaves the triangle in place. The presentation would be

$$D_3 = \langle r, f : f^2 = 1, r^3 = 1, rf = fr^2 \rangle.$$

Here $r$ is a rotation of 120 degrees around the center of the equilateral triangle and $f$ is a flip along one of the medians. This presentation says that if one flips a triangle twice, one will end up at the same starting place as one started. If one rotates a triangle three times, one will end up at the starting point, and a rotation followed by a flip is equivalent to first flipping the triangle and then rotating it twice. Notice that for the case of an equilateral triangle, each permutation of the vertices is actually given by a symmetry. So, $D_3$ is the same as $S_3$, the group of permutations on 3 symbols. For any regular $n$-gon with $n > 3$, this is no longer true. For example, if $n = 4$, $D_4$, the symmetries of a square have order 8, while $S_4$, the symmetric group on the four vertices, has order twenty-four.

The next example deals with free groups.

**Example 2.3.** Consider the free group of rank 3. Here, we suppose that a group is free on three elements, $a, b$ and $c$ hence rank 3. The presentation of this group is then

$$F_3 = \langle a, b, c; \rangle.$$

Later on, we will see that the overall idea behind the graph of this group is that coming out of the identity, there will be six types of edges, $a, b, c, a^{-1}, b^{-1}$ and $c^{-1}$.

Now we will define what a Cayley Color Graph is.

**Definition 2.2.** Let G be a given finite group with $X = g_1, g_2, \ldots, g_k$ a set of generators for $G$. It is common not to include the identity in a generating set. We describe a graph (actually a diagraph) called the **Cayley Color Graph of** $G$, denoted by $D_X(G)$. (The symbol $X$ is included in the notation since the Cayley color graph ordinarily depends on the set of generators chosen. In general, several Cayley color graphs may exist for a given group, depending on the generating set.) We associate a vertex of $D_X(G)$ with each element of the group $G$. With each generator $g_i$ of G we associate a color, say color $i$. Suppose $G = \{x_1, x_2, \ldots, x_p, \ldots\}$ and suppose that vertex

$v_i$ of $D_X(G)$ corresponds to $x_i$. Then there is an edge colored $i$ from $v_j$ to $v_k$ if and only if $x_j = g_i x_k$. This gives us the Cayley color graph of $G$

We will now draw the corresponding Cayley graphs for the group presentations given in examples 1 through 3 above. For example 1, to draw the corresponding Cayley color graph, since there are two generators in the set, $X$, the graph will contain two colors and there will be two edges coming out of each vertex, a $g$ edge and a $g^3$ edge. Since the order of $\mathbb{Z}_4$ is four, the graph will have four vertices.

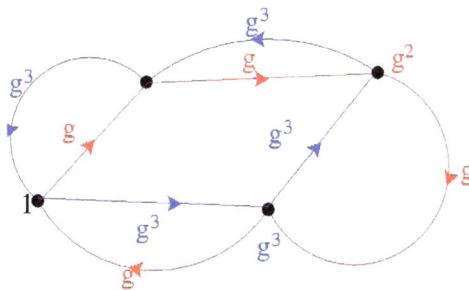

As we can see from the Cayley graph of $\mathbb{Z}_4$, the order of $g$ and $g^3$ are each four, which can clearly be seen in the graph since the edge of each takes four trials to get back to the identity, which is one.

For example 2, the Cayley graph of this group is:

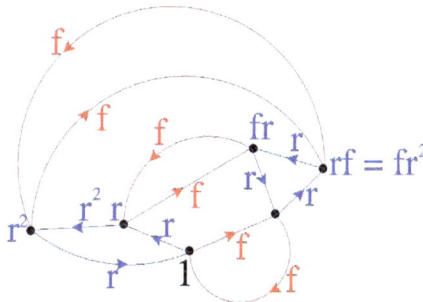

Notice from the first two examples, this graph follows the same kinds of patterns found previously. Going on an $r$ edge from the identity, it takes us three edges to get back to the identity, hence the order of $r$ is three. Going on an $f$ edge from the identity, it takes two edges to get back to the identity, hence the order of $f$ is two. There are six vertices, which corresponds to the fact that the order of $S_3$ is six, and going on an $r$ edge from $f$ is the same as going on an $f$ edge from $r^2$, hence, $rf = fr^2$.

For example 3, the Cayley graph of this group is:

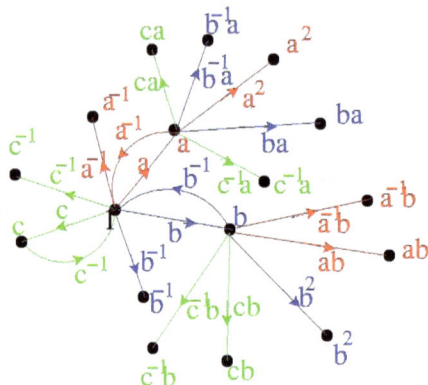

As we can see, what we have is a tree like graph and coming out of each point there will be five different edges because the inverse part will lead back to the previous vertex. In this case, each point has valency of six, which is the number of edges coming in or leaving each vertex. We say that a graph has valency $n$, if this is the number of edges that come in and out from every vertex.

## 3. Sabidussi's Theorem

Sabidussi's Theorem gives conditions under which a graph is the Cayley graph for some group. Before stating Sabidussi's Theorem, we must first define what a group acting means, define an automorphism of a graph, define transitive on a graph, and define free action on a graph. In addition, we will state a lemma, which will be used in the explanation of Sabidussi's Theorem.

**Definition 3.1.** Let $X$ be a set and $G$ a group. An **action** of $G$ on $X$ is a homomorphism of $G$ onto the permutation group on $X$. That is a map $\alpha : G \to S_X$ such that

(1) For each $g \in G$ the element $\alpha(g)$ is a permutation on $X_1$.
(2) For each $g_1, g_2 \in G$ we have $\alpha(g_1 g_2) = \alpha(g_1)\alpha(g_2)$.

**Definition 3.2.** An isomorphism from a graph to itself is called an **automorphism** of the graph. The set of all automorphisms of a graph G under the binary operation of composition is always a group, called the **automorphism group**, or simply the group of G, and is denoted by Aut(G). In other words, Aut(G) = the set of all automorphisms of G.

**Definition 3.3.** A group $G$ acting on a set $X$ is **transitive** on $X$ if for each $x_1, x_2 \in X$ there exists a $g \in G$ such that $g(x_1) = x_2$. In other words, a group $G$ acting on a set $X$ is transitive on $X$ if we can map one element of $X$ to another via the permutations in $G$.

**Definition 3.4.** A **free action** of a group $G$ on a set $X$ is one where no element of $G$ has a fixed point. That is, $g(x) \neq x$ for any $g \in G$ and $x \in X$.

**Lemma 3.1.** *A group $G$ acts on any of its Cayley graphs by* **left multiplication**. *On the directed Cayley graph $\Gamma$ the left multiplication is a colour preserving automorphism.*

**Theorem 3.1.** (*Sabidussi's Theorem*) *A graph $X$ is a Cayley graph of a group $G$ if and only if there is a transitive and free action of $G$ on $X$.*

What Sabidussi's Theorem tells us is that a graph $X$ is a Cayley graph if and only if the automorphism group of $X$ contains a subgroup $G$ acting regularly on the vertex set of $X$. It characterizes when a graph can be a Cayley graph. Cayley graphs have to be vertex transitive because one can multiply an element by a group and get an action on the group, which means a mapping. If one maps a group $G$ in this manner then all the other elements of the group are mapped onto it. This action then sends a vertex to a vertex, hg, and an edge to that h. The action being transitive means that one can map any element of a group onto any other element with this map. So, the graph has to be vertex transitive meaning that there has to be an automorphism of the whole graph that takes any vertex to any other vertex. The next section will talk about random elements and how one can choose a random element on a Cayley graph by performing a random walk on the graph.

## 4. Choosing Random Elements in a Group, Random Walks on the Cayley Graph

Let us begin by giving a definition of a random walk on a graph $\Gamma$.

**Definition 4.1.** A **u-v walk** in a graph $\Gamma$ is an alternating sequence of vertices and edges of $\Gamma$, beginning with u and ending with v such that every edge joins the vertices immediately preceding it and following it. [C]

The purpose of these walks relevant to this investigation is to randomly choose an element in a group, we need to take a random walk on its

Cayley graph. In addition, each Cayley graph defines a random walk and thus a Markov Process. These random walks will be used in the proof of the theorem about free groups. The next section will deal with Markov Processes, Markov Chains, transition matrices, and the long-range vector.

## 5. Markov Processes

Our goal in this section is to describe Markov Processes and how they relate to Cayley graphs and random walks on groups. First we must explain and define both a Markov Process and Markov Chain.

A Markov Chain or Markov Process is a process in which what happens in the future is totally dependent on the present state of the system. At any stage, a Markov Process is in one of a number of states, with the next stage of the experiment consisting of movement to a possibly different state. The probability of moving to a certain state depends only on the state previously occupied and does not vary with time. A convenient way of displaying these probabilities $p_{ij}$ is through a matrix which is called the **transition matrix** of the Markov process.

**Definition 5.1.** A **Markov Process** is a stochastic process characterized as follows: There are a finite or countable infinite number of possible states $\{s_1, \ldots, s_n \ldots\}$ that a system can be in. At discrete time steps, the system can move from one state to another. We denote by $C_k$ the state of the system at time $k$. The **state sequence** up to time $N$ is then is $\{C_0, \ldots, C_N\}$. It is a Markov Process if $P(C_k|C_0, \ldots, C_{k-1}) = P(C_k|C_{k-1})$. That is the probability of changing states depends only on the present state, which is called the **Markov Property**. The probability of being in state $i$ at time $k$ is denoted by $p_k(i)$. The **state vector** is $p_k = (p_k(0), \ldots, p_k(n), \ldots)$.

For the most part, with the exception of our characterization of free groups, we will only consider Markov Processes with finitely many states.

**Definition 5.2.** A **Markov Chain** is a sequence of experiments each of which result in one of a finite number of states that we label as $\{s_1, \ldots, s_m\}$. The probability that a given state is entered depends only on the state previously occupied. As before, we let $P_{ij}$ denote the probability of moving from state $i$ to state $j$. In a Markov Chain with $m$ states, the transition matrix $P = (p_{ij})$ is the $m \times m$ matrix giving the transition probabilities.

A Markov Processes and a Markov Chain apply to the evolution in time of the state of a random phenomenon. In general in a Cayley graph of a group, the states are the points and the transition probability is the probability of going to another point (state) on one edge. Hence, for a group with $n$ generators, there are at most $2n$ states that it can go to on one transition. However, the Cayley graph is connected so from any state, we can eventually get from any state to any other state and so, the Markov Process is **ergodic**. A Markov Process is **regular** or **ergodic** if any two states communicate, that is given any state $s_i$ there is a positive probability of moving to state $s_j$ in a finite number of steps. For ergodic Markov Processes there is a long-range equilibrium vector describing the long-range state vector.

If $P$ is the transition matrix for a Markov process then it can be proved that $P^n$ represents the n-step transition probabilities.

An $n$-vector $(x_1, \ldots, x_n)$ is a **probability vector** if the entries add up to one. Each row of a transition matrix of a Markov Process is a probability vector. A matrix with this property is called a **stochastic matrix** and hence any transition matrix of a Markov Process is a stochastic matrix.

**Theorem 5.1.** *Let $P$ be the transition matrix of a regular Markov Chain. Then*

(1) *the matrices $P^n$ approach a fixed matrix $T$ as $n \to \infty$,*
(2) *the rows of $T$ are all identical and equal to a probability row vector $t$,*
(3) *the equilibrium vector $t$ is the unique probability vector that satisfies $tP = t$.*

The significance of this long-term equilibrium vector is that it will give us what the long-term probabilities will be in terms of the states.

Relative to this paper, consider the following question. If one were to randomly walk around a Cayley graph for a group $G$ indefinitely, why would any element of a group have a particular probability or be preferential over another? We prove that this is true. First we give an example.

Consider the finite abelian group $\mathbb{Z}_2 \times \mathbb{Z}_2$, which has the presentation of $\langle a, b : ab = ba, a^2 = b^2 = 1 \rangle$. Since the order of $\mathbb{Z}_2 \times \mathbb{Z}_2$ is four, there will exist four states in the corresponding Cayley graph. The corresponding transition matrix is given by

|      | 1             | $a$           | $b$           | $ab$          |
|------|---------------|---------------|---------------|---------------|
| 1    | 0             | $\frac{1}{2}$ | $\frac{1}{2}$ | 0             |
| $a$  | $\frac{1}{2}$ | 0             | 0             | $\frac{1}{2}$ |
| $b$  | $\frac{1}{2}$ | 0             | 0             | $\frac{1}{2}$ |
| $ab$ | 0             | $\frac{1}{2}$ | $\frac{1}{2}$ | 0             |

Suppose that the long-range vector is $t = (t_1, t_2, t_3, t_4)$. By using $tP = t$ and $t_1 + t_2 + t_3 + t_4 = 1$ we obtain the following the system of equations:

$$-2t_1 + t_2 + t_3 = 0$$
$$t_1 - 2t_2 + t_4 = 0$$
$$t_1 - 2t_3 + t_4 = 0$$
$$t_2 + t_3 - 2t_4 = 0$$
$$t_1 + t_2 + t_3 + t_4 = 1$$

Solving this system we find that $t_1 = t_2 = t_3 = t_4 = \frac{1}{4}$.

As we can see, since this is random, there is no preference for any of the states. This means that if one were to wonder around long enough on the Cayley graph 1/4th of the time one would end back at the identity, 1/4th of the time at $a$, 1/4th of the time at $b$, and 1/4th of the time at $ab = ba$. The long-range vector verifies the fact that there no preference for any group element in an infinite random walk on the Cayley graph.

This example is actually a prototype for any finite group. For the transition matrix for the Cayley graph of a group, each column must also be a probability vector. A Markov Chain that has every row and every column as a probability vector is called **doubly stochastic**. It follows that the transition matrix for the Markov Process defined on any Cayley graph for the group is doubly stochastic. For a finite Markov Chain with $n$ states, it can be proved that if it is doubly stochastic then the long-range equilibrium vector is $(\frac{1}{n}, \ldots, \frac{1}{n})$. Hence we have the following theorem.

**Theorem 5.2.** *Given a finite group $G$ of order $|G|$ the long-range equilibrium vector of the associated Markov process on any of its Cayley graphs is given by*

$$v = \left( \frac{1}{|G|}, \ldots, \frac{1}{|G|} \right)$$

## 6. Markov Processes and Free Groups

A group presentation defines a Cayley graph, which in turn defines a random walk and a Markov Process. Conversely, the Cayley graph in principle determines the structure of the group defined by the presentation. For free groups this relationship is particularly straightforward. In fact a free group is characterized by its corresponding Markov Process.

**Theorem 6.1.** *A group G is a free group on n generators if and only if its corresponding Markov Process has the following properties*

(1) *the properties that make the corresponding graph a group and*
(2) *at each point there are exactly $2n$ connected points and the transition probability to each is exactly $\frac{1}{2n}$.*

**Proof.** Suppose that $F$ is a free group with free basis $\{x_1, \ldots, x_n\}$. Then the Cayley graph is a tree with valency exactly $2n$ at each point. Further, since the graph is a tree, there are no loops and hence there are unique paths between any two points. It follows that the probability of moving from a point $P$ to any connecting point is exactly $\frac{1}{2n}$. Hence the corresponding Markov Process constructed from the Cayley graph has the desired properties.

Conversely, suppose that we have a Markov Process with the desired properties. Since it satisfies Sabadussi's Theorem, it is the Cayley graph of a group. Suppose the probability is exactly $\frac{1}{2n}$ moving from one point to any other point. If the resulting group was not free then the graph could not be a tree and hence there would be a loop. Therefore for at least two points $P_1, P_2$ there would be two paths connecting them. It follows that moving out of $P_1$ there must be an edge with transition probability less than $\frac{1}{2n}$ contradicting the assumption on the Markov Process. Therefore the group must be free. □

### References

[B]   G. Baumslag, *Topics in Combinatorial Group Theory*, **Birkhauser** Lectures in Mathematics, ETH Zurich, 1993.

[C]   C. Chartrand, *Introductory Graph Theory*, **Dover Publications**, 1985.

[J]   D. Johnson, *Presentations of Groups*, London Math. Soc. Student Texts 15, 1990.

[K]   B. Kron, *Introduction to Ends of Graphs*, Ver. 22 The Unviersity of Sydney School for Mathematics and Statistics NSW 2006.

[LS]  R. Lyndon and P. Schupp, *Combinatorial Group Theory*, **Springer-Verlag**, 1977

[MKS] W. Magnus, A. Karass, D. Solitar, *Combinatorial Group Theory*, **Wiley Interscience** New York, 1966.

[MS]  M. Mizrahi and M. Sullivan *Finite Mathematics with Applications for Business and Social Sciences*, **Wiley**, 1988.